COUNTING FEMINICIDE

COUNTING FEMINICIDE

DATA FEMINISM IN ACTION

CATHERINE D'IGNAZIO

The MIT Press
Cambridge, Massachusetts
London, England

The open access edition of this book was made possible by generous funding from the MIT Libraries.

The MIT Press would like to thank the anonymous peer reviewers who provided comments on drafts of this book. The generous work of academic experts is essential for establishing the authority and quality of our publications. We acknowledge with gratitude the contributions of these otherwise uncredited readers.

This book was set in ITC Stone Serif Std and ITC Stone Sans Std by New Best-set Typesetters Ltd. Printed and bound in the United States of America.

Library of Congress Cataloging-in-Publication Data

Names: D'Ignazio, Catherine, author.
Title: Counting feminicide : data feminism in action / Catherine D'Ignazio.
Description: Cambridge, Massachusetts : The MIT Press, [2024] | Includes bibliographical
    references and index.
Identifiers: LCCN 2023027064 (print) | LCCN 2023027065 (ebook) | ISBN 9780262048873
    (hardcover) | ISBN 9780262378000 (epub) | ISBN 9780262378017 (pdf)
Subjects: LCSH: Feminism. | Feminism and science. | Big data—Social aspects.
Classification: LCC HQ1190 .D5735 2024 (print) | LCC HQ1190 (ebook) |
    DDC 305.42—dc23/eng/20230828
LC record available at https://lccn.loc.gov/2023027064
LC ebook record available at https://lccn.loc.gov/2023027065

10   9   8   7   6   5   4   3   2   1

This book is dedicated to all of the feminicide data activists—the political collectives, data journalists, small nonprofits, individuals, and academics—across the Americas and around the globe. Thank you for your time, labor, and dedication to caring for the women and people no longer with us, to restoring the balance of justice for families and communities, and to envisioning a world where gender-related violence has been eliminated.

CONTENTS

Warning: Readers are advised that this book contains the names and images of people who were killed.

LAND ACKNOWLEDGMENT

I would like to acknowledge the Wampanoag, Massachusett, and Nipmuc peoples on whose lands I resided as a guest while undertaking the research and writing for this book.

I would like to acknowledge I am part of an institution, MIT, that has played a role in the genocide of Indigenous peoples and the theft of their land, through its leaders like Francis Amasa Walker playing a public role in justifying the reservation system, through occupying territories in present-day Cambridge, and through profiting from the sale of Native lands through the Morrill Act. We can do better to acknowledge the painful truths of our past, to build relationships with Indigenous communities, and to support Native students, staff, and faculty. I am a part of a growing number of voices on campus, including some voices of senior leadership, who are working together toward what those better relations might look like.

I would like to acknowledge that *settler colonialism*, the term that describes these harmful practices, is ongoing and plays a causal role in reproducing the gendered violence that this book is about. Indigenous feminist thinkers teach us that land sovereignty and body sovereignty are interlinked.

I would finally like to acknowledge the grace and patience and teachings that Indigenous colleagues, collaborators, and students have offered to me. Thank you.

I DATA AND FEMINICIDE

INTRODUCTION

In 2017, a colleague sent me an article from the citizen media outlet Global Voices. The headline read: "One Woman Is Behind the Most Up-to-Date Interactive Map of Femicides in Mexico."[1] This was my first encounter with the work of María Salguero, and I was both stunned with sadness and painfully curious at the same time. Since 2016, this Mexican activist has spent hours a day scanning news reports, digging into government websites, and investigating crowdsourced tips. Her sole focus is logging *feminicides*—gender-related killings of women and girls, including cisgender and transgender women. For each feminicide, Salguero plots a point on the map and logs up to three hundred fields in her database, everything from name, age, whether the victim was transgender, and relationship with the perpetrator to mode of death, the case status in the judicial system, and whether organized crime was involved. She also includes the full content of the news report where she sourced the information. Salguero's map is called "Yo te nombro: El Mapa de los Feminicidios en México" (I name you: The map of feminicides in Mexico) because one of her many goals is to show that each murdered woman or girl had a name and a life and a place and a community (figure 0.1a, b).

Salguero estimates that it takes her three hours a day, every day, to scan her daily sources, verify information, and organize it into structured database fields: "It's just that it's a mess of information from all the tweets and all the things I get sent."[2] She takes breaks for mental health, but the work is still excruciating and massive in scale. In 2019, she logged 2,900 feminicides, an average of eight killings per day. Meanwhile, the Mexican government logged 1,006.[3] What accounts for this significant discrepancy?

It's hard to know, because despite the fact that Mexico has a law defining and criminalizing feminicide, which additionally guarantees women's rights to live a life free of violence, the official data are not publicly available in any disaggregated form.[4] As a

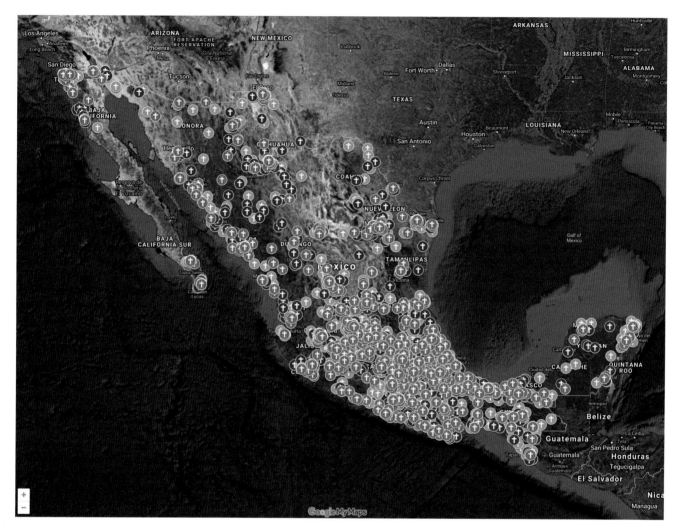

FIGURE 0.1

María Salguero's map of feminicides in Mexico (2016–present). (a) Map extent showing the whole country. (b) A detailed view of Ciudad Juárez with a focus on a single report of an anonymous transfeminicide. Courtesy of María Salguero.

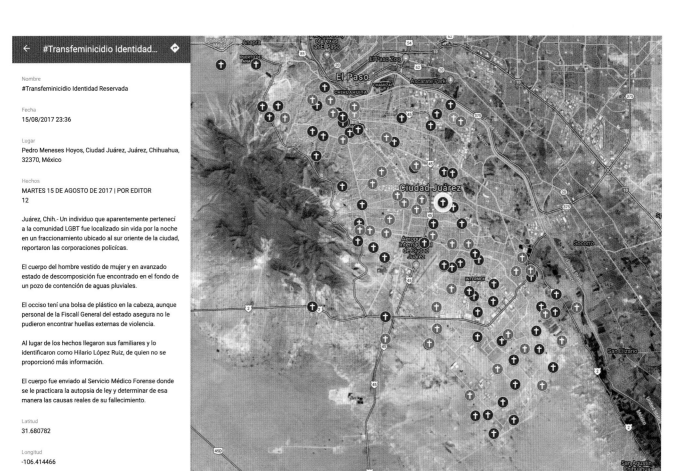

FIGURE 0.1 (continued)

result, Salguero's unofficial database has become the largest publicly accessible dataset of feminicide in Mexico. Her map has been used to help find missing women, and Salguero herself has testified multiple times before Mexico's Congress about the findings of her work. She has made her data available to journalists, activists, and nonprofits to support their efforts. Parents of victims have called her to give their thanks for making their daughters visible. Salguero is honest about the broad aims of her work: "This map seeks to make visible the sites where they are killing us, to find patterns, to bolster arguments about the problem, to georeference aid, to promote prevention and try to avoid feminicides."[5]

But the question remains: Why is this work left to one person?

In *Data Feminism*, Lauren Klein and I frame data about feminicide as *missing data*—data that are neglected by institutions, despite political demands that such data should be collected and made available. Missing data may include data that are entirely absent but also data that are sparse, poorly maintained, infrequently updated, contested, removed, and/or underreported. In the case of feminicide, the state and its institutions systematically ignore the phenomenon, neglect to count cases, and often neglect to conceptualize gender-related violence in such a way that it could even be counted precisely—and thus it does not count. When the state and its institutions fail to collect important information, activists and journalists are increasingly stepping into those data gaps and producing their own *counterdata*—data that are produced by civil society groups or individuals to challenge structural inequality, to protest state bias and inaction, to galvanize media and public attention, to work toward policy change, and/or to help heal wounded communities. As Alice Driver notes in the case of Mexico, "The most accurate records of feminicide are still kept by individuals, researchers, and journalists, rather than by the police or a state or federal institution."[6] This is true for Mexico, and it also applies to countries in the Global North such as the United States and Canada that keep no records of feminicide because, in contrast to the vast majority of Latin American countries, they have no legal formulation of the concept (see table 1.1).

This is a book about transnational grassroots data activism in relation to feminicide and gender-related killing. It documents the creative, intellectual, and emotional labor of data activists across the Americas. Their work with data focuses on care, memory, and justice. In so doing, it challenges the hegemonic and extractivist logics of mainstream data science. Thus, this is a book about the potential for undertaking *restorative/transformative data science*—an approach to working with systematic information with the twin goals of restoration and transformation. *Restoration* involves restoring rights, dignity, life, living, and vitality to the individuals, families, communities, and larger publics harmed by structural inequality. *Transformation* involves work to dismantle and shift the structural conditions that produced such violence in the first place. It is both visionary and preventative. As we will see, the concept of restorative/transformative data science was developed inductively and collaboratively—from observing and being in community with grassroots activists working with data about feminicide. Indeed, one of the main arguments of this book is that it is not computer scientists, think tanks, philosophers, critical data studies scholars, engineers, or policy people who are developing the most sophisticated approaches to the ethical use of data for public benefit. Rather it is grassroots feminist data activists, predominantly from Latin America, who are at the forefront of data ethics in the service of justice.

María Salguero's work has received an immense amount of media coverage. She has even become something of a folk hero in Latin American feminist movements, which have been mobilizing fiercely against feminicide for the past decade. On March 8, 2021, International Women's Day, activists took to the streets in Mexico City to cover existing street names with names of notable women, victims of feminicide, women's groups, and feminist events. At least two separate streets were named for Salguero, and the results were showcased on social media with the hashtags #TomaLasCallesNoCalles and #LasCallesTambienSonNuestras (figure 0.2).[7] These translate to "take to the streets, don't be silent" and "the streets are also ours (women's)."

Salguero's work deserves to be celebrated. But her efforts are not alone. Sociologist Saide Mobayed found that there are at least twenty projects in Mexico that register feminicide.[8] Not all of them attempt to do it at the scale of the country; some are focused on specific states, and others are focused on cities like Ciudad Juárez and Mexico City. Beyond Mexico, as part of the Data Against Feminicide project, my colleagues and I have cataloged more than 180 grassroots projects around the world producing data to challenge feminicide and gender-related killings. Activists compile their databases by painstakingly scanning news reports and social media feeds, triangulating with official data, talking with families and friends of loved ones, and logging details of people's murders into digital maps, spreadsheets, and databases. This book is an exploration of these grassroots efforts and how their data practices challenge conventional ideas about data science in the world today. It makes the case that feminicide data activists provide a powerful model for how data may be used in the service of justice.

WHAT IS FEMINICIDE?

What are we talking about when we use this word?[9] *Feminicide* (or *femicide*) is the misogynous and gender-related killing of women, including transgender and cisgender women and girls. It is used to denote domestic or intimate partner violence that is fatal, and also murders perpetrated where a woman's gender, and her gender subordination, are part of the motivation for the crime. Both feminicide and femicide are evolving concepts. Their exact definition and scope shifts across contexts, and you will see this surface in the various definitions of what "counts" as feminicide used by activists in this book. Yet despite this variation, feminicide and femicide are increasingly mobilized in legislation, national statistics and activism. As we will explore further in chapter 1, the term *femicide* emerged from the feminist work of Radford and Russell. Building on this work, Latin American activists and scholars introduced the term *feminicidio* as a way to capture the role of the state in enabling violence against

FIGURE 0.2
On International Women's Day, March 8, 2021, activists temporarily renamed several streets to honor the work of María Salguero. Courtesy of Fátima Araujo.

women through either omission, negligence or complicity. In recent years, feminicidio has traveled back into English as *feminicide*, as a way of capturing the public nature of such violence—basically implicating states for widespread failure to ensure women's basic human right to life. Feminicide, in other words, asserts that this violence is not a "domestic" or "private" problem, but a public problem in which the state is complicit. In their 2010 work *Terrorizing Women*, Fregoso and Bejarano define feminicide as (1) the murders of women and girls founded on a gendered power structure; (2) both public (implicating the state) and private (implicating individual perpetrators); and (3) intersecting with other inequalities. They write, "The focus of our analysis is not just on gender but also on the intersection of gender dynamics with the cruelties of racism and economic injustices in local as well as global contexts."[10] This is the formulation of feminicide that I will use throughout this book.

Activism on the issue of feminicide from families, feminist groups, and women's movements, particularly across Latin America, has played a fundamental role in raising awareness and promoting policy change globally. Most notably, mobilization around the disappearances and murders in Ciudad Juárez, Mexico, since the 1990s, along with the Ni Una Menos (not one less [woman]) uprisings in Argentina and across the region since 2015, have brought worldwide attention to the issue of feminicide. The vast majority of Latin American countries now have a law defining feminicide or femicide (see table 1.1), and dozens of grassroots efforts to monitor the phenomenon have proliferated in the past decade.

Because Latin American feminists and women's groups are leading this work, the geographic center of gravity in this book is focused on Latin American data activism on the topic of feminicide. As you will see, the book also intentionally weaves in feminicide data activism from groups based in the United States and Canada for three reasons. First, there is an equally long legacy of activism in North America organizing against femicide, fatal violence against Black women, and missing and murdered Indigenous women and girls and two-spirit people (MMIWG2). Among many other efforts, we can see this in the #SayHerName campaign and in the Women's Memorial March, which started protesting and memorializing the killings of Indigenous women in Vancouver in 1991 and continues to this day. Second, as I began to speak publicly about this work, I realized that there is a risk of people from Anglo-America dismissing feminicide as a problem "down there," south of the US-Mexico border. This reaction upholds racialized and colonial stereotypes about Latin America. It is both ignorant of and incorrect about the current realities of feminicide, MMIWG2, and gender-related violence in North America. High-profile cases have made headlines—Breonna Taylor, Rita Hester, Savanna LaFontaine-Greywind, Gabby Petito—but too many other cases have not. So one reason this book focuses on the Americas is as a corrective to the gringo exception-alism that tries to invisibilize feminicide in the Global North.[11]

Third, the book focuses on data activism in the Americas broadly because feminicide data activists themselves are increasingly in dialogue with each other across national and regional borders. They participate in networks like the Interamerican Anti-Femicide Network (RIAF); they set up one-on-one meetings to share tips and strategies; and they draw inspiration from the concepts of each other. This very book emerges from a South-North participatory action research project called Data Against Feminicide. This way of working is what the editors of the *Translocalities/Translocalidades* anthology frame as *translocal*: an effort to create feminist and antiracist bridges, connections, and move-ments across grassroots efforts in the Americas.[12] Instead of formal movement building at the scale of the state, this approach emphasizes the construction of relationships at the scale of the community. These relations are born from shared struggle but respect local differences and variations in culture, context, and history. This is how feminicide data activists are in relation to each other.

COUNTING FEMINICIDE

This leads us to the book's title and its central questions: Why count such fatal vio-lence? What may be gained by systematically tabulating, monitoring, documenting, aggregating, analyzing, and visualizing cases of feminicide and gender-related killing? How is it possible to seek justice through databases and spreadsheets, charts and statis-tics? What does counting do and who do we count for?

According to estimates by the United Nations, in 2017, eighty-seven thousand women were intentionally killed across the world. Nearly 60 percent of them were murdered by intimate partners or other family members.[13] In the United States, reports state that around three women are killed every day by their current or former partners, though research has shown that these numbers are underreported.[14] Ten feminicides happen every day in Mexico.[15] In Latin America and the Caribbean (LAC), every two hours a woman is killed in incidents related to her gender.[16] Many killings of women are perpetrated by intimate partners and family members. But other feminicides are perpetrated by mass shooters, organized crime groups, sex traffickers, and the state itself in the form of overpolicing, systemic economic violence, the social exclusion of LGBTQ+ people, and racialized violence visited on low-income communities. Beyond the loss of lives, these murders have deep ripple effects on the mental health and liveli-hoods of relatives and communities. Not only do they cause and perpetuate different forms of intergenerational trauma, but because women are so often providers of care and/or income, these murders leave their loved ones more vulnerable to food, health, and housing insecurities.

Yet as terrible as these existing statistics are, they are terribly incomplete. Official government data on gender-related violence and feminicide are often absent, infrequently updated, contested, or underreported. This is true despite the fact that a 2015 report from the UN Special Rapporteur on Violence Against Women called for the establishment of a femicide "watch" or monitoring effort in every country in the world.[17] The reasons for these data gaps range from stigma and victim-blaming to matters of legal interpretation and pervasive patriarchy. We will explore the phenomenon of missing data about feminicide at length in chapter 2. Although numerous reform efforts exist to address these gaps from national and international perspectives—through the creation of standardized definitions, data standards, and indices—the central focus of this book is on grassroots efforts that count *from within* specific places and communities. Citizen data monitoring of feminicide has stepped into the data gaps that exist in many countries and regions around the world, often through the model of an *observatory*: an ongoing monitoring effort, versus a one-time research project, that works to document human rights violations. Individuals and groups compile cases of feminicide from government and media sources; they verify and analyze those data; they circulate visualizations, reports, and statistics. Yet this work is not only about producing and analyzing information. Feminicide data activists also accompany families through the justice system; they memorialize and humanize victims; and some provide direct services and emotional support to wounded communities.

Many feminicide data activism efforts have taken shape in the past decade using digital technologies, but they follow in a predigital activist tradition of using media reports to monitor gender-related and racialized violence. The earliest use of such methods that I have encountered is from the legendary Ida B. Wells, who meticulously compiled statistics about lynching from white, mainstream media reports in *A Red Record. Tabulated Statistics and Alleged Causes of Lynchings in the United States, 1892-1893-1894.* The monitoring of feminicide from news reports can be traced back decades to work in Boston, in Ciudad Juárez, in Ontario, in Uruguay, and other places that we will explore in chapter 1. These projects used fax machines, pamphlets, newspaper clippings, and handwritten accounting ledgers to document feminicides and gender-related killings, to protest and memorialize, and to fight for policy change and justice. Scholars Fregoso and Bejarano place these data practices in a long history of *observatorios comunitarios* (community-based observatories) in Latin America, which have served to engage communities in monitoring state agencies and using collective action strategies to hold them accountable.[18]

In this book, we will meet contemporary groups working in this tradition and explore their motivations and their methods for producing counterdata. These efforts are mainly initiated and sustained by women—especially Indigenous women and Black women—as well as queer people. Most projects are explicitly aligned with intersectional feminist values, feminist and antiracist movements, and/or movements for Indigenous sovereignty. There is variation across the projects: they count and classify different things, their reasons for counting are diverse, and the uses to which the data are put are diverse. While the majority of activists use feminicide as their organizing concept, those that focus on gender-related violence as it intersects with white supremacy and/or settler colonialism and/or cisheteropatriarchy name that violence with more intersectional specificity. These include missing and murdered Indigenous women, girls, and two-spirit people, Black women killed in police violence, or fatal violence against transgender people. Indigenous women and Black women are at the forefront of grassroots efforts that monitor racialized feminicide. While they sometimes coordinate with broader antifeminicide movements, they repeatedly critique the same for sidestepping the role of racism and colonialism in producing gender-related violence. Despite this variation, all of these groups are joined in their determination to count and document one of the most egregious violations of human rights in the world today: the violation of the right to life, liberty, and security of person. As the Combahee River Collective wrote on a large banner in their march to protest Black women's murders in Boston in 1979: "3rd World Women: We Cannot Live Without Our Lives."[19]

One thread common to all feminicide data activism discussed in this book is that counting is never neutral; it is a deeply political act. For that reason, I will mainly refer to counterdata *production* in this book versus counterdata *collection*, to keep at top of mind that the work being undertaken is not about a neutral observation of what is already present in the world, but about the deliberate crafting of political visibility and advancement of concepts like feminicide that can work to name and challenge structural political problems.[20] I want to thank my friend and collaborator Helena Suárez Val for our conversations drawing out this point. This is the basic starting point that underlies all data activism and reformative/transformative data science: data are always political, and data are always produced.

DATA FEMINISM IN ACTION

If counting is not neutral, which other assumptions about data and data science might we need to challenge in order to bend it toward justice? Which other standard operating procedures of Western Anglo white rich cis men's data science might need to be disrupted and decolonized? In *Data Feminism*, Lauren Klein and I outlined what a feminist approach to data science might look like. We drew from intersectional feminist theory, activism, and writing to outline seven principles for working with data in a feminist way: examine power, challenge power, elevate emotion and embodiment, rethink binaries and hierarchies, embrace pluralism, consider context, and make labor visible (see the full list in the toolkit in chapter 8 of this book).

These principles were designed to challenge what one might call *hegemonic data science*. The use of *hegemonic* here refers to philosopher Antonio Gramsci's conception of power, which outlines how the ruling class of society manipulates social and cultural norms and values, as well as economic structures, in order to benefit itself and with the result that minoritized people are excluded and subordinated.[21] Hegemonic data science can be defined as mainstream data science that works to concentrate wealth and power, to accelerate racial capitalism and perpetuate colonialism, and to exacerbate environmental excesses and social inequality.[22] This is the data science critiqued in pathbreaking works of public scholarship like *Weapons of Math Destruction*, *Indigenous Statistics*, *Race after Technology*, *Automating Inequality*, *The Age of Surveillance Capitalism*, *Algorithms of Oppression*, *The Costs of Connection*, and *Artificial Unintelligence*.

Hegemonic data science can be directly implicated in feminicide and a number of other public emergencies if we draw from formulations of technical "glitches" by Ruha Benjamin and Meredith Broussard.[23] Glitches tend to be considered temporary interruptions in a system, but what if what they actually do is expose standard operating procedure? What if glitches actually "illuminate underlying flaws in a corrupted system"?[24] In this model, a feminicide is not an aberration but an exposure of the violent inner infrastructure that racial capitalism and patriarchy require in order to persist. Artist Erin Genia, from the Sisseton-Wahpeton Oyate and Odawa tribes, has a name for our moment. She says that this is a *cultural emergency*.[25] Emergencies, glitches, crises— including feminicide as a crisis—these are not one-off, temporary accidents. They are the entirely predictable outcomes of the way that our present political-economic system is organized to favor the lives, livelihoods, and thriving of those at the top of the economic order and wage slow violence on the rest of us as the disposable middle and bottom. Hegemonic data science is the technocratic and ideological infrastructure of this extractive system: racialized databases, biased algorithms, discriminatory machine learning models, and unintelligent AIs enable and exacerbate economic, environmental, and physical violence with devastating consequences. Related to MMIWG2, geographer Annita Lucchesi (Cheyenne) discusses hegemonic data practices as constitutive of *data terrorism*.[26] Many theorists of feminicide and racialized violence call our attention to who is disposable, ignorable, and "disappearable" under neoliberal and extractive modes of governance—and it turns out to be the majority of us.[27]

Yet, as media scholar Paola Ricaurte asserts in her call for decolonial approaches to data science, there are "possible alternative data frameworks and epistemologies that are respectful of populations, cultural diversity, and environments."[28] Some alternative epistemological approaches are being created and advanced in academia. I would situate data feminism here along with emerging and exciting work on feminist data refusal, decolonial AI, Indigenous data sovereignty, queer data, Black data, and emancipatory data science (see the toolkit in chapter 8 for more on these approaches). But academia cannot and should not do this work alone. New media scholars Stefania Milan and Lonneke van der Velden call attention to the important role of data activists who function as "producers of counter-expertise and alternative epistemologies, making sense of data as a way of knowing the world and turning it into a point of intervention. They challenge and change the mainstream politics of knowledge."[29] Data activists are inventing and enacting new ethical visions and expanded notions of citizenship in the digital age. They are forging alternative *data epistemologies*—ways of knowing things about the world with data. We see these at work in the Indigenous data sovereignty movement, Data 4 Black Lives, the antieviction data activism taking root in the United States; in new communities that have emerged to connect data study and data practice such as Tierra Común; and, of course, in the central subject of this book: feminicide data activism.

As I learned more about María Salguero's work, and that of other feminicide data activists, I was struck by how their ways of working resonate with data feminism's principles. The role of emotion, care, and lived experience is omnipresent (*elevate emotion and embodiment*). The approach to sourcing and verifying information is almost always pluralistic (*embrace pluralism*). Data activists work in coalition and collaboration—though not always without conflict. The labor of counting is not always publicly visible, but it is at the center of the internal work of activists (*make labor visible*). For me, different data feminism principles appeared to be more and less salient at different stages of a feminicide data activism project. Data activists also navigate ethical issues not addressed by data feminism, such as consent to name someone in a database, which points to concerns that have been elevated by other important contributions to feminist data science.[30]

Since *Data Feminism* came out, readers have repeatedly asked Lauren and I how to realize our principles in action. For me, working on feminicide and in community with feminicide data activists over the past years has sparked a great deal of reflection on these principles. Which activist data science practices resonate with data feminism? Which practices challenge or point out gaps in the principles? As all practitioners know, practice is messy and rarely adheres cleanly to pleasing principles. Throughout the book, I will highlight resonances and tensions between activist data practices and the principles of data feminism. As such, this book attempts to reflect on data feminism's principles as I, Catherine, see them surfacing in data activism about feminicide. It can be thought of as an extended action-reflection on the frictions that arise when moving from speculative ethics to real-world relations. Here I wish to be careful about *imposing* data feminism on groups from the outside, from a Global North positionality, or from a white academic perspective. For that matter, there is also a risk of imposing feminism itself. Though the majority of groups in this book do use an explicitly feminist frame, not all of them do. This is due to the ways in which feminism itself has been used to uphold racial hierarchies and exclude the lived realities, work, and leadership of women of color. Thus, it would be tautological and misguided to think that I went into "the field" and Lo! I discovered data feminism! Instead, I will focus throughout this book on the idea of resonances and tensions in practices and principles and on drawing out lessons that can be learned from grassroots activists that move us closer to using data science in the service of restoration and transformation.

The stakes are high. There is currently enormous investment in the AI ethics and "data for good" space. Stanford and MIT are aiming to fundraise on the order of billions of dollars, foundations are creating funds in the millions, and Big Tech corporations are assembling large AI ethics research teams (and then firing their Black and white women leaders who raise questions about potential harms).[31] But perhaps these powerful institutions are looking to and investing in the wrong places for such ethical guidance. At best, these monies are funding work by well-intentioned people who need a lot more training in feminism, critical race theory, Indigenous studies, ethnic studies, queer theory, community organizing, post-colonial thought, social work, and other frameworks and fields of practice that rigorously tackle inequality and oppression. At worst, these are cynical efforts at ethics-washing.[32] Grassroots data activists at the margins— real-world people who are using data science in the service of real-world struggles for justice—have expertise to offer to those people who are trying to craft data frameworks and practices in the service of justice. Areas like health, housing, urban planning, policing, transportation, and education are characterized by durable and extreme structural inequalities based on patriarchy, white supremacy, classism, colonialism, and more. To work with data in these domains is, inevitably, to confront inequality and oppression. Grassroots organizations and movements know this and live this. They are far better equipped to deal with how power infiltrates datasets in these domains than today's hegemonic data scientists who are trained at elite institutions or in the elite institutions' money-making MOOCs. Activists and movements are increasingly challenging these inequities using data science as one tactic in a larger struggle. This book makes the case that feminicide data activists are at the forefront of a data ethics that rigorously and consistently takes power and people into account.

HELLO, READER

Before we go further together, I would like to take this moment to introduce myself to you and explain how I got to the point of writing a book about feminicide and data activism. I often describe myself as a "hacker mama." I'm a hacker because I spent the early years of my career as a freelance software developer and database programmer. I am also a hacker in sensibility and method, where *hacking* means the "clever or playful appropriation of existing technologies or infrastructures or bending the logic of a particular system beyond its intended purposes or restrictions to serve one's personal, communal or activism goals."[33] Fundamentally, I have never believed that tech is deterministic. That is to say that just because technology is birthed from war and militarism and colonial violence doesn't mean that it cannot be employed toward serving as resistance to and liberation from the same. Over the course of my careers as a software developer, artist, and professor, I have sought to point out the power imbalances in tech, but also to find and use the cracks; to bend and borrow the tools; and to support the flourishing of alternative, feminist, *humble* visions for the role of data, information, and technology in working toward social justice.

The "mama" part of hacker mama is because I am mother to three children and three cats, and that takes up a whole lot of my time. *Mama* is also a relational term, existing in a relation of care with people and pets and places. These interconnections manifest as collaborations, and I rarely work alone. This is my quiet but persistent refusal of the relentless individualism of academia, of the art world, and of Western society generally. I am also a mama in the sense of its gendered political identity. I care deeply about those experiences of mamas that are systematically overlooked and undervalued: birth, breastfeeding, and reproductive justice; care work; and political work toward healthy, safe, thriving communities. In 2014 and 2018, I co-organized the Make the Breast Pump Not Suck Hackathon with a collective of awesomely wonderful women whom I continue to admire. And though that work was about the beginning of life, and the present book is about lives cut short by violence, there are many resonances between them. Indeed, in much of the activism against feminicide and gender-related violence, mothers are out front—and they have been organizing collectively for decades to demand systemic change.

Following a long period of freelancing—as an artist, software developer, and educator—I have been full-time in academia since 2014, and I am currently at MIT in the

Department of Urban Studies and Planning. There I run a lab called the Data + Feminism Lab, where our mission is to use data and computational methods to work toward gender and racial justice, particularly as they relate to space and place. As a white settler, I have been on a many-years-long journey to understand how I may contribute to racial justice and Indigenous sovereignty. At the current stage of my journey, I believe this is through establishing meaningful and ongoing relationships with individuals and organizations, diverting financial and social capital back to where it belongs, creating joyful and collaborative educational spaces for minoritized students, and playing a supporting role in projects led by women of color and queer people of color.

While I researched María Salguero for *Data Feminism*, I became deeply inspired by both her work and the many-pronged efforts of Latin American feminist movements. On the Italian immigrant side of my family, my grandfather's cousins emigrated to South America. These family ties led me to live in Argentina for several years, both as a young person and a middle-aged person, and to develop my Spanish, which I speak (imperfectly) to my children and cats. This is part of the story of how I came to the work of data activism about feminicide. My partner and I moved our family to Argentina in 2019, where I was able to attend gatherings, interview activists and data journalists, and make connections to feminist groups. I enter the topic of feminicide with a distinctly North American perspective, and you will see that surface in this book. I enter with appreciation for everything that the framing of feminicide as a feminist concept has been able to accomplish in legal, activist, policy, and intellectual circles. And I enter with questions about the boundaries of that frame and about whose killings may be sidelined or made invisible within our current formulation of feminicide—to name a few, Indigenous women or sex workers or trans women or Black women killed in police violence. Patricia Hill Collins teaches us that violence is a *saturated site*—a place where intersectional forms of domination are rendered most visible.[34] Obviously violence against women involves patriarchy and sexism, but how do these intersect with settler colonialism? With anti-Blackness? With cissexism? With the extractive logics of neoliberal capitalism? I hold this appreciation and curiosity about feminicide as a concept together with a commitment to intersectional thought and action. It turns out that many of the data activists working on feminicide also ask these questions.

DATA AGAINST FEMINICIDE

In May 2019, I took a bus across Buenos Aires to meet Dr. Silvana Fumega, who was, at that time, research and policy director of the Iniciativa Latinoamericana por los Datos Abiertos (ILDA; Latin American Initiative for Open Data).[35] I didn't know quite what to expect, just that three people in my network had told me that if I was interested in learning about feminicide and data, I must meet Silvana. From this very first meeting, which actually turned into a brainstorming session, Silvana and I drank coffee, shared interests, and discussed ideas for using machine learning to detect news articles about feminicide. Since 2017, Silvana had been leading the development of a Latin American data standard for feminicide data collection. At the end of our meeting, she suggested we invite Helena Suárez Val into our emerging collaboration. Helena is a feminist activist, researcher, and writer who has been producing data about feminicide in Uruguay since 2015. In the midst of her doctoral project, she told us about the joy and connection she found by convening a group of feminist data activists to discuss their work. Remarkably, Helena, Silvana, and I have formed a tight and thriving collaboration (and friendship) despite the fact that, mainly due to COVID, all three of us did not meet together in person until the spring of 2022.

Our collaboration evolved to be named *Data Against Feminicide*.[36] It is a bit of a sprawling project, as are our conversations and brainstorms, but it's perhaps best characterized as a South-North action-research collaboration across activism, academia, and civil society. We outlined three goals for the project, each stemming from an area that one of us had been working on:

1. To foster an international community of practice around feminicide data

2. To develop digital tools to support activists' production of feminicide data from media sources

3. To support efforts to standardize the production of feminicide data where appropriate

The Data Against Feminicide project *does not* collect or aggregate activist data.[37] Our three goals are focused on supporting and sustaining the already existing practices of activists who care for feminicide data in their own contexts. Since 2020, we have organized annual convenings around the International Day for the Elimination of Violence against Women. This day of awareness falls on November 25 and was designated by the UN as a remembrance for the three Mirabal sisters—"las Mariposas"—from the Dominican Republic who were assassinated by order of dictator Rafael Trujillo in 1960. Our Data Against Feminicide virtual events have featured panels with data activists and public sector officials, workshops and hands-on learning sessions, brainstorming and network building opportunities, and short talks by academics and graduate students.[38] Hundreds of participants have attended these events, from thirty-eight countries and five continents. The primary language of the events is Spanish, with live interpretation in English and Portuguese to support participation from across the Americas. Between annual events, we host a network directory, email list, and Slack channel

FIGURE 0.3
Silvana Fumega, Helena Suárez Val, and I worked together for three years on the Data Against Feminicide project before we were able to meet in person in Montevideo in spring 2022. Courtesy of the author.

so that participants may connect with each other. In spring 2022, we experimented with creating a more formal learning structure and, through ILDA, offered a nine-week online course in Spanish about feminicide and data.[39]

In collaboration with my lab and students at MIT, we have also been co-designing and piloting digital tools and technologies to support the data production work of activists and civil society organizations. Data Against Feminicide has released two free tools for activist use, and the process of creating these is the subject of a case study in participatory design in chapter 7. The way I think about our community events and our digital tools is that in both cases we are building infrastructure. This infrastructure is always *sociotechnical*—consisting of both technical and social relations at the same time. And in both cases—community events and digital tools—the infrastructure we craft is deliberately small-scale, noncentralizing, and consists of actions and relations that support and sustain (not outsource, not automate) the difficult labor of activist data production.

To make the past years of community-building and tool-building possible, we needed to better understand activist data practices: How widespread are practices of counterdata production about feminicide and gender-related violence? Why do groups and individuals begin their monitoring projects? What sources do they use for information? How do they classify and categorize cases? How do they publish and circulate their data and with what impacts and effects? What challenges do they face in collecting, analyzing, and using the data they produce?

Helena already had answers to these questions for herself. Since 2015, she has been logging cases of feminicide in Uruguay into a Google spreadsheet and publishing them on her website, Feminicidio Uruguay. The project started during a period of intense feminist organizing in Uruguay; there were mobilizations around the country to prevent the overturning of the country's law legalizing abortion and to increase attention to feminicide. Helena, at the time a member of the Coordinadora de Feminismos del Uruguay movement, remembers that they were collectively outraged by a series of murders of young women in late 2014 and by the sensationalist, victim-blaming media coverage of their killings. The group started going out into public space to protest each killing, and Helena began the spreadsheet primarily as a record of protests. Eventually, the collaboration dissolved, but Helena kept recording each new case as it was reported in the press. For each violent death, she logs details in a spreadsheet and on a map, and then also publishes those to Twitter and Facebook. The work—of reading misogynist news reports about women who were beloved to their families or communities—is exhausting, and she has often contemplated ending the monitoring—but she has not.

To understand Helena's experiences in relation to others in our emerging community, we began first to keep a list of all the feminicide data activist groups we could find. We started with Helena's prior research into other feminicide mapping projects.[40] As we read research papers or hosted community events, each time we learned about a monitoring group we would do research about them online and add them to our list. This informal spreadsheet of feminicide data activists has now been formalized and expanded to include more than 180 efforts from around the world. It includes groups who document and produce data about cases of feminicide, femicide, MMIWG2, LGBTQ+ killings, Black women killed in police violence, and other forms of gender-related killings. We also started, in 2020, to conduct in-depth interviews with data activists in the Americas. At first we interviewed ten groups. Then we began appreciating how truly widespread across the Americas this form of data activism was. Ten interviews became twenty, twenty became thirty, and to date, we have interviewed more than forty individuals or groups.

We have published about Data Against Feminicide individually and collectively, with an eye toward supporting publications led by each other and supporting our students and project partners to take the lead on publications as well.[41] With the blessing of my collaborators, I proposed documenting our learnings from the interviews and larger list of data activist projects, and this book is the form that has taken. Throughout the book, you will see work by Silvana, Helena, and Data Against Feminicide surface many times as an object of reflection and discussion. Their prior work and our collaborative work together has been formative in shaping this book, and indeed the book could not exist without the learnings that I have gained from our conversations and friendship and without its grounding within the Data Against Feminicide project and community as a whole.

In fact, this book draws significantly from *two* collaborative projects—Data Against Feminicide and *Data Feminism*. Throughout the book, I will be in dialogue with these projects and my collaborators, who I will refer to by their first names: Silvana, Helena, and Lauren. To do right by them, as well as by the feminicide data activists who we interviewed and who we continue to be in community with, I set up an Academic-Community Peer Review Board for this book. Members of this board include Silvana, Helena, Lauren, Annita Lucchesi from Sovereign Bodies Institute, Debora Upegui-Hernández from the Observatorio de Equidad de Género Puerto Rico (Gender Equity Observatory of Puerto Rico), and Paola Maldonado Tobar and Geraldina Guerra Garcés from the Alianza Feminista para el Mapeo de los Femi(ni)cidios en Ecuador (Feminist Alliance for Mapping Femi[ni]cide in Ecuador). They participated in reviewing and providing detailed feedback on the manuscript. While traditional peer review is designed to provide expert scholarly feedback, the purpose of the Academic-Community Peer Review Board, for me, was to invite feedback from the people that I am in community with in the process of doing this work. They are the ones who I feel most accountable to, as well as the ones who are best positioned to correct, question, or challenge my interpretations or usage of our shared work. Appendix 2 describes this board and our process in more detail.

OVERVIEW OF THE BOOK

Counting Feminicide is divided into three major sections. The first section provides background on some of the major concepts in the book. In chapter 1, we will explore how the concepts of femicide and feminicide evolved through multiple South-North border crossings. Counting feminicide is not a new activist practice, and we will visit several historical examples of documenting cases and situate them in relation to literature on data activism. Chapter 2 describes a case study in Puerto Rico as a way to examine missing data—why official information about feminicide is often absent, sparse, incomplete, unreliable and contested. This chapter introduces the idea of counterdata, characterizes the data activists who produce counterdata about feminicide, and introduces the concept of restorative/transformative data science.

The second part of the book is called the Process of Restorative/Transformative Data Science. The four chapters in this section are empirical, meaning that they come out of our research team's in-depth interviews with more than forty data activist groups and individuals. From our qualitative analysis, the themes of *resolving*, *researching*, *recording*, and *refusing and using data* surfaced as common workflow stages for restorative/transformative data science projects about feminicide and gender-related killing. Each of these stages of work is the focus of a chapter in this section, and the collaged diagram in chapter 2 (figure 2.4) links these stages together. "Resolving" (chapter 3) addresses activists' motivations for starting a database of cases about gender-related killings. "Researching" (chapter 4) is the process of seeking information and discovering relevant cases of feminicide or gender-related killing to add to groups' databases. The stage of "Recording" (chapter 5) relates to the process that data activists use for information extraction—typically moving from the unstructured text of press reports, social media, or personal exchanges into structured datasets—the verification and management of those data, the classification of cases according to diverse typologies, and the management of data. While some groups use the laws in their country to determine if a case is feminicide, others deliberately count feminicide using a more expansive conception of the phenomenon.

"Refusing and Using Data" (chapter 6) is the final stage of a restorative/transformative data science project about feminicide and refers to where counterdata go and who uses them and toward what ends. This chapter surfaces five major ways that activist data are used to contest feminicide. Activists that engage in *repair* work use their data to provide direct support and services to relatives and communities who have lost beloved members. There are a variety of ways that counterdata production serves activist efforts to *remember* and memorialize killed people. Many activists are focused on narrative change efforts to *reframe* feminicide. They use data to communicate about lives lost on social media or through data journalism. Counterdata are also used to push to *reform* existing institutional practices around feminicide, and there are examples of collaboration and communication between government and activist groups around the production of feminicide data. In these cases, counterdata and official data begin to mix in fascinating and sometimes uneasy ways. Finally, there are efforts to *revolt*—to use counterdata to support large-scale mobilizations, usually in conjunction with social movements tied to specific political demands.

Each of the chapters in part II focuses on the grassroots data practices of feminist, women, and queer activists to produce and circulate data about feminicide and gender-related killing. These practices, and the goals and values and relations that motivate them, are often radically different from those of hegemonic data science. Thus, at the end of each of these chapters, I try to draw out key lessons for the development of restorative/transformative data science: how more people and groups may undertake data science grounded in the *restoration* of life, living, vitality, rights, and dignity and the *transformation* of the structural conditions that produce inequality in the first place. This is a contribution to urgent contemporary conversations about data and AI ethics: how we can craft alternative epistemological approaches to data science that are grounded in healing and liberation.

In the third and final section of the book, Action-Reflection, I model some ways to carry forward and apply what our research team learned from our interviews with data activists. In chapter 7, I present a case study about how our Data Against Feminicide team has used these findings along with the principles of data feminism to undertake participatory design of digital tools and machine learning systems together with data activists. We have developed and piloted two tools in four languages with the goal of supporting and sustaining data activism across the Americas and beyond. This chapter describes our design process and how we used the principles of data feminism as guideposts, putting our work in dialogue with other justice-oriented work in human–computer interaction (HCI) and participatory design (PD). It also surfaces key unresolved tensions that principles alone are insufficient to address and details how we navigated some of the intersectional shortcomings of our first designs.

The final chapter in this section and in the book provides a toolkit for restorative/ transformative data science (chapter 8). While this book is focused on data activism about feminicide, there are many other domains that are mobilizing community-produced data in order to work toward social justice. These include monitoring other forms of human rights violations such as police killings or LGBTQ+ hate crimes, tracking and contesting evictions, monitoring government spending, logging voting rights violations, auditing monuments, and tabulating extractive environmental harms, to mention just a few. From our in-depth interviews with feminicide data activists, it was clear that while some aspects of their work are unique to the topic of gender-related violence, other aspects have relevance in other domains—for example, the concept of triangulating information across multiple formal and informal sources in order to verify cases, the emotional labor involved in cataloging stories of human trauma, and the positionality of the data producer to the communities about whom data are produced. This is to say that people who are interested in using data for social justice can learn a great deal from the process and labor of feminicide data activists. Thus, chapter 8 is geared toward practitioners and researchers who aspire to start or join restorative/transformative data science projects.

Now that I've told you what the book will do, I want to state various worthy endeavors that are outside its scope and point you to the people who have already done excellent work in that space. I also want to detail some explicit political decisions I have made about how I will enter this topic and be in relation to the already-existing communities that work in this space. First, from a geographic perspective, this book has a transnational focus primarily on the grassroots data activists and data journalists challenging feminicide in the Americas. It is not an in-depth focus on feminicide in one country or place and will not explain specific country histories, government policies, and cultural attitudes that lead to gender-related violence there. There are excellent place-based works in this vein, including Lorena Fuentes's dissertation on femicide in Guatemala, Julia E. Monárrez Fragoso's work on Ciudad Juárez, Lagarde y de los Rios's monumental work on feminicide in Mexico, and so many more.[42] It is also not a study about the use of data, statistics, and quantification by international institutions like the UN to compare violence on a global scale or to quantify mass human rights violations. For this, I refer the reader to sociologist Sally Merry's excellent book *The Seductions of Quantification: Measuring Human Rights, Gender Violence and Sex Trafficking*, as well as the anthropological study *Who Counts? The Mathematics of Life and Death after Genocide* by Dianne M. Nelson about the role of numbers in post–civil war Guatemala, and the collaborative book from Sandra Walklate et al. about intimate femicide, *Towards a Global Femicide Index: Counting the Costs*, among other resources.[43]

The object of study, reflection, design, and discussion in the present book is transnational, grassroots data activism about feminicide. As you travel through this book you will meet feminist collectives, data journalists, academics, nonprofits, artists, and others working on the issue and learn about how they do data science and work with information. This point of view is purposefully situated. I will not be placing myself or you, the reader, in the god trick position of looking down from above to try to determine the "best" international policy intervention or the universal data standard that captures feminicide in a data schema, or to determine what "we" (the unmarked "we" of academia, typically signifying a white, liberal, academic, elite, Global North establishment) should be doing to address feminicide. This is an intentional political choice because grassroots, women-led, Indigenous-led, feminist-led, Black-led, Global South–led, queer-led groups are sidelined in many international conversations about data and gender-related violence. As Merry puts it regarding measurement of gender violence, "There is typically little opportunity for input from those who are being measured."[44]

Likewise, the book is not an assessment of the "success" or "failure" of feminicide data activism as a method of social change. As I have begun speaking about this work publicly, some more policy-minded listeners have pressed me to detail cases of concrete policy change so that we can pass judgment on whether data activism is "successful." I found this surprising because I do not consider the purpose of this work to evaluate the effectiveness of activist practices. But I began to see these requests as related to how academics typically enter a topic—as critics, as technical experts, and as evaluators. I do not enter in these roles. For my part, I enter as a documenter, a supporter, an organizer, and a person deeply touched and inspired by the tremendous labor of feminicide data activists. I enter in the middle, not from above, and I am firmly aligned with the political goals of data activists, despite the fact that this may produce some discomfort and definitely requires ongoing critical reflection. This is what Firuzeh Shokooh Valle calls "solidarity as a method."[45]

These cases of policy change that people are requesting do exist, and you will learn about some of them in this book (see chapter 2), but this is also a narrow and limiting frame that misses the full scope and goals of feminicide data activism. While many groups monitor feminicide in order to reform policy and hold the state accountable for its failure to prevent gender violence, this book does not presume that policy change is the only goal of activists, nor that policy change is a metric of success to be judged by. In fact, data activists working within the framework of Indigenous sovereignty actively reject such a positioning of the work with the settler state at the center. Counting feminicide and gender-related killing has value because many diverse human rights defenders are doing it, full stop. They count for diverse reasons and use their data in diverse acts of refusal of the gendered status quo. These acts are valuable whether or not they can be told as the type of stories that academia likes: cause and effect stories, linear stories, measurable outcome stories, progress stories, or even those delicious "everything is bad and there is no hope" stories. Data activists' informatic acts and practices are valuable in and of themselves. The smallest and most intimate acts with information are sometimes the most monumental.

Finally, while the book is focused on the work of activists and human rights defenders, it is important to acknowledge the humanity of people who were killed and the true grief and outrage and loss experienced by the communities they leave behind. I purposefully do not focus directly on the stories of killed people and their families in this book because I strongly believe that these are not my stories to tell. Too often, stories are used in ways that can perpetuate the violence they represent—to sell sensationalism, misogyny, dehumanization, and misery porn. For an excellent and thoughtful exploration of the ethics of narrative and artistic representations of feminicide, I refer the reader to Alice Driver's book *More or Less Dead: Feminicide, Haunting, and the Ethics of Representation in Mexico*.[46] And Amber Dean explores various strategies of artistic commemoration that have given visibility, sometimes problematically, to the missing and murdered Indigenous women in Vancouver's Downtown Eastside in the book *Remembering Vancouver's Disappeared Women: Settler Colonialism and the Difficulty of Inheritance*.[47] We will return to some of these works when we discuss how feminicide data circulate publicly and politically and emotionally in chapter 6.

CONCLUSION

My hope is that this book can work to link data activists to each other and share grassroots data science and communication practices. I also see it as a first step toward the study and practice of restorative and transformative uses of data science and technology design, and specifically toward those uses that support the creation of intersectional, antiracist, antiableist, decolonial, abolitionist, Black, Indigenous, queer, trans, and feminist futures. The essential tension of counting and producing counterdata about feminicide is that we—and here *we* means myself and my collaborators, you the reader, and the activist groups and individuals that I profile in this book—seek nothing less than the complete eradication of the phenomenon we are counting. We want a world where no one counts gender-related killings because our political economy ceases to produce gender-related violence as an inevitable and completely predictable outcome of its white supremacist, extractivist, patriarchal, settler colonial economic logic. That is the world to fight for: the world where counting is not necessary and serves no purpose. We will hold this tension throughout.

1 A SHORT GENEALOGY OF FEMINICIDE AND DATA ACTIVISM

On May 11, 2015, Argentine journalist Marcela Ojeda tweeted in frustration and outrage at the news of two separate murders from different parts of the country. Suhene Carvalhaes Muñoz and Chiara Paez, both young women, were beaten to death by their partners.[1] Ojeda tweeted: "Women actors, politicians, artists, business women, social activists . . . women, all of us . . . are we not going to raise our voices? THEY ARE KILLING US."[2] The call to action circulated through her network and far beyond and led to a massive uprising in the streets of Buenos Aires and across the country. Hundreds of thousands of people turned out at the seat of Argentina's national government and in more than 120 other cities in the country in early June 2015 (figure 1.1).[3] They chanted and raised signs and placards that said "#NiUnaMenos" (not one less [woman]). The hashtag and the protests in public spaces traveled virally across Argentina, Latin America, and the world. In ensuing years, public demonstrations spread rapidly to Uruguay, Chile, Mexico, Perú, El Salvador, Paraguay, Bolivia, Brazil, Spain, and beyond, and were increasingly linked to gender rights issues such as street harassment, the wage gap, the legalization of abortion, and *machista* (sexist) culture. Recognizing murder as the culmination of many forms of gendered violence, the movement also took up the slogan #VivasNosQueremos, which translates to "we (women) want to stay alive" or "we (women) love being alive."

These massive mobilizations were first and foremost about challenging feminicide. That the protests traveled across Latin America and beyond echoes the fact that the concepts of femicide and feminicide have a long history of crossing South-North borders, undergoing translation, transmutation, and adaptation to local feminist contexts.

The first documented use of the term *femicide* in English occurred in an 1801 book and referred to the killing of a woman. Linked to gender oppression, *femicide* was first used

FIGURE 1.1

Ni Una Menos mobilization in Buenos Aires in 2015. Courtesy of Fabian Marelli and La Nacion/
Argentina.

by the South African activist and scholar Diana Russell in testimony to the International
Tribunal on Crimes against Women in Brussels, Belgium, in 1976, to assert that the
homicides of women and girls took a distinctly misogynistic form.[4] In her subsequent
coedited book with British scholar Jill Radford in 1992, the two authors provided a
more explicit definition of femicide as "the misogynist killing of women by men."[5] That
is to say, femicide constituted a crime motivated by the victim's subordinated gender
status as a woman (or girl). Because of that, these crimes took distinctly different forms
than the murders of men (or boys).[6] Men are not frequently violated and killed in their
homes, for example, and men's bodies are not typically desecrated in brutal and sexual-
ized ways. Included within Radford and Russell's formulation of femicide were explicitly
misogynist acts of murder against women and girls such as honor killings and female
infanticide, as well as more widespread but equally extreme forms of violence such as
the killing of a woman by her intimate partner. Femicide as a concept seeks to reframe
fatal violence against women from a "private" matter to a structural phenomenon.

    This chapter provides some background on feminicide and data activism as a base for
reading the rest of the book. It traces a short genealogy of the terms *femicide* and *feminicide* and describes other activist, feminist, Black feminist, and Indigenous framings of

structural violence against women. *Data activism*—understood as the production and circulation of data for the purposes of social change—has long been integral to challenging gender-related violence. This includes data activism that preceded the widespread use of digital technologies, when records were kept as paper files, newspaper clippings, and ledgers. In the years following the Ni Una Menos uprising, there has been a surge in the practices of feminicide data activism, as well as growing scholarly attention on data activism.

FEMICIDE TO FEMINICIDIO TO FEMINICIDE

Just as Radford and Russell's book about femicide was published, the North American Free Trade Agreement (NAFTA) was signed and the now infamous disappearances and murders of women in the US-Mexico border town of Ciudad Juárez started to become publicly known. From 1994 until 2001, the homicide rate in Juárez increased by 300 percent for men, while it increased 600 percent for women.[7] Esther Chávez Cano, resident of Juárez and an accountant by training, was one of the earliest people to document the killings in Juárez by physically clipping articles and obituaries from newspapers and saving them in physical files. In 1993, she cofounded the Grupo 8 de Marzo women's rights group and, in 1999, went on to found Casa Amiga, the first women's shelter in Ciudad Juárez.[8]

The antifeminicide activist ecosystem in Juárez was robust and involved many actors from diverse perspectives, often working in coalitions. Grupo 8 de Marzo and Casa Amiga came to the topic from feminism and women's rights. Other groups, like Nuestras Hijas de Regreso a Casa (Bring our daughters home), were primarily comprised of family members who were consistently denied access to justice for their daughters' murders. Nuestras Hijas de Regreso a Casa started building on Chávez Cano's documents. They began keeping records, lists of names, and newspaper clippings. They organized protests and lobbied the press and international community to pay attention and take action.[9] Various groups banded together to form a social movement called Ni una más. The coalition took its name from a 1995 phrase written by poet and activist Susana Chávez Castillo, depicted in figure 1.2a, who was from Ciudad Juárez. She wrote, "Ni una mujer menos, ni una muerta más"—Not one less woman, not one dead [woman] more.[10] You can see this slogan in action in figure 1.2b, a poster published by Nuestras Hijas de Regreso a Casa protesting the feminicides in Chihuahua, the Mexican state where Ciudad Juárez is located. Sadly, outrageously, Chávez Castillo herself was the victim of feminicide in 2011 and never lived to see how her words were taken up four years later by the global Ni Una Menos movement.

FIGURE 1.2

(a) A poster protesting feminicides in Juárez published by family-led activist groups dated some-time between 1999–2006. The text reads, "Not one dead [woman] more! Not one less woman in Chihuahua!" Courtesy of the Esther Chávez Cano Collection, New Mexico State University Library, Archives and Special Collections. (b) "Ni una mujer menos, ni una muerta más" are words drawn from the poetry of Susana Chávez Castillo who was from Ciudad Juárez and who was killed in a feminicide in 2011. Courtesy of Zerk, CC BY-SA 3.0, via Wikimedia Commons.

While the murders in Juárez have often been depicted with sensational, "true crime" types of storytelling that paints such violence as the exceptional and pathological work of serial killers, feminist investigations have pointed to the structural conditions that have fomented such gender-related violence. One of the earliest works to try to use data to document the scope of the murders of women in Juárez was the 1999 book *El Silencio Que La Voz de Todas Quiebra* (The silence that the voice of all [women] breaks). The authors did not yet use the terms *femicide* or *feminicide* to describe the problem, but they did scour official data and news articles to undertake a statistical analysis of deaths and disappearances of women in Juárez between 1993 and 1998. The writers assert that the problem is the public culture of "silence, self-censorship, complicity, and negligence" that permits the murder of women and girls.[11] This climate invisibilizes and normalizes gender-related violence: it is not seen, it is not named, it is simply part of the fabric of everyday life.

Mexican sociologist Julia E. Monárrez Fragoso, from the Colegio de la Frontera Norte in Ciudad Juárez, was the first academic to theorize and develop the concept of feminicidio in relation to the murders in her city.[12] Drawing from the work of Radford and Russell, Monárrez Fragoso expanded the idea to mean not only the gender subordination of women by men, but also the role of institutions such as the state and the church in creating a climate of impunity.[13] Monárrez Fragoso also pointed to the intersection of the Juárez murders with neoliberal economic policy, the rise of the *maquiladoras* and the feminization of their labor force, the migration of rural women, the ongoing presence of intimate partner violence (still accounting for at least 30 percent of murders in Juárez between 1993 and 2007), and the predominating culture of machismo and subordination.[14] Building on these connections to culture, Mexican scholar Marcela Lagarde y de los Ríos wrote, "Feminicidal violence flourishes under the hegemony of a patriarchal culture that legitimates despotism, authoritarianism, and the cruel, sexist—macho, misogynist, homophobic and lesbophobic—treatment reinforced by classism, racism, xenophobia and other forms of discrimination."[15] Indeed, Mexico was no stranger to a climate of impunity around violence. In the Mexican Dirty War, a Cold War–era regime of state violence supported by the United States, thousands of citizens, young people, and students were disappeared and executed by the state.

Lagarde y de los Ríos, an anthropologist by training, took up the issue of feminicide when she ran for a seat in Mexico's federal legislature in 2003. She was elected and then proceeded to undertake one of the most thorough government-sponsored country-level studies of feminicide to date, a fourteen-volume report issued in 2006 that eventually led to the codification of feminicide in Mexican law in 2012. In the process of her legislative work, Lagarde y de los Ríos built on the significant theoretical shifts introduced by Monárrez Fragoso in which *feminicidio* means both the killing of a woman or girl for gender-related reasons and also the linking of those killings to human rights violations and to the climate of impunity created by state inaction. Lagarde y de los Ríos thus framed feminicide as a crime of the state, an assertion that has been upheld in international courts of law.[16] This is to say that the state was systematically failing to ensure the most basic human right for more than half of its citizens.

This formulation has laid the groundwork for drawing these murders out of the private and interpersonal sphere of life (typified by misogynist media framings, which often depict them as "crimes of passion") and demanding public action and public accountability for the widespread, systematic discrimination they represent. Whether a gender-related killing happens as part of domestic violence in the home perpetrated by an intimate partner or from a sexual assault perpetrated by narcotraffickers in public space, Lagarde y de los Ríos would say that both constitute *feminicidal violence*. Both crimes violate women's human rights because they are gender-related crimes; both are made possible because of state negligence and because of widespread, systematic discrimination, including gender inequality and unequal access to economic opportunity.

Monárrez Fragoso's and Lagarde y de los Ríos's contributions also laid the groundwork for more intersectional elaborations of the concept of *feminicidio* and for its adaptations into contexts outside of Juárez and Mexico. Countries such as Bolivia, Paraguay, and Colombia adopted legislation defining and criminalizing feminicidio in the 2010s. Yet the theoretical shift around feminicidio—as a crime that implicates the state—is not a settled matter. Before and during this work, other Latin American scholars such as Montserrat Sagot Rodríguez and Ana Carcedo Cabañas had translated *femicide* as *femicidio* and undertook important investigations of violence against women in the context of Central America. *Femicidio* is also used by activists and state officials in other Latin American countries such as Venezuela and Argentina, and many countries passed laws codifying femicidio (see table 1.1 for a full list).[17]

In the past decade, activists, journalists, and academics based in the United States and Canada have taken note of the work on femicide and feminicide by their Latin American counterparts and traveled some of these concepts back into the English language. For example, in 2010, Rosa-Linda Fregoso and Cynthia Bejarano published an important edited volume called *Terrorizing Women: Feminicide in the Americas*, which summarizes some of the Latin American theoretical innovations and brings them to an English-speaking audience. Fregoso and Bejarano land on *feminicide* as a term to mean the murders of women and girls founded on a gender power structure and intersecting with racism and economic injustice. Other scholars writing in a North American context have worked to further the understanding of femicide and feminicide back into English.[18] Canadian scholar Paulina García-Del Moral has written about *feminicidio* as a resonant frame for transnational and local activism, as well as argued for a more decolonial and intersectional conception of femicide in the interest of understanding the violence that Indigenous women in Canada experience.[19] Significant transnational exchanges on feminicide have been led by Indigenous women. For example, in 2020, the Enlace Continental de Mujeres Indígenas de las Américas organization (ECMIA; Continental Network of Indigenous Women of the Americas) organized multiple sessions at their annual gathering about the feminicide of Indigenous women, including one on data production and one on legal frameworks.[20] These were led by activists from North and South America, exchanging diverse conceptions of gendered colonial violence as well as data practices and legal strategies to challenge it.

The rising tide of global feminist activism sparked by Ni Una Menos in 2015 has also helped Latin American conceptions of femicide and feminicide, as well as Latin American feminisms more broadly, to travel into the English-speaking world. While Ni Una Menos may have appeared to some people to arise suddenly, it was built on Argentina's long history of human rights and feminist organizing. This includes the prominence of the Abuelas and Madres de la Plaza de Mayo, who, since 1977, have silently protested the disappearance of their relatives in Argentina's Dirty War. It includes the rising tide of the Buen Vivir movement (1990s–present) linking gender rights to Indigenous rights and led by Indigenous groups such as Movimiento de Mujeres Indígenas por el Buen Vivir, as well as three decades of national *encuentros*—annual conferences and organizing events focused on convening women, trans people, nonbinary people, and *travestis*.[21] The transnational spread throughout the Latin American region was bolstered by decades of continent-wide organizing like Encuentros Feministas de América Latina y Caribe (EFLAC), Encuentros Lésbicos-Feministas de América Latina y Caribe (ELFLAC), las Cumbres Continentales de Mujeres Indígenas de Abya Yala, and more. As Maria Florencia Alcaraz writes, there was "fertile ground" for Ni Una Menos, which consisted of many social relations—physical and digital—and relational infrastructure laid over decades.[22] This powerful transnational infrastructure is what Gago and Gutiérrez Aguilar are referring to in their essay "Women Rising in Defense of Life." They write, "We are part of the hope for an internationalized insubordination-in-action."[23] While such a phenomenon might appear as a sudden flash, this is a case where decades of popular and community-based feminist organizing from the South laid the groundwork for global, ongoing impact.

Ni Una Menos was also responsible for placing calls for better data and information at the center of political debates about feminicide. One of the five main demands of Ni Una Menos in Argentina in 2015 was to "create a single Official Registry of victims of violence against women. Produce official and updated statistics on femicides. We can only design effective public policies by understanding the scope of the problem."[24] This demand was only partially met in the form of the National Registry of Femicide from the Argentina Justice System, created in response to movement demands in 2015, but activists criticize it for the length of time it takes to publish information, the lack of inclusion of trans women and travestis, and the duplication of efforts across the executive and judiciary branches of government. Moreover, the Ministry of Women has struggled to retain public servants who are charged with implementing more comprehensive data collection of gender-related violence.[25] As we will see, the most complete national registries of feminicide in Argentina, as elsewhere, continue to be produced by civil society and not by governments.

#SAYHERNAME, #MMIWG2, AND FEMINICIDE ACTIVISM IN NORTH AMERICA

In the United States and Canada, there are other important framings of gender-related violence that have been and continue to be the subject of intense organizing and data activism. Indigenous and Black women's groups have long been organizing against fatal gender-related violence, as well as linking this violence to multiple forces of structural domination, including patriarchy, white supremacy, settler colonialism, and economic violence (see the glossary in chapter 8 for precise definitions of these terms).

For example, following a spate of murders of Black women and girls in Boston in 1979, the Combahee River Collective organized a march of hundreds of people to memorialize their lives, protest their deaths, and challenge the lack of justice families had received, both from the judicial system and from the media. The collective kept records on the women murdered and published a pamphlet for self- and community defense (figure 1.3). Here we see one of the earlier examples of "counting feminicide." In this marked-up draft with notes, the numbers of Black women killed increase from six, to seven, to eight, and finally eleven. While the numerical precision is important, the authors end the pamphlet by saying that it is not only about these eleven women but also about the "1000s and 1000s of women whose names we don't even know. As Black women who are feminists we are struggling against all racist, sexist, heterosexist and class oppression. We know that we have no hopes of ending this particular crisis and violence against women in our community until we identify *all* of its causes, including sexual oppression."[26] Here, the Combahee River Collective insisted on an intersectional approach to understanding the root causes of gender-related violence. They refused a single-axis analysis that would attribute the violence in their community to *either* sexism *or* racism. Counting was central to this campaign, and yet the

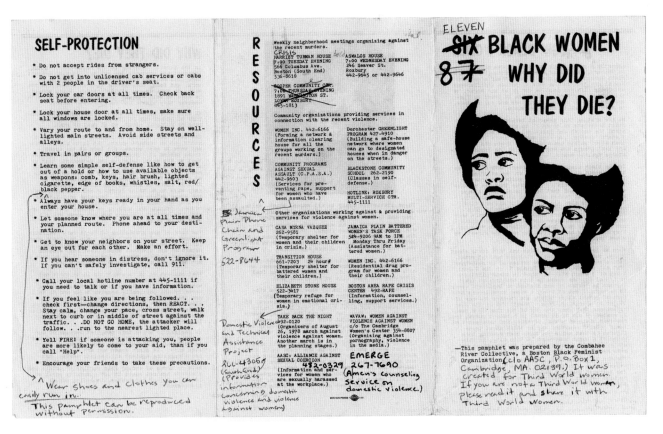

FIGURE 1.3

Combahee River Collective pamphlet, *Eleven Black Women: Why Did They Die?* (1979). The collective circulated more than eighteen thousand copies of the pamphlet in both Spanish and English. Courtesy of the Barbara Smith Collection at the Lesbian Herstory Archives. Pamphlet originally published by Red Sun Press.

goal was not to produce a single comprehensive number. Rather, the pamphlet used counting to open a conversation about the countless women whose names we don't even know.

This early organizing in support of Black women finds an echo in contemporary campaigns like #SayHerName, which seeks to uplift the stories of Black women and gender nonconforming people who are killed in police violence in the United States. Their lives and deaths are often invisibilized by the larger narrative around anti-Black police killings, which focuses on men, as well as pervasive underreporting in the media due to long-standing racism that constructs Black women victims as less worthy.[27]

In her work on Black femicide and the Black Lives Matter movement, political theorist Shatema Threadcraft traces the #SayHerName campaign back to "the United States' most iconic necropolitical warrior, one Ida B. Wells."[28] In the *Red Record*, published in 1895, Wells collected reports of Black lynchings published in the white press to systematically present the scope and scale of these brutal racialized killings and to challenge the white narratives that circulated about them.[29] Here *necropolitics* means mobilizing the politics of death, especially as they intersect with state power—who is targeted for death, who kills, and, most importantly, what their deaths mean.[30] This is a concept from Cameroonian political theorist Achille Mbembe that focuses on how death chances are distributed by the state—on how people from subordinated groups are marked for death and the ways in which their deaths are normalized. Necropolitics, then, involves not only the empirical fact of the disproportionate deaths of specific groups of people but also the performative and discursive element of interpretation around what those disproportionate deaths *mean*. Another way to think about necropolitics is to ask the question, "In a particular society, whose deaths matter?" And, following geographer Lorena Fuentes, what are "the discourses that visibilize, differentiate, and/or obscure the bodies of victims"?[31]

While Threadcraft describes the Black Lives Matter movement's necropolitical achievements in reframing Black police killings as unjust, she outlines tensions for #SayHerName to be able to do the same for Black women: "Activists concerned with stemming Black femicide should reflect on the fact that the movement has relied on amplifying the spectacle of death in a context in which Black women suffer from a severe spectacular violent death deficit. . . . what can stand in the place of this public spectacle, when what they are dealing with are a greater number of wholly private murders. What, then, will motivate people to rally around the bodies of our Black female dead?"[32]

Even studying the phenomenon is challenging because of missing data, underreporting due to media bias, and because Black women are killed in less public and therefore less visible ways than Black men.[33] As Kimberlé Crenshaw and colleagues write, "There is currently no accurate data collection on police killings nationwide, no readily available database compiling a complete list of Black women's lives lost at the hands of police, and no data collection on sexual or other forms of gender- and sexuality-based police violence."[34] In response, advocates like Black Femicide US, Black Girl Tragic, the African American Policy Forum and Andrea Ritchie have created important databases of Black women and gender nonconforming victims of violence that they use in advocacy, support for families, consciousness-raising, and movement building.

MISSING AND MURDERED INDIGENOUS WOMEN, GIRLS, AND TWO-SPIRIT PEOPLE

During the same time period that women's groups were bringing the feminicides in Ciudad Juárez to public attention, activism led by Indigenous women was surging in Downtown Eastside Vancouver, the unceded territories of the xʷməθkʷəy̓əm (Musqueam), Sḵwx̱wú7mesh Úxwumixw (Squamish), and səlilwətaɬ (Tsleil-Waututh) Nations in present-day Canada. Since the late 1980s, Indigenous women had been gathering in the Downtown Eastside to protest missing and murdered women in their community by holding signs with photographs of their relatives. Among them was prominent MMIWG2 activist Mona Woodward of the Cree, Lakota, and Saulteaux peoples. She began demonstrating following multiple relatives being affected by violence, including her sister, Eleanor "Laney" Ewenin, who was murdered in 1982 and whose case has still not been brought to justice.[35]

On February 14—Valentine's Day—1991, a small number of people, including Woodward, gathered to mourn the death of a beloved Coast Salish woman who had been sexually assaulted and murdered. Her mother and other family members staged a small memorial to her life.[36] This gathering became annual and has grown to encompass many thousands of participants in Vancouver as well as spread to more than twenty cities in Canada and the United States. It is known as the Women's Memorial March and its annual theme is Their Spirits Live within Us (figure 1.4). Each year, the march moves slowly through the Downtown Eastside neighborhood and makes stops at sites where Indigenous women have last been seen or where they were killed. The march honors individuals, and family members often speak. But since the beginning, it has also challenged the structural conditions and cultural representations that produce such violence. This has involved keeping a count and publishing a list of disappeared and murdered women in the Downtown Eastside area. More than 970 names have been added to the list since the march started, and seventy-five names were added just in the year 2019.[37] A pamphlet passed out at the 2001 Women's Memorial March read: "WE ARE ABORIGINAL WOMEN. GIVERS OF LIFE. WE ARE MOTHERS, SISTERS, DAUGHTERS, AUNTIES AND GRANDMOTHERS. NOT JUST PROSTITUTES AND DRUG ADDICTS. NOT WELFARE CHEATS. WE STAND ON OUR MOTHER EARTH AND WE DEMAND RESPECT. WE ARE NOT HERE TO BE BEATEN, ABUSED, MURDERED, IGNORED."[38]

The Women's Memorial March, along with many other Indigenous-led efforts, laid the groundwork for the growing movement across North America now known by its hashtags—#MMIW, #MMIWG2, #MMIP—and by its slogan, No More Stolen Sisters.

FIGURE 1.4
The Women's Memorial March, pictured here in 2017, has been held every year on February 14 since 1991 and it has spread to more than twenty cities across the United States and Canada. Courtesy of Vancouver Is Awesome (https://vancouverisawesome.com). Photo by Dan Toulgoet.

Indigenous organizers and scholars have been unequivocal in linking this violence to settler colonialism. Mohawk scholar Audra Simpson states plainly, "Canada requires the death and so called 'disappearance' of Indigenous women in order to secure its sovereignty."[39] Others have challenged the gender binary itself as a settler colonial construction of gender—a toxic ideological import from Europe—that has been wielded as a tool to disrupt Indigenous family relations and plunder Native communities for generations.[40] These scholars definitively establish the links between gender-related violence, the disruption of Indigenous kinship, and the dispossession of Indigenous land.

Contemporary Indigenous feminist scholars like Sarah Deer of the Muscogee (Creek) Nation build on this legacy of linking gender-related violence to colonial violence and speak of the paths to multiple forms of sovereignty—land sovereignty, body sovereignty, and soul sovereignty: "It is impossible to have a truly self-determining nation when its members have been denied self-determination over their own bodies."[41] In parallel, Latin American Indigenous leaders have led the development of the concept of *cuerpo-territorio*. This is an analysis from Lorena Cabnal, Mayan Q'eqchi'-xinka heal-ers, and other Indigenous feminists and community defenders that links violence on the body (*cuerpo*) to violence on the land (*territorio*) because so many Indigenous women are killed in defending their land or protecting their water from destruction by man camps, extractivist industries, and toxic public-private partnerships.[42] The group Coordinadora Nacional de Mujeres Indígenas (National Coalition of Indigenous Women, CONAMI), based in Mexico, calls these deaths *ecofeminicidios*. These concepts show how deeply gender-related violence against Indigenous women is intertwined with *colonial dispossession*—past, present, and ongoing. Feminicide is a kind of final dispossession in which all rights have been stripped and stolen—the rights to land, sovereignty, language, and traditions, along with the very right to life itself.[43] Michi Saagiig Nishnaabeg scholar and artist Leanne Betasamosake Simpson describes how Indigenous bodies are themselves political orders. This means that the body sover-eignty of Indigenous women and two-spirit people is a direct threat to the authority of the settler state and thus becomes a target.[44] In 2023, Indigenous activists and femi-nicide scholars came together to assert that the mainstream formulation of feminicide is characterized by "a failure incorporate a focus on colonialism and the systemic rac-ism/discrimination that shape the experiences of violence of Indigenous women and their peoples."[45] These scholars and activists demonstrate how the gender-essentialist, single-axis frame of Russell and Radford's definition of femicide as "women killed by men" is not expansive enough, nor precise enough, nor historical enough to appropri-ately describe this violence.

Much of what is demonstrated in scholarship is corroborated by reports from national inquiry commissions and advocacy groups, such as the prevalence of gender-related violence against Indigenous women at sites of resource extraction.[46] Still, the crisis persists and official data about MMIWG2 is widely known to be missing, poor quality, fragmented, misclassified, or purposefully shielded from public view by the state. This has prompted a growing number of Indigenous-led activist and civil society efforts to collect the data, such as Sisters in Spirit (Canada), the Safe Passage project (Canada), Sovereign Bodies Institute (North America and beyond), and Emergencia Comunitaria de Género (Mexico). Geographer and cofounder of Sovereign Bodies Insti-tute Annita Lucchesi (Northern Cheyanne) has written that negligent and discrimi-natory institutional practices cause Native women "to disappear not once, but three times—in life, in the media, and in the data."[47] She has always carried with her the date of the first Women's Memorial March: "The march was in February 1991 and I was born in May 1991. There was never a time in my lifetime where my people weren't advocat-ing for my right to live a life free from violence as an Indigenous woman. My right to not be murdered or go missing. As the keeper of an MMIP database now, that means something to me. I'm a part of an intergenerational commitment to ending this crisis that existed before I was born."[48]

KEY TERMS

I have narrated a short genealogy of some key terms that describe and bring to light fatal gender-related violence: femicide, feminicide, MMIWG2, the Say Her Name movement, and Black women killed in police violence. All of these are insurgent necropolitical concepts aiming to reframe fatal gender-related violence, to move it out of the private sphere and place it into a structural, systemic, and public context. In this book, I will use *feminicide* most of the time for two reasons. First, as I stated in the introduction, the center of gravity of this book is feminicide data activism in Latin America. Feminicide is the framing concept that the majority of the data activists and civil society groups that we interviewed and worked with use. Second, as elaborated by Largarde y de los Ríos and built on by Latin American feminist theorists as well as transborder scholarship like that of Fregoso and Bejarano, it is a concept that, in contrast to femicide, can hold more intersectional consideration of how and why the violent killing of women happens and persists. But my choice to use *feminicide* is not to castigate those who use *femicide* or any other framing of gender-related killing as politically incorrect. Grassroots groups and individuals know best what framing concepts to use for their work. When I speak about the data activism of a specific group, I will name the violence as they name it and draw from their framing concepts and motivations to describe the work.

Women in this book is a political category that includes cisgender and transgender women. This is true for the vast majority of activists that we interviewed who produce data about feminicide. This is in contrast to legislation about feminicide in which governments mostly use *women* to refer to cisgender women and exclude transgender women (see table 1.1). When I speak about gender-related violence that includes genders other than women, such as nonbinary people, trans men, two-spirit people, travestis, or others, I will use *feminicide and gender-related killing*.

You may have also noticed that I am using the term *gender-related violence* instead of the more common term *gender-based violence*. This decision comes out of discussions with my collaborators on the Data Against Feminicide project. We decided to use this term to indicate that understanding feminicide requires an intersectional perspective. It is a term that is also favored by Rashida Manjoo, former UN Special Rapporteur on Violence against Women.[49]

Finally, throughout this book I will frequently reference the interlocking systems of power that result in gender-related violence and inequality. These include terms like *settler colonialism*, *cisheteropatriarchy*, *white supremacy*, and *racial capitalism*. If these are new concepts for you, see the glossary in chapter 8, which contains short definitions of each.

LAWS AND OFFICIAL DATA ABOUT FEMINICIDE

Feminicide is gathering energy in national and international law, policy frameworks, and governance. It has increasingly featured in public and policy debates, especially in Latin American countries. In the past two decades, public pressure from feminist and women-led movements has led to the passage of legislation that criminalizes feminicide or femicide in nearly all Latin American countries. Table 1.1 lists all of those countries (and one territory) in the Americas with laws about femicide or feminicide.[50]

Table 1.1 shows that eighteen countries and one territory have passed laws codifying and criminalizing feminicide or femicide since 2007. All are part of Latin America where feminicide activism, advocacy, and public uprising has been strongest. Despite legal advances, feminicide laws differ in definition and scope. In some countries, such as Nicaragua and Argentina, feminicide has been defined to consist of only intimate partner violence, which means that gender-related murder perpetrated by a stalker, a family member, a client, or a stranger would not be covered. Depending on the context, this could be leaving out a large proportion of feminicide cases. Moreover, the vast majority of feminicide laws do not protect the rights of transgender women. Legislation in only four countries and one territory explicitly mentions trans women, gender identity, and/or transfeminicide. This means that fifteen countries adopt an implicit definition of *women* to mean only cisgender women. That said, activists in many countries are challenging this exclusion—in the streets, in the media, and in the courts.[51]

At the bottom of table 1.1 is a list of countries in the Americas that do not have laws about femicide or feminicide. The regional differences are striking, with legal advances on feminicide being led by the Latin American region. Despite extremely high rates of female homicide, countries in the Caribbean and Northern America largely do not have legal definitions, frameworks, and protections in place for one of the most basic human rights: the right to life, liberty, and security of person.[52] In the United States, murder represents a leading cause of death for women under forty-five, higher than diabetes, stroke and heart disease across races.[53] Homicide is the second leading cause of death for Black women and girls under twenty, responsible for more than 15 percent

Table 1.1

List of countries and territories in the Americas that have laws about femicide or feminicide

Country	Femicide/ Feminicide	Year	Covers more than intimate partner violence?	Includes trans women?	Government publishes official data in some form?
Argentina	Femicide	2012	N	Y	Y
Bolivia	Feminicide	2013	Y	N	N
Brazil	Feminicide	2015	Y	N	N
Chile	Femicide	2010	Y	Y	Y
Colombia	Feminicide	2015	Y	Y	Y
Costa Rica	Femicide	2007	Y	N	Y
Dominican Republic	Feminicide	2014	N	N	Y
Ecuador	Femicide	2014	Y	N	PARTIAL
El Salvador	Feminicide	2011	Y	N	Y
Guatemala	Femicide	2008	Y	N	Y
Honduras	Femicide	2013	Y	N	N
Mexico	Feminicide	2012	Y	N	PARTIAL
Nicaragua	Femicide	2012	N	N	N
Paraguay	Feminicide	2016	Y	N	Y
Peru	Feminicide	2013	Y	N	Y
Panama	Femicide	2013	Y	N	Y
Puerto Rico	Feminicide	2021	Y	Y	PARTIAL
Uruguay	Femicide	2017	Y	Y	PARTIAL
Venezuela	Femicide	2014	Y	N	N

Countries in the Americas with *no legislation* about femicide or feminicide include:

- **Northern America:** Canada, Greenland, and the United States
- **Central America:** Belize
- **South America:** Guyana and Suriname
- **Caribbean:** Cuba, Haiti, Jamaica, Trinidad and Tobago, Bahamas, Barbados, Saint Lucia, Grenada, Dominica, Saint Kitts and Nevis, St. Vincent and Grenadines, Antigua, and Barbuda

Source: Reprinted with permission of the authors. For detailed notes about each entry in the table, consult the original publication.

of deaths in 2018.[54] Asian women who were spa workers were targeted in a mass killing in Atlanta in 2021. White, rich, and/or elite women are not exempt, and their cases often gain more media attention. While the United States and Canada do not have laws, they have begun to strengthen legal protections for related phenomena like intimate partner violence and MMIWG2. In the United States, the most expansive federal law is the Violence Against Women Act (VAWA), which outlines provisions for services and programs to prevent various forms of gender-related violence but stops short of defining feminicide as a separate crime or the murder of women as a violation of their civil rights. In fact, this is one of the main challenges of legal reform: writing legislation that recognizes, defines, and provides redress for feminicide as a structural, public problem rather than an interpersonal, private problem; a violation of civil rights and human rights, not (only) a personal dispute or "domestic" or "intimate" matter.

Many of these legal reform efforts include mandates for improved collection of data and publication of statistics. Often these are due to demands from advocates specifically requesting such provisions, as we saw in the case of Ni Una Menos in Argentina in 2015. For example, a 2021 law passed in Puerto Rico ordered the systematic collection of data about feminicide and transfeminicide, stating: "If there are no reliable and comparable data collection mechanisms for a certain type of crime, there will be no appropriate ways to understand it nor effective strategies to combat it."[55] All recent laws regarding missing and murdered Indigenous people in the United States have provided extensive provisions for improving data collection and statistics, though the implementation leaves much to be desired from the standpoint of grassroots groups, advocates, and families.[56]

But legal mandates get complicated when they run up against funding challenges like the neoliberal austerity measures increasingly being implemented in Latin America or bureaucracy challenges like the fragmentation of agencies, each addressing a part of the issue or collecting part of the data with no agency fully responsible for data production. For example, despite numerous provisions for data collection around MMIP, federal legislation like Savanna's Act in the United States is out of compliance with its own directives and criticized by the Indigenous families and communities whom it was meant to serve.[57] Research by the Iniciativa Latinoamericana por los Datos Abiertos (ILDA; Latin American Initiative for Open Data) in Argentina that explored government data collection around feminicide corroborates this: "The strong fragmentation of this space in institutional, legal and occasionally political terms, makes it difficult to coordinate who is responsible for reporting and in what way on this particular type of crime."[58] In addition, there is almost universally a lack of training and knowledge about the gendered and racialized nature of violence for public sector employees who handle and classify cases on the ground, such as police and medical examiners. MMIWG2 reports highlight the persistent racial misclassification of Indigenous women by the state.[59] Moreover, there is the added informational complexity of what needs to be known in order to determine whether a murder was motivated by gender—namely, the relationship between victim and perpetrator, the motive for the murder, and the types of prior violence that the person may have suffered. In a 2022 analysis and report called *Datos Para la Vida* (Data for life), the Mexican nonprofit Data Cívica found that authorities did not properly record data fields that they were required to collect by law in order to determine whether a violent death constituted a feminicide. The report found that the government also failed to record variables that would help disaggregate violence faced by women on the basis of racism, ableism, heteronormativity, and other intersectional power dynamics.[60] The fact that these data prove so challenging for agencies to accurately record, adding a time and resource burden to each case, has led the UN and some government officials to suggest that the violent death of a woman should be considered *by default* to be motivated by gender, thus shifting the burden of proof onto officials to prove otherwise.[61] For example, in August 2020, Alejandro Gertz Manero, attorney general of Mexico, recommended that all murders of women be investigated as feminicides.[62] However, there is the distance between what authorities say and what actually happens on the ground. Sociologist Mariana Mora studies feminicide in Costa Rica and states that the official party line is that state investigators should assume that every killing of a woman is a feminicide and later rule it out, but this is not actually followed in practice.

As nations grapple with these difficulties, regional and international efforts have stepped in to try to help provide both data standardization and technical guidance to governments (and, in some cases, such as Mexico's, to provide international accountability in the face of government inaction[63]). For the past four decades, transnational networks of feminist groups, largely led by Latin American women, international NGOs, and supranational agencies like the World Health Organization (WHO) and United Nations (UN) have held conferences and released reports on the topic of femicide/feminicide. These have framed the topic as a global issue of gender inequality and invariably highlight the lack of reliable, comparable data. Surveying this work, sociologist Sandra Walklate and colleagues have posited the potential value and challenges of working toward a *global femicide index*, a set of standard methods to count, map, and measure the prevalence of femicide across countries.[64] In 2015, Dubravka Šimonović, UN Special Rapporteur on Violence against Women, called for establishing a *femicide watch*—an observatory effort to count, collect, and monitor data about gender-related killings—in every country.[65]

In 2014, amid this intense and growing international coordination on feminicide, the UN published an important technical guidance document about feminicide. Titled *Latin American Model Protocol for the Investigation of Gender-Related Killings of Women (Femicide/Feminicide)*—and from here on out I'll just call it the Latin American model protocol—it outlines a baseline set of definitions, data fields, and procedures for official government data collection about feminicide for use by, among others, police, medical examiners, forensic experts, public sector data analysts, and social workers.[66] It is called the Latin American model protocol because the UN developed it mainly with professionals in the Latin American region.

Some key contributions of this document include a working definition of femicide/feminicide as gender-related killings that are rooted in the structural subordination of women. It also elaborates two key categories of femicide (here I use *femicide* because this is what the report primarily uses). *Active or direct femicides* include intimate partner femicides, misogynist murders, so-called honor killings, female infanticide, hate crimes against lesbians, and more. *Passive or indirect femicides* include deaths from unsafe abortions, maternal mortality, deaths linked to organized crime, and deaths linked to negligence. In the latter category, the state is implicated due to its role in creating an unsafe environment for women that diminishes their life chances, reduces their reproductive autonomy, and creates a climate of impunity where women are disproportionately killed and justice is not served. The Latin American model protocol also outlines a typology of fifteen kinds of femicide that the authors found to be common in the Latin American experience. The categories of femicide in the Latin American model protocol have been influential for federal governments and, as we will see in chapter 5, for data activists as well. But coming in at two hundred pages of technical documentation, the protocol is dense and there are numerous institutional barriers to governments just jumping in and adopting it.

As public conversation around feminicide in Latin America surged following the Ni Una Menos uprising, ILDA started investigating how they could contribute to addressing some of the institutional barriers from a public sector data perspective and from a Latin American regional perspective. ILDA often works with government officials on open data and information systems and sought to understand whether a common femicide data standard would be useful to governments and what such a standard might involve. In 2017, they developed a project around femicide data standardization that sought to analyze how femicide data are constructed in different countries, what variables are included, and how open the data are. From this work, they released a preliminary data standard that described a baseline for how public agencies can register femicide cases.[67]

ILDA piloted the data standard in workshops with federal and provincial officials in Argentina and Uruguay and, drawing on their feedback, drafted a fifteen-page guide that builds on the Latin American model protocol with a more specific protocol for standardizing the registration of femicides for public agencies.[68] The ILDA guide has flowcharts for data collection and case registration processes, plus a series of sixty-six variables recommended to be collected about each case. ILDA recommends that government agencies collect key information that, following legal definitions, will help determine whether a murder constitutes femicide, such as the relationship between perpetrator and victim, whether there was sexual aggression, and whether there were prior complaints filed against the perpetrator. As ILDA's work progresses, they publish updates to their guide and recommended registration processes, such as the flowchart in figure 1.5 that serves as a visual tool to help identify a crime as femicide and advises adhering to a quarterly and annual schedule for reviewing cases. Shifting processes and protocols in one country—not to mention a whole region—is slow work, but former research director Silvana Fumega was cautiously optimistic: "We're seeing some governments taking what we put out as an ideal structure and comparing to what extent they have that in place or not, which is a good sign that at least it's being taken as a reference. We know that what we're suggesting is an ideal."[69]

Despite legal advances in Latin America, technical guidance, and persistent political pressure on feminicide, governments are still unable to produce reliable public data about the phenomenon. In the final column of table 1.1, our research team attempted to catalog which countries with legislation on feminicide publish any form of official data about the phenomena. We did not judge the completeness of the data, just whether or not the government publishes anything. Even so, only around half of the countries represented publish any official data about feminicide. In Mexico, sociologist Julia Monárrez Fragoso notes, "national statistics do not document the reason for the murder, the relationship between victim and victimizer, nor the various types of violences that the women suffered prior to being murdered. [. . .] In the face of such absences, it is necessary to find alternative means to understand femicides with greater precision."[70] And one of the main findings in Canada's multiyear MMIW national inquiry is that "there is no reliable estimate of the numbers of missing and murdered Indigenous women, girls, and 2SLGBTQQIA persons in Canada."[71]

DATA ACTIVISM AND FEMINICIDE

Indeed, the vast majority of scholarly works, commissions, and advocacy reports about feminicide lead with the fact that data about this violence are either completely absent

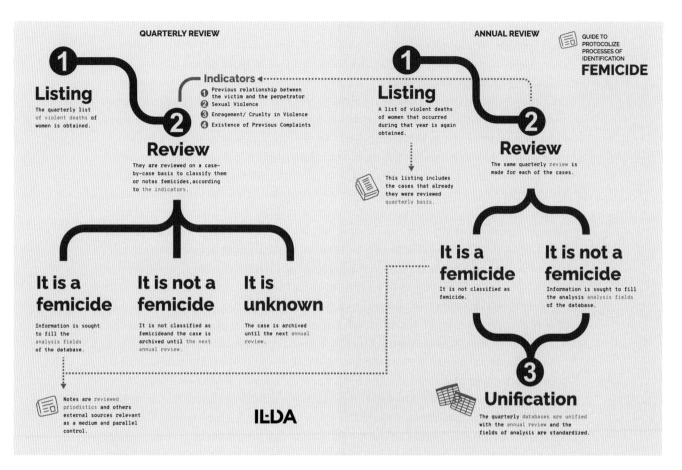

FIGURE 1.5

ILDA's flowchart to identify femicides. They propose that organizations collecting data undertake a quarterly review and an annual review of all cases and then unify the results. Courtesy of ILDA. Graphic design by DataSketch. Graphic adaptation for this book by Wonyoung So.

or else sparse, unreliable, unavailable, untimely, neglected, misclassified, or not public. This is a phenomenon called *missing data* that we will explore further in the next chapter. An increasing number of grassroots activists and civil society groups, particularly in Latin America, have turned to the production and circulation of *counterdata* not only to fill in missing data or to counter official data, but also to challenge the hegemonic systems that produce gender-related violence in the first place: cisheteropatriarchy, white supremacy, colonialism, and more. They use their data to confront state bias and inaction, to galvanize media and public attention, and to help heal wounded communities.

This work fits under the broader umbrella of *data activism*, a concept that foregrounds the use of data and software to pursue collective action and exercise political agency. Critical data studies scholar Stefania Milan characterizes data activism as an emerging social movement tactic and draws parallels between data activism and media activism. In both cases, citizens and people are taking advantage of widely available digital tools of production to "uncover stories of injustice or change."[72] In Milan's framework, codeveloped with scholars Lonneke van der Velden and Miren Gutiérrez, there are two types of data activism: reactive and proactive.[73] *Reactive data activists* resist datafication and data surveillance practices. *Proactive data activism*, on the other hand, may have any number of objects of struggle. What proactive data activists share is that they "create, mobilize, solicit, appropriate, or crunch data in view of supporting alternative narratives of the social reality, questioning the truthfulness of other representations, denouncing injustice and advocating for change."[74] Grassroots counterdata production about feminicide falls into the latter category and undertakes all of these actions with data, with a particular emphasis on the act of *creating* datasets and databases that document cases of feminicide. These datasets are often manually crafted (row by row, case by case, person by person) and constitute data produced outside of—and often in opposition to—mainstream counting institutions like governments and corporations.

Indeed, prior work has shown that one of the key goals of data activism is to challenge whether and how institutions see, count, measure, and evaluate particular phenomena. The 2016 report *Changing What Counts* describes numerous case studies of citizen data action that attempt to shift official institutional measurement of various things: police killings, air pollution, water access, and government pardons.[75] Civil society actors in these cases produce their own data as well as collate and combine data from diverse sources in order to challenge public sector numbers and also to challenge public sector *narratives* around these phenomena. That is to say that datasets do not operate as a bucket of disembodied facts but are in fact mobilized by activists as "a vector for the circulation of affective and emotional bonds."[76] Scholars Kathleen Pine and Max Liboiron (Red River Métis/Michif) have discussed data activism as the deployment of *charismatic data*, data whose dramatic or spectacular nature impels different stakeholders to take action toward social change.[77] This means that datasets are—or can be strategically deployed as—acts of rhetorical and political communication. The act of registering diverse instances of gender-related killings into a dataset about feminicide is an act of counting *and* an act of classification *and* a rhetorical act to assert feminicide as a valid concept and a valuable thing to count. Sociologists Aryn Martin and Michael Lynch frame this as *numeropolitics* and discuss how acts of counting are also always acts of classification. To count feminicide is also to select events from the world and to assert them as feminicide and also to assert that feminicide *matters* as a concept. As Pine and Liboiron state succinctly, "measurements make things."[78]

Recent work has examined the intersection of data activism and grassroots data practices around feminicide. In *Terrorizing Women*, Rosa-Linda Fregoso and Cynthia Bejarano place feminicide counterdata production practices in Latin America in a long history of observatorios comunitarios (community-based observatories), which have served to engage communities in monitoring state violence and using collective action strategies toward accountability and justice.[79] Helena Suárez Val—interdisciplinary scholar, activist, and colead on the Data Against Feminicide project—has advanced the concept of *strategic datafication* to describe the motivations behind activist production of feminicide data. Activists strategically mobilize the perceived legitimacy of data and numbers to draw attention to feminicide at the same time as they refuse the positivist epistemology of hegemonic data science; they refuse the colonial and military histories of their tools; and they refuse to have women's and people's lives and bodies be reduced into rows and columns.[80]

In a 2019 case study on #NiUnaMenos, Jean-Marie Chenou and Carolina Cepeda-Másmela described how activist demands for data constitute both an appropriation of hegemonic technology and the production of alternative imaginaries around big data—namely, that data could and should be used in the service of gender justice.[81] In studying how feminicide data circulate on social media, Helena has asserted them as *affect amplifiers*, digital cartographies that seek to translate feminist grief and rage into public action.[82] This resonates with Lucchesi's scholarship around using an Indigenous decolonial approach to mapping MMIWG2, in which maps and stories are used in the service of "resilience and resurgence and not just of loss."[83] Lucchesi outlines how data gathering and mapping projects run the risk of replicating *data terrorism*, "the use of data to terrorize a population into submission for political, ideological, or social gain."[84] Data terrorism describes projects that collect data and produce maps on gender-related violence against Indigenous women and girls and two-spirit people that *reproduce* and *enact* the colonial violence that they purport to describe. This is to say that such projects use data and statistics to construct deficit narratives of Indigenous women as being in need of help; they ignore the role of the settler state and its institutions in perpetuating the violence; and they create the conditions for further policing and regulation of Indigenous women's bodies. The path toward addressing data terrorism is not to accumulate more data but rather to center Indigenous women's data sovereignty.

These works emerge as calls for decolonizing and de-Westernizing data scholarship are growing. Movements around Indigenous data sovereignty emphasize that colonization—in the form of settler colonialism—has not ended. Conventional uses of data participate in ongoing *epistemicide*, defined by Stephanie Russo Carroll and coauthors as the suppression and appropriation of Indigenous knowledges and data systems.[85] Stefania Milan, Emiliano Treré, Mohan J. Dutta, and other scholars in critical data studies situated many of the harmful effects of contemporary datafication as a continuation of Western European colonization and their (our) extractivist, violent knowledge regimes.[86] Numerous important concepts have been advanced for naming these regimes—among them, surveillance capitalism, the New Jim Code, automated inequality, data extractivism, data terrorism and data colonialism.

These concepts have been essential to make sense of the current moment. And yet critical data studies has the potential to do more than retroactively describe hegemonic data science practices. In his essay "Seeking Liberation: Surveillance, Datafication, and Race," information studies scholar Roderic Crooks challenges the field of critical data studies to do more than merely describe various forms of data injustice: to actively participate in "seeking liberation" for minoritized people from those harms.[87] Building on this vision, critical data studies can be a generative site for developing liberatory approaches to data, knowledge, and power.

In her 2019 paper "Data Epistemologies, the Coloniality of Power, and Resistance," media scholar Paola Ricaurte analyzes how present regimes of data power are violent. They are colonial in their methods for capturing and extracting value from human life. They pose a threat to humans, biodiversity, climate, and life on Earth itself. Yet Ricaurte leaves open the possibility that, through collective resistance, "we can reverse extractive technologies and dominant data epistemologies in favor of social justice, the defense of human rights and the rights of nature."[88] She discusses María Salguero's map of feminicides in Mexico as an example of such resistance, enacting the use of grassroots data to further justice, memory, and human rights in the face of economic and patriarchal violence. For Ricaurte, feminicide data activism represents a step toward the development of "alternative data frameworks and epistemologies that are respectful of populations, cultural diversity, and environments."[89] Such alternative epistemological approaches are flourishing in both scholarship and activism, including data feminism, feminist data refusal, emancipatory data science, decolonial AI, Indigenous data sovereignty, queer data, and more (see chapter 8 for more on these data epistemologies).

As I stated in the introduction, throughout this book I will seek to draw out resonances between data feminism and data activism about feminicide. In our book *Data Feminism*, Lauren and I described Salguero's map and counterdata production work as an example of the *challenge power* principle—appropriating hegemonic data science and mapping tools to visibilize (*visibilizar*) violence that is systematically invisibilized by the state and its institutions. One of the underlying assumptions of this book is that the real-world, already-existing practices of data activists have much to offer those of us who seek to support and sustain alternative, feminist, anti-racist, Indigenous, queer, Black and/or decolonial data epistemologies. This case is aligned with Milan's recent paper outlining what critical data studies as a field may learn from data activism as it was practiced during the COVID-19 pandemic.[90] It also builds on the case made by feminist scholar Aristea Fotopoulou that scholars should shift their object of research from data, algorithms, and platforms themselves toward the human practices of acquiring, analyzing, and using data, so that we may "reinstate the materiality of data, to think about laboring bodies, invisible human practices, and social relations and activities."[91]

CONCLUSION

Grassroots data activism about feminicide invites us to imagine a data science, epistemology, and ethics that rigorously takes power and people into account; that understands how structural inequality produces missing and flawed data and develops creative strategies to mitigate that; that views data science not as a technosolutionist panacea but, first, as an intimate act of care, witnessing, and memory justice; and, second, as a vector for transformative social change. This is not to romanticize the very difficult and fraught labor of feminicide data activists, which we will be exploring in more detail in the rest of this book. Rather, it is simply to offer that mainstream data practitioners and critical data studies scholars have much to learn from these reflective practices, particularly those of us who wish to mobilize data science in solidarity with movements for social justice.

2 OFFICIAL DATA, MISSING DATA, COUNTERDATA

In 2019, two nonprofit organizations released a somber report called *La Persistencia de La Indolencia: Feminicidios En Puerto Rico 2014–2018* (The Persistence of Indolence: Feminicides in Puerto Rico 2014–2018).[1] The report compared news reports and official death registry records about feminicide with police records of murdered women and found that, in any given year, police data were missing between 11 percent and 27 percent of the year's feminicide victims. That is to say, the police did not know about, or have on record, the murders of more than a quarter of the women killed in Puerto Rico in a given year. Thus, the "indolence" in the title of the report referred to the state's response to the problem of feminicide on the island. This indolence had grown in the years leading up to the report and was woven together with larger shifts: the Office of the Attorney General of Women was increasingly politicized and rendered ineffective; neoliberal measures gutted public services; and the island experienced increasing environmental shocks.[2] All of these were backgrounded by the colonial political context. As a territory of the United States, Puerto Rico has struggled for autonomy for more than a century. The report was undertaken by Proyecto Matria, an organization that works toward economic justice for women, in collaboration with Kilómetro Cero, a nonprofit that monitors human rights and police violence in Puerto Rico. As Mari Mari Narváez, founder of Kilómetro Cero, explained to our team, "Our interest initially was to collect data for us to be able to further expand our work on the state's response, particularly that of the police, in regards to gender-related violence. There really was a very basic data problem in Puerto Rico. In other words, it was simply not known how many feminicides there were."

This chapter explores the official data published by the state in relation to the twin phenomena of missing data—exemplified by the many murders of women in Puerto

Rico that are inexplicably absent from official reports—and counterdata—those cases painstakingly compiled by individuals and feminist organizations in Puerto Rico and assembled for analysis in the report. Missing data and counterdata are central concepts for data activism about feminicide, so this chapter provides more background on both, including literature review, practical definitions, and further theoretical elaboration. The chapter also introduces the idea of *restorative/transformative data science*—a concept that aims to encompass the motivations, process, and impacts of undertaking data activism about feminicide, as well as in other domains characterized by durable structural inequalities. We will see how all of these concepts are at work in this specific case of the *Persistence of Indolence* report and its impact in Puerto Rico. The production of counterdata is not a simple task, particularly in an information ecosystem characterized by institutional inaction—a will to *not know* that invisibilizes feminicide and legitimizes impunity. As Alice Driver observes, there is a disquieting parallel here with the disappearances that often mark the murder of women: missing bodies are accompanied by unrecorded violence.[3] And no one can be held accountable for what is not known.

When the two nonprofit groups decided that they wanted to study feminicide in Puerto Rico, they had to figure out how to get the data. At the time, there was no official data explicitly about feminicide because no legislation had been passed that outlined a legal framework for the concept. The closest thing to official data was the police data about deaths due to domestic violence, which only came in aggregated form. That was how, in 2018, Proyecto Matria and Kilómetro Cero made a visit to an *égida*—a senior living facility—in San Juan where Carmen Castelló lives on her state pension. Castelló retired from a career in social work in 2010 and began caring for her grandniece, baby Alba (figure 2.1). While the baby slept, she would watch the news on TV and read newspapers, and she was alarmed at what appeared to be increasing rates of feminicide and sexual assault. She started applying skills she had learned on the job, "Because we, as social workers, when we serve people, we open a case file and follow up on it, right?"

Castelló learns about new cases through the news and logs each case of feminicide that she finds in a Word document. She monitors the news every day but tries not to do it early in the morning because it affects her emotionally to start her day with violence. Castelló has become an expert on the media ecosystem in Puerto Rico and in particular looks to lesser-known news sources from small towns and villages for details about each case. As the case develops, she uses the media reports on it to follow it through the justice system and updates the Word document with the new details, carefully noting the sources of the information. At the same time, she also publishes all cases and updates on her Facebook page: Seguimiento de Casos (Case tracking). Castelló also compiles

FIGURE 2.1
Carmen Castelló is a retired social worker who produces the most reliable data on feminicide in Puerto Rico while she cares for her young relatives. Courtesy of Ana María Abruña Reyes / Todas (https://todaspr.com).

separate databases about sexual abuse and missing women. When Proyecto Matria and Kilómetro Cero approached her about their study, Castelló was happy to work with them; she shares her files freely with any organizations working to address the issue.

Castelló's data became the starting point for the *Persistence of Indolence* report. Kilómetro Cero, an organization led by journalists, fact-checked and verified each case in Castelló's database. Says Debora Upegui-Hernández, a data analyst who joined the Observatorio de Equidad de Género Puerto Rico in 2020, "So far [Castelló's data] has been the most reliable and most up-to-date source I have seen."

Proyecto Matria and Kilómetro Cero then did something quite creative—but, as we will see, also quite common across feminicide data groups—to try to assess the undercounting of feminicides. They *triangulated* data gathered from press reports with data from official records. Triangulation across information sources helps verify and substantiate the details about each case. Data activists explained to our team that press reports often have incorrect information, especially the first reports that come out about a case. These often get basic details—such as name, age, and method of death—

incorrect. In the case of Proyecto Matria and Kilómetro Cero, they combined the data from Castelló with open data from the federal demographic registry that logs all births, deaths, and marriages across the country. Notably, these data were only made public because of a lawsuit won by the Centro de Periodismo Investigativo (Center for Investigative Journalism) in 2018. This lawsuit had stemmed from another high-profile dispute around numbers: the insistence of the Puerto Rican government that there were only sixty-four deaths following Hurricane Maria in 2017, versus investigative reports and research studies showing thousands more. The final estimate was pegged at 2,975 and accepted by the government only after significant domestic and international pressure.[4]

Triangulating Castelló's data from news reports and the demographic registry for the time period of 2014 to 2018, the researchers found 156 cases in both sources, 68 only noted in news reports, and 33 cases only in the demographic registry (figure 2.2). They were then able to compare these totals with aggregate numbers published by the police to demonstrate that agencies of the state had failed in "rigorously documenting the situation of feminicides in Puerto Rico and in disclosing it to the public."[5] The report concluded that the numbers published by the police significantly underreported feminicide. With these techniques—of counterdata production, triangulation across multiple data sources, and systematic case logging—Proyecto Matria, Kilómetro Cero, and Castelló were able to quantify missing data in order to build a case for institutional neglect and make demands for visibility, resources, and better measurement. They wrote in the report, "Collection and analysis of information on feminicide are fundamental tools to determine its magnitude, to understand its patterns and trends, and to establish international comparisons that may serve as an instrument to assess the successes and failures of prevention efforts."[6]

FEMINICIDE AND MISSING DATA

The data sourcing, analysis, and triangulation techniques used by Castelló and the *Persistence of Indolence* report are without doubt creative and rigorous. But it raises the earlier question that I asked about María Salguero: Why is this work being undertaken by individuals, journalists, nonprofits, and feminist collectives? Not surprisingly, these groups are very aware of the irony that, as Marta Pérez of the Argentine group Mujeres de Negro stated to our team, they "are doing work that belongs to the State." According to Irma Lugo Nazario, Coordinator of the Observatorio de Equidad de Género Puerto Rico (Gender Equity Observatory of Puerto Rico), the government has reduced funds and dismantled services for gender-related violence over the last decade, and so

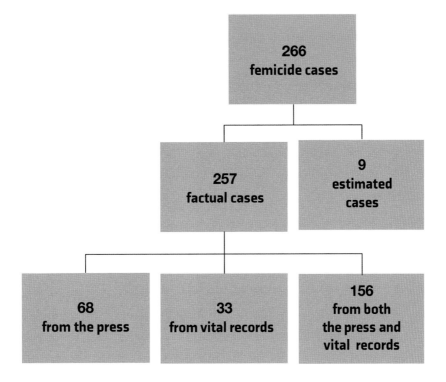

Femicide cases 2014–2018
N=266

FIGURE 2.2

A diagram from *The Persistence of Indolence: Feminicides in Puerto Rico 2014–2018* illustrates how the authors triangulated their case counts across multiple sources of information. The report can be downloaded in English at https://www.kilometro0.org/informes. Courtesy of Kilómetro Cero and Proyecto Matria.

feminist organizations in Puerto Rico have stepped in to fill that void, even though they continue to demand that the government take responsibility for such work.

Indeed, political demands for data and information are a central aspect of *missing data* as I am discussing it in this book. There is more than a century of literature about missing data in statistics, including theories of missing data and techniques for handling such missing values in the data analysis process. Craig K. Enders has an excellent applied introduction to this work and, in 2019, Sara Fernstad explored how data visualizations should represent missing data.[7] This literature has a typology of three mechanisms that lead to missingness, which include missing completely at random, missing at random, and missing not at random. Feminicide data, in this typology, can be considered *missing not at random* data; this is to say that there are underlying patterns to the missingness (such as patriarchy) that are not random and that are dependent on variables not recorded in the dataset. We will explore some of these informatic patterns leading to missingness in chapter 4.

In general, the statistical literature on missing data does not generally try to explain—in any causal or contextual way—*why* observations or values or whole datasets do not exist. Missing data in this literature is a technical enterprise rather than a political proposition. That said, there are some interesting case studies that take more of an applied statistics approach to feminicide. In a 2021 paper, statistician Maria Gargiulo outlines two main forms of missing data related to femicide data: deaths that are recorded but information is missing to classify them as femicide, and deaths that are not recorded at all. When official or activist datasets about femicide do exist, they represent convenience samples influenced by a number of factors, including selection bias (e.g., it is easier to document violence in urban areas versus remote rural areas) and reporting bias (media consistently underreport cases about racialized women). Gargiulo concludes with a set of recommendations for mitigating some of these biases for researchers undertaking analysis of femicide data.[8] And in a 2021 sociology paper on the global measurement of femicide, scholars Myrna Dawson and Michelle Carrigan review various sex-/gender-related indicators that try to distinguish femicides from other homicides. These include aspects such as evidence of a prior relationship between the victim and the perpetrator, sexual violence, mutilation of the body, and existence of prior complaints, among others. They examine these in relation to existing government datasets and find a high proportion of missing and inaccessible data that impede efforts to design public policies that prevent feminicide.[9] This is to say that governments often do not collect the most important variables that would permit the determination of femicide in a given case.

Other contextual explanations of missing data have come from scholars in the social sciences, international development, humanities, and the arts. For example, there is extensive feminist literature about women's invisibility in economic and labor data due to their relegation to the domestic sphere. Since the 1970s, social scientists have been pointing out how these systemic biases result in missing data: "Quite simply, governments and international agencies produce data only for those aspects of social life they deem important."[10] Geographer Joni Seager has spoken about the *data voids* that result because of institutional neglect of women's lives, a perspective recently popularized by Caroline Criado Perez's book *Invisible Women: Data Bias in a World Designed for Men*.[11] And through a 2016 artwork and accompanying essay, Mimi Ọnụọha posited the idea of *missing datasets*—whole collections of data that are disregarded, overlooked, and discounted. In her project, missing datasets include such topics as "people excluded from public housing because of criminal records" and "trans people killed or injured in instances of hate crime."[12]

Ọnụọha links these missing datasets to the study of ignorance: "That which we ignore reveals more than what we give our attention to. It's in these things that we find cultural and colloquial hints of what is deemed important. Spots that we've left blank reveal our hidden social biases and indifferences."[13] In fact, there is a whole body of work called *agnotology*—the study of ignorance. Whereas the fields of philosophy and sociology tend to think deeply about how people come to know things, work on the epistemology of ignorance examines how people and societies come to *not know* things. For philosopher Charles W. Mills, who brought critical race theory to bear on this work, this is exemplified by the ignorance of white people about the effects of their system of racial domination in North America.[14] Feminist philosopher Nancy Tuana applies this line of thinking to knowledge about women's health, where she develops a taxonomy of types of ignorance, including "willful ignorance"—where missing data and missing knowledge about gender-related violence would be situated. This category entails "a systematic process of self-deception, a willful embrace of ignorance that infects those who are in positions of privilege, an active ignoring of the oppression of others and one's role in that exploitation."[15] Indeed, in a reflection on lessons to be learned from Audre Lorde's life and work, health researcher Lisa Bowleg stated that "epistemological ignorance is one of the master's most formidable tools."[16] In these framings, ignorance is not an individual question—knowledge that a single individual does or does not possess. Rather, ignorance is a social and a political phenomenon whereby dominant groups and dominant institutions render the experiences of minoritized groups invisible—outside the realm of knowledge. Through her framing of *data silences*, Indigenous studies scholar Bronwyn Carlson renders this process of ignorance even more active: she links the production of not-knowing directly to gendered colonial violence whereby the settler state actively silences accounts of violence perpetrated on Indigenous bodies.[17] There are also cases when the state has a strategic interest in denying the existence of not only a widespread phenomenon but also certain types of human beings. For example, in her book *The Uncounted*, Sara Davis outlines how governments try to fight the AIDS epidemic while at the same time denying the very existence of the people most vulnerable to AIDS: sex workers, men who have sex with men, drug users, and transgender people.[18]

While Donna Haraway taught us that knowledge is situated, Tuana writes about how ignorance, too, is situated.[19] This can lead towards Carlson's more active framing of data silences: the absence of data is not merely due to oversight, accident, or "unconscious bias." Missing data are actively produced as part of the systemic suppression of the experiences of women, Indigenous people, racialized people, Global South people, and trans people that is necessary to maintain a racialized, colonized, patriarchal, global economic order. This resonates with what activists refer to as the *invisibilization* of feminicide and to what literary scholar Sayan Bhattacharyya calls *epistemically produced invisibility*.[20] Thus, the absence of data can be part of an active and ongoing *production* of invisibility and *maintenance* of invisibility, requiring loads of labor to simply *not* know and *not* understand and *not* see a systemic phenomenon and *not* count it.

MISSING DATA AND POWER

The working definition of missing data that I am using in this book derives from these social, political, and structural considerations on the presence and absence of knowledge. Missing data are those data that are neglected to be prioritized, collected, maintained, and published by institutions, despite political demands from specific groups that such data *should be* collected and made available. As Ọnụọha explains, to frame data as missing is to make a normative assertion: "It implies both a lack and an ought: something does not exist, but it should."[21] Data are never missing in any absolute sense. Rather, data are missing because there is a political demand that such data should exist. Moreover, missing data includes not only the total absence of data but also the production of sparse, incomplete, unreliable, misclassified, inaccessible, contested, removed, and untimely information.

Missing data, therefore, is a concept that is situated, relational, and political. In the case of official data about feminicide, there is missing data precisely because grassroots feminist, Indigenous, Black, queer and/or women's groups, journalists, and social movements make demands that such data be produced. Thus in relation to missing data we must ask, for whom are the data missing? Which groups are making the political demand for data collection and knowledge production? Who is named as the entity responsible for (or negligent in) data collection? The answers to these questions matter for understanding how power is at work throughout the sociotechnical environment of missing data. Indeed, the first two principles of data feminism call on us to *examine power* and to *challenge power* in relation to data.

Yet what exactly do we mean by *power*, especially in relation to data produced about feminicide? Lawyer and trans activist Dean Spade makes a strong case that power is decentralized: there is no one person or institution responsible for administering and distributing systemic oppression. And conversely, to challenge power one cannot only look to one person, nor even one institution such as the law. He writes, "This way of understanding the dispersion of power helps us realize that power is not simply about certain individuals being targeted for death or exclusion by a ruler, but instead about the creation of norms that distribute vulnerability and security."[22] Sociologist Patricia Hill Collins developed a Black feminist conceptual model for this multisited operation of power called the *matrix of domination*. This refers to the uneven allocation of *privilege*— increased life chances, opportunities, access, care—for some groups, and *oppression*— decreased life chances—for other groups. The matrix of domination is a conceptual model that, among other things, outlines a taxonomy of four domains of oppression: structural, disciplinary, hegemonic, and interpersonal (table 2.1).[23] This model can help to pinpoint some of the diffuse and decentralized operations of power that result in the *nonproduction* of feminicide data. Missing data about feminicide is produced across these four domains, and activists have also taken action in these four domains, as we will explore further in chapter 4.[24] This is a high-level framework because it attempts to grasp broad patterns in the nonproduction of feminicide data across contexts. It is one step toward answering the question "Why and how does official data about feminicide and gender-related violence come to be missing?"

Two domains in the matrix of domination can be framed as the purview of the state—*laws and policies* and their *implementation*. The first is where oppressive laws are enacted or where, conversely, lack of or inadequate legislation reinforces oppression of certain groups. As discussed in chapter 1, thanks to persistent efforts by feminist and women-led and family-led movements, many countries have passed legislation criminalizing femicide or feminicide or strengthening protections for MMIWG2 and fatal violence against women. But this legislation often retains a narrow definition of

Table 2.1
The four domains of the matrix of domination

Laws and policies	Implementation of laws and policies
The structural domain organizes oppression and inequality.	The disciplinary domain manages oppression and inequality.
Media and culture	**Interpersonal**
The hegemonic domain circulates oppressive ideas.	Individuals experience oppression in the interpersonal domain.

Source: Based on concepts introduced by Patricia Hill Collins in *Black Feminist Thought: Knowledge, Consciousness, and the Politics of Empowerment.* Partially adapted from *Data Feminism* with permission of the authors.

the phenomenon and only in rare cases includes transfeminicide (see table 1.1). Until recently, for example, Puerto Rico only counted cisgender women's murders in the context of domestic violence—what is known as *intimate feminicide*—but lumped in all other gender-motivated killings into the category of homicides.[25] In other places, such as Mexico, definitions of feminicide vary across states, which obfuscates national comparisons. The lack of adequate legal definitions and categories thus underpins missing data; without these the state simply does not recognize a systemic form of violence, let alone count it or ensure justice for those who experience it.

The *implementation of laws and policies* happens in the disciplinary domain of Collins' matrix, and here many governments have failed to implement adequate information collection, failed to devote resources to accurately classify deaths, and failed to create supportive climates for survivors and families to report violence. This is primarily where the *Persistence of Indolence* report was directed: "The Police force is indolent in its lack of diligence to renew its practices of collecting, analyzing, interpreting and disclosing statistics of murdered women in accordance with widely used international standards."[26] Their main finding was the systemic undercounting of women's murders. This came to be through various factors: different state agencies were responsible for different types of violence, so there is an office for sexual assault, an office for domestic violence, and another office for homicide. Each was counting differently and their numbers on feminicide didn't correspond with each other. This institutional fragmentation is not unique to Puerto Rico; it was echoed in ILDA's workshops with public sector employees in Argentina and Uruguay.[27] There are also errors in those data that do exist. For example, reports about MMIWG2 in North America have consistently highlighted the racial misclassification of Native women.[28] In Puerto Rico, as the *Persistence of Indolence* team was going through the demographic registry, there were more than a dozen cases of women murdered in a single year, which were inexplicably and incorrectly marked as "suicides." These bureaucratic incompetencies constitute some of the more pernicious ways that the disciplinary domain contributes to missing data—through what Menjívar and Walsh have called "state acts of omission and commission," wherein the state either indirectly or directly contributes to underreporting and allows violence to go unpunished.[29] This is also why Marcela Lagarde y de los Ríos stressed the role of the state in the definition of feminicide: by repeatedly not taking action, the state creates a climate of impunity that permits the persistence of mass human rights violations.

The implementation of laws and policies can also see the active suppression of data and information. High numbers of feminicides make the police and the state look bad. As Castelló recounts, "[The government] was telling us they couldn't do that [collect data about feminicide]. They themselves said: 'we cannot do that, because people will think that the relevant agencies are not doing their job.' And, we told them: 'Well, exactly. You are not doing your job.'" Members of the Puerto Rican team also described how feminicide data requested from police came only in aggregated form and were irregularly reported—sometimes reported by week, sometimes by month, sometimes by year. Often there were no updates for months at a time. And Upegui-Hernández expressed great skepticism at the official reports published about domestic violence in Puerto Rico because they were so "supremely consistent over the years." For more than ten years, the official data about domestic violence complaints appeared to be decreasing with fewer and fewer complaints each year, a trend she found extremely hard to believe given her work with grassroots survivor-led groups.

A third domain is the purview of *media and culture*: the hegemonic domain is where stereotypes and harmful cultural ideas circulate, and this too plays a role in producing missing data. Media coverage of feminicide tends to reproduce gender and racial stereotypes, blames victims for their own deaths, reinforces stigma, and ultimately both spectacularizes and normalizes gender-related violence. Renowned anthropologist Rita Segato has described how the voyeuristic media coverage of feminicide glamorizes and reenacts the violence it purports to describe, leading to a kind of social contagion of the phenomenon.[30] It is typified by coverage that emphasizes especially sexual or brutal aspects of the case or paints intimate partner violence as a "crime of passion" or relates details about how victims were dressed or which industries they work in or how late they were out at night or how drunk they were or their prior history of incarceration. This creates stratified categories of worthy victims (typically whiter, richer, settlers, gender-conforming, heterosexual, students, or professionally employed) and unworthy victims.[31] These articles reproduce messages that justify and naturalize (some) feminicides, contributing to the overall climate of impunity. In Guatemala, anthropologist Sarah England found that newspaper coverage of feminicides typically provides "mainly speculation on the part of the people interviewed, the police, and the reporter."[32] In Brazil, a study found that news articles provided little information on the context of women's deaths, and often relied on common tropes such as "jealousy" and "violent emotion"—an approach that deemphasizes the structural and systemic character of gender-related violence.[33] Media often directly import the framing and perspective of the police and thus can function as a kind of cultural arm of the state, something we will explore further in chapter 4. For example, in the case of Puerto Rico, when the police reported out their feminicide data, one of the categories they used is "crime of passion," a category that authors of the *Persistence of Indolence* rejected as "offensive and obsolete."[34] Finally, media often misgender transgender victims of violence, resulting in what Sofia Vanoli Imperiale and Eloi Leones have separately called a *double murder*, in which the person's lived identity is murdered again in the media coverage of the event.[35]

When media circulate harmful stereotypes that dehumanize victims and normalize gender-related violence, this further reduces political and public will for (1) seeing fatal gender-related violence as a systemic violation of human rights; (2) measuring and counting fatal gender-related violence; and (3) preventing gender-related violence in the first place. But an equally harmful way that media contribute to missing data is in their unequal coverage of killed people. For example, it is so well documented that murders and disappearances of white women in the United States receive more media attention than those of women of color that there is a name for the phenomenon— *missing white woman syndrome*—created by Black journalist Gwen Ifill in 2004.[36] A case in point was Gabby Petito, a young white woman from the United States who went missing in 2021 and was later determined to have been murdered by her fiancé in Wyoming. There was extensive media coverage throughout the period of her disappearance, drawing criticism from many groups who put forward that more than seven hundred Native women had disappeared or been murdered in the same state with virtually no public attention.[37] As a case in point, the *New York Times*, a US-based newspaper of record, has published a total of nineteen articles that mention Gabby Petito from September 2021 to the date of this writing, whereas they have published eighteen articles to date that make any mention of the entire phenomenon of missing and murdered Indigenous women in the United States over all time.[38] The point is not that Petito didn't deserve public attention and justice, but rather that Native and Black women who experience violence are being systematically denied that public attention and justice. The domain of media and culture thus also contributes to missing data through systematically declining to pay attention to the killings of certain groups of women. They fail to "ignite any form of public outrage," something that Carlson and other Indigenous scholars have linked to symbolic annihilation—cultural invisibilization and erasure through nonrepresentation.[39] This prevents the public from knowing more and demanding more from its institutions. The bias in coverage also poses a significant problem for counterdata producers like Castelló and other data activists who rely heavily on media sources to compile cases that are not being tracked elsewhere, as we will explore in chapter 4.

Finally, the interpersonal domain is where individuals experience oppression directly as discrimination and violence. In relation to feminicide, the violence extends beyond the act itself, leaving families devastated, demoralized, and reluctant to report violence. In a climate of bureaucratic ineptitude (whether intentional or not), open hostility, and victim-blaming, families and communities may understandably want to seek as little involvement with the state as possible, perpetuating the cycle of missing data by not reporting. In places where there are high levels of corruption and organized crime, many police officers may be on criminal payrolls, so families themselves can face retribution and violence simply for filing a report. For example, political scientist Mneesha Gellman has read many US asylum cases where women have to state their reasons for not reporting gender-related violence in their home country: "Reasons include the police being either gang-involved themselves or colleagues or friends with the abuser. Other reasons were that the police would either further the abuse themselves, or dismiss the abuse as to be expected given the cultural context of machismo."[40]

As is evident, missing data are enmeshed in power relations. Missing data are actively produced in the domains of law and policy; the implementation of law and policy; media and culture; and the interpersonal realm. Indeed, the production of silence, ignorance, and invisibility across multiple domains is central to the maintenance of the matrix of domination itself. The takeaway here is that the silence is structural, the ignorance is social, and the invisibility is political. Latin American activists often talk about "visibilizing" feminicide, and it is precisely these multisited absences they are up against.

FEMINICIDE AND COUNTERDATA

When the state and its institutions fail to render justice (as well as to produce even basic information) and when the media reproduce oppressive narratives and when families are demoralized and disempowered, activists and journalists are increasingly using informatic strategies to produce their own counterdata in order to *challenge power*, the second principle of data feminism. Just as missing data are produced across all four domains of the matrix of domination, data-driven acts of resistance and refusal happen across these four domains as well. We will explore these at length in chapter 4, which maps activists' information-sourcing practices across the matrix.

First, a short genealogy of counterdata as a concept. The term *counterdata* was introduced in 2014 by geographers Craig Dalton and Jim Thatcher, who drew from *counter-mapping* to describe acts of resistance that use data to upset dominant power relations. They call for examining both speculative and existing practices: "We must ask what counter-data actions are possible? What counter-data actions are already happening?"[41] As such, counterdata can be viewed as part of the larger universe of data activism practices discussed in chapter 1. Case studies have built on this concept of counterdata. For example, geographer Morgan Currie and colleagues described a Hackathon on

Police Brutality conducted in Los Angeles in which participants worked with official and citizen-collected data about police killings as a way of interrogating what actions data do and do not make possible.[42] Amanda Meng and Carl DiSalvo demonstrate that grassroots counterdata production around housing in Atlanta is rooted in a long history of Black community organizing.[43] In their case study on a community group in Los Angeles that documents police killings, information scholar Roderic Crooks and Currie place the group's work in a larger field of *agonistic data practices*. These practices constitute efforts by communities to use data to document harm, but the group they profile also uses their data in affective and narrative ways—"to motivate people to act on their passions and imagination."[44] In *Data Feminism*, Lauren and I discussed numerous examples of counterdata projects emerging from academia, data journalism, community organizing, and activism.

Counterdata production is an apt term for describing the work that most (but not all, as we will see) groups do to compile spreadsheets and databases about feminicide.[45] But what is counterdata countering? Is it a response to missing data? Is it a counterpoint to inadequate or erroneous official data? Is it countering the patriarchal, colonial state? Here I would answer "yes, and . . ." to all of these questions because each is only part of the story. Drawing from Latin American thought and its long history of engagement with the work of Antonio Gramsci, counterdata means *counterhegemonic*. Gramsci's conception of hegemony laid the groundwork for a multisited conception of power in which dominant groups not only control legal and institutional structures but also manipulate social and cultural norms and values in order to enrich themselves at the expense of minoritized groups. For Gramsci, counterhegemony describes the ways that people work together to critique and dismantle these oppressive structures in multiple domains. In Latin America, counterhegemonic approaches are also linked to anticolonial and anti-imperialist struggles—for example, resisting a global neoliberal order that extracts labor and resources, destabilizes democratically elected leaders, and spreads violence and inequality to enrich the North. In relation to counterdata, this means that it is broader than filling in the gaps in official data or reforming the informatic practices of the state. Indeed, as described in the prior section, the state is only one part of the multisited matrix of domination that produces missing data about feminicide. Counterdata, then, is about using data to counter those structural, hegemonic forces of power that cause feminicide in the first place, specifically cisheteropatriarchy, white supremacy, racial capitalism, colonialism, extractivism, and more (see the glossary in chapter 8 for definitions of these terms). The goal is nothing less than the elimination of feminicide, so this requires the elimination of the structural conditions that produce it.

Specifically in relation to feminicide, Segato has developed the idea of *counterpedagogies of cruelty*. She argues that whereas feminicide and its media representations habituate the public to a pedagogy of cruelty, counterpedagogies work to teach and spread practices of life, living, and vitality.[46] Counterdata production about feminicide can be situated as one component of such a counterpedagogy of cruelty. As Geraldina Guerra Garcés from the Alianza Feminista para el Mapeo de los Femi(ni)cidios en Ecuador related to me, "Registering feminicides is a job in defense of life. Even when it seems that we work with death. In reality, it is about life, so that it does not happen again, to amplify the collective voice and achieve the prevention of feminicides in our countries."[47] Activists told us over and over again, in different ways and with different words, how their counterdata production was not about quantifying the dead but about defending life itself.

Counterdata, countermapping, counterpedagogies: all these counterhegemonic concepts directly acknowledge—and then attempt to counteract—asymmetrical power relationships. They often appropriate hegemonic cultural forms such as the database, the spreadsheet, or the map, and shift them into the service of minoritized groups and subjugated knowledge. In this sense, counterdata production, like data activism more broadly, is a citizenship practice. It is an informatic form of enacting democratic dissent, prompting protest, and insisting on political engagement.

For example, in the case of Puerto Rico, Proyecto Matria and Kilómetro Cero developed a collaboration with Castelló and carried their project from data collection through analysis and public distribution. While the police had claimed there were no feminicides in 2018, their report demonstrated that in fact a feminicide occurred every seven days on the island.[48] When the team published its results in the *Persistence of Indolence* report, it framed feminicide as a structural public health problem that was being systematically neglected by the state. The report included background and definitional information, infographics, and narrative and statistical analysis. It demanded that the state formulate a definition of feminicide and that it include transfeminicides, cases under investigation, and feminicides relating to drugs and organized crime. The report's six recommendations targeted changes in law and policy (broader definition of feminicide; more regulation for firearms); the implementation of law and policy (more training on gender-related violence for police, courts, and health professionals) as well as media and culture (promote a culture of nonviolence).[49] On its publication, the report received wide press coverage in English and Spanish, including a feature from the Intercept and television coverage on Telemundo. In August 2021, the governor signed Senate Bill PS 130 into law, which was the first legislation in Puerto Rico to define feminicide and transfeminicide. The text of the legislation specifically named and quoted the report as a motivation for the necessity of the law and outlined feminicide and transfeminicide as crimes of first-degree murder. It also ordered the creation of a national data collection system on feminicide and transfeminicide that was slated to launch in 2023. The Observatorio de Equidad de Género Puerto Rico sits on the advisory committee for the development and implementation of this system, and it is being designed to follow the Latin American model protocol outlined in chapter 1. In this case, the production and circulation of counterdata about feminicide in Puerto Rico had a significant impact on media, law, and policy.

COUNTERDATA CAVEATS

Data activists know that counterdata, especially on its own, is not sufficient to produce emancipatory outcomes. Counterdata production should be regarded as one tactic among many in the action repertoires of activists and social movements to effect structural change. Not all groups working on feminicide and gender-related violence work with data. Many grassroots groups focus on family or survivor support, accompany families through the justice system, develop programs focused on prevention, or utilize legal strategies to push for policy change. Even among those data activists who do produce counterdata, there is variation. For some, like Castelló and many individually led efforts, the production and circulation of datasets is the sole focus and goal, and they are happy when others use their data (with consent) for media reports, policy change, or to secure grants for prevention programs. But for other monitoring groups, particularly those with more people and financial resources, the production of data is one tactical aspect of their larger strategic efforts to visibilize feminicide and gender-related violence as well as work toward justice, healing, and prevention. For example, while Sovereign Bodies Institute began as a research and data-production effort, it now provides direct services to families, produces reports in collaboration with tribes and Indigenous public health programs, and works to empower Indigenous-led governance around MMIWG2/MMIP. In addition to running a national feminicide observatory, the Red Feminista Antimilitarista in Colombia also works in the space of prevention and organizes community protection circles for women facing intimate partner violence. For these groups, counterdata play a supporting role in a larger constellation of efforts. It is important to underline that in none of these cases—absolutely zero—do activists think that more data alone can lead to social change.

Indeed, there are certainly plenty of reasons to be skeptical of counting and counterdata production. These come from the data activists themselves, many of whom have told our research team "no somos números"—that is, "we [women] are not numbers"—a tension explored further in chapter 6. They also assert that a fundamental contradiction of this work is that producing data about feminicide should be someone else's job. Why does it fall to a school nurse in Texas or a geophysicist in Mexico or a feminist collective in Argentina or two nonprofit organizations and a retired social worker in Puerto Rico to produce the most reliable national statistics about feminicide? In the face of a retreating neoliberal state, why are women, and often Indigenous and Black and queer women, left to pick up the pieces and sustain their communities and defend their collective right to life in the face of structural violence?

Skepticism also comes from scholars. In 2015, legal scholar Sarah Deer—citizen of the Muscogee (Creek) Nation of Oklahoma—published an award-winning book: *The Beginning and End of Rape: Confronting Sexual Violence in Native America*. The first chapter reflects at length on the benefits and drawbacks of data. Deer outlines what is known, such as federal statistics according to which one in three Native women in the United States will be raped in their lifetime, and what is simultaneously known and unknown, such as grassroots advocates' assertions that the actual prevalence of rape is much, much higher. While Deer affirms the importance of statistics about rape as rhetorical devices that have helped usher in reforms in federal law, she raises questions about fixating on numbers: "But do we need more data in order to move forward?" Her answer is clear: "A continued emphasis on the aggregate data about the rate of rape committed against Native women may serve to eclipse long-term victim-centered solutions."[50] At what point does a fixation on data just divert resources and attention from taking action? How much evidence is "enough" and for whom? Especially since Native women and two-spirit people and Indigenous advocates already understand the scope and scale of the problem—because they live it.[51]

Feminist sociologists also offer a powerful critique of the way in which reducing violence against women into counts, ratios, and indicators—particularly at the global scale—works against a rich understanding of the circumstances of violence. In her book *The Seductions of Quantification*, Sally Merry draws from science and technology studies (STS) to compare four frameworks for producing global indicators about gender-related violence. She finds that because each framework is based on a different conceptual background, each would lead to very different policy actions and "none of these approaches is comprehensive. Each one is insufficient without detailed qualitative studies that reveal the social and cultural context of the violence."[52] Sociologist Saide Mobayed draws from Merry's work to elucidate the situation about feminicide data in Mexico specifically, and how the War on Drugs has led to more public feminicides and more feminicides with firearms. This narrative is only visible when combining government data with activist data like that of María Salguero and Data Cívica. Thus, Mobayed argues that data about feminicide should not be viewed as micro- versus macroscales but rather as an "interconnected web."[53]

Criminologist Sandra Walklate and her colleagues share Merry's concerns about decontextualization and outline further risks of counting femicide specifically. They assert that a serious risk of creating counts is that the available data are so poor, with so many femicides left out—especially those of women at the intersection of forces of domination—that they may possibly do harm by appearing to be objective while obscuring the deeply underreported reality.[54] A paper by statistician Patrick Ball aligns with this argument. An expert in using statistical methods to determine the scope of wide-scale human rights violations, Ball challenges the use of convenience data—data collected from what is readily available such as press reports rather than a representative sample. "After more than 20 years creating statistics for human rights analysis," he writes, "I have come to believe that descriptive statistics of convenience samples are worse than no statistics." In his experience, decision-makers will give more weight to deeply flawed numbers, often bypassing grounded, qualitative accounts from insiders, no matter how many caveats and disclaimers the statisticians themselves may make about those numbers.[55]

In relation to the utility of data and quantification for queer and trans people, technology scholar Os Keyes finds fundamental dissonance: "Trans existences are built around fluidity, contextuality, and autonomy, and administrative systems are fundamentally opposed to that. Attempts to negotiate and compromise with those systems (and the state that oversees them) tend to just legitimize the state, while leaving the most vulnerable among us out in the cold."[56] In other words, helping the state better understand and represent queer and trans lives might help the state better target and oppress queer and trans people.[57] And, moreover, participating in the datafication of trans lives risks legitimizing the very knowledge methods (quantification, counting) that have led to structural violence in the first place. This represents a kind of epistemological complicity with the matrix of domination. As Walklate and colleagues write, "When we focus our attention and activism on counting the killing of women, we then become part of these knowledge projects that have so successfully maintained the invisibility of gendered violence of all types."[58]

Finally, in the first writing of this book manuscript, I made a case for considering the full spectrum of feminicide data activism as *counterdata science*—a term that could encompass all of the data science practices that groups use to produce, verify, secure, systematize, analyze, publish, and circulate their data. But in dialogue with members of the Academic-Community Peer Review Board (see appendix 2), we explored the limitations of the concepts of counterdata and counterdata science to name and describe feminicide data practices. In particular, I am grateful that Indigenous-led organizations and scholars, along with a Latin American group that focuses on community defense, explained to me during the review process for this book that counterdata and counterdata science do not resonate as a term for their work because they do not see their work as existing in counterpoint to the settler state. Annita Lucchesi from Sovereign Bodies Institute described that for their organization the work is emphatically not about resistance or building counterpower. It is about sovereignty and kinship. It is for Indigenous people and communities first and foremost, not about producing alternate numbers to hold the state accountable. These conversations spurred me to think about the ways in which all counterhegemonic approaches—whether to mapping, storytelling, or counting—still center the powerful people and institutions responsible for producing structural violence in the first place.

Perhaps these "counter" concepts set up a powerful and too easy binary. On the one hand there is power—conceived as monolithic, and often associated with state power—and then, on the other hand, there is counterpower—conceived as its opposing force, often also framed as monolithic. As we will see in part II, these simple binaries do not resonate with the complex ways that activists themselves engage with state power, sometimes themselves becoming state officials. Nor does it leave space for how public sector officials work with activist data and collaborate with civil society organizations. In this chapter's case study about Puerto Rico, for example, the Observatorio de Equidad de Género now advises the implementation of the data collection system for the government's new law about feminicide and transfeminicide. Moreover, having been in community with data activists, journalists, and public sector officials for the past four years on feminicide data, this led me to think about ways in which "counterdata science," with its emphasis on agonistic relations, did not encompass the relations of care, restoration, and healing that many groups are engaged in through their data practices. Thus, the community review process for this book led me to use the term *counterdata* in a more specific way and then to set aside *counterdata science* and seek an alternate frame developed in dialogue with activists themselves—*restorative/transformative data science*—that can hold more of the complexity, variation, and vision of feminicide data practices.[59]

Following the data feminism principle of *rethinking binaries and hierarchies*, this book will continue both to work with and to trouble the interrelated concepts of official data, missing data, and counterdata as it works toward elaborating a restorative/transformative data science. Not all data that society does not have is missing. As I stated previously, *missing data* are constituted by the political demand that such data should exist. The political demand comes from specific groups and is directed toward specific institutions. Naming official feminicide data as missing—as virtually all activism, reports, guides, and policy recommendations about the topic do—is in line with feminist efforts to tie such violence back to the state's involvement in enabling human rights violations against significant groups of its own citizens and residents. "Feminicide is a state crime," Lagarde y de los Ríos states plainly.[60]

But Indigenous studies scholars such as Audra Simpson (Mohawk), Eve Tuck (Unangax̂ and), and K. Wayne Yang, among others, highlight the viability and urgency of refusal: witholding knowledge, witholding data, and withholding consent.[61] For Indigenous data activists, producing missing data—data that the settler state wants but is purposefully withheld by Indigenous communities from its view—might in fact be the most emancipatory path forward. This perspective is grounded in *sovereignty*—the right for Indigenous peoples to refuse extractive relations with the settler state and/ or its knowledge systems. Sometimes dominant groups and their institutions do not deserve to know things about others because they have proved themselves unworthy of that knowledge. Some knowledge is sacred. Some communities must actively organize to obfuscate institutional knowledge about themselves because they are so profoundly targeted by those same institutions. Not all data gaps should be filled.

This brings us back to the *who questions* that Lauren and I referenced often in *Data Feminism*: Who collects data? About whom? Who benefits (and who is overlooked or actively harmed)? Whose values guide the process?[62] It is not a given that "missing data" are a "bad thing." It is also not a given that counterdata are emancipatory. It is not a given that counterdata production will lead to desirable social change (or any change at all). And it is not a given that producing counterdata will not, in fact, harm communities it may intend to help. Navigating these questions is crucial for people who are exploring the utility of data science for their efforts to shift power, and for this reason I included the final chapter of this book, "A Toolkit for Restorative/Transformative Data Science," which supports teams to ask and answer hard questions about their current or future data science projects.

TOWARD A RESTORATIVE/TRANSFORMATIVE DATA SCIENCE

As is clear from the *Persistence of Indolence* report, feminicide activists and journalists do much more than produce counterdata. Proyecto Matria and Kilómetro Cero did not produce a dataset, upload it to the internet, and consider their work done. Data activists' and journalists' work comprises more than simply creating counts and tallies of fatal violence. They verify and triangulate information; they undertake exploratory and explanatory analysis; they draw from data to produce reports, artworks, infographics, social media campaigns, family support systems, and protests in public space; they develop new theoretical concepts from their data analysis. This is why this book makes an extended case for considering *restorative/transformative data science* as a concept that encompasses the care, rigor, and systematization that grassroots groups enact as they work to count feminicide and gender-related violence.

Restorative/transformative data science is an approach to working with information with the twin goals of restoration and transformation. Drawing from approaches to restorative justice, *restoration* involves restoring rights, dignity, life, living, and vitality to the individuals, families, communities, and larger publics harmed by structural violence. Here I am drawing inspiration from the words of Rita Segato. In her book *Contra-pedagogías de la crueldad*, she talks extensively about the importance of "lo vivo y lo vital"—the living and the vital—and those practices that can support and defend life, living, and vitality from the cruelty of structural violence.[63] Restoration seeks first and foremost to reduce harm, repair relationships, and heal the material and spiritual wounds left by violence. It also seeks the restoration of rights—the right to live a life free from violence, for example. That said, a person's life, once taken by patriarchal violence, can never be fully restored. This has led Saide Mobayed to draw parallels between data activists' work and the Japanese method of *kintsugi*—mending broken pottery pieces with colored glue so that the fractures are visible.[64] The bowl publicly wears its history of rupture and repair.

Building from thinking around transformative justice, *transformation* involves work to dismantle and shift the structural conditions that produced the violence in the first place. It is both visionary and preventative. Transformation, as a goal, involves a conceptualization of the event or problem in its systemic, historic, intergenerational, political, and social context. For example, to address feminicide as a public problem, we cannot always remain at the scale of an individual case. We must understand how one case reflects a larger pattern that must be disrupted and transformed to prevent similar cases in the future. This is one of the reasons that the production of systematized information—data—is a useful tactic in producing a structural understanding of an issue and working towards structural transformation.

As dual goals, restoration looks to the past, to healing and repairing, and operates closer to the scale of the individual, the family, and the community. Transformation looks to the future and to systems, to envision and struggle for the world where violence and inequality no longer exist, and operates closer to the scale of the society, culture, or nation. The multiscalar nature of these goals when working with data and information can produce tensions, as outlined in chapter 6.

As this book will demonstrate, activists' and journalists' work to produce, triangulate, verify, and circulate feminicide data constitutes an approach to data science grounded in restoration and transformation. But why call it *data science*? There are two reasons. First, if we return to the Latin roots of the word *scientia*, these relate to knowledge, knowing, expertness, and experience. Feminicide data activists and journalists are building and sharing systematic knowledge using alternative data epistemologies grounded in, variously, feminisms, women's rights, Indigenous data sovereignty, Black liberation, and others (see the toolkit in chapter 8 for an explanation of various data epistemologies). Activists' production and circulation of data easily meets the definition of science as a social practice—the process of "people seeking, systematizing and sharing knowledge."[65]

The second reason to name this work as data science is that those who currently call themselves data scientists stand to learn a great deal from the alternative data epistemologies and ethical data practices described in this book. Feminicide data activists and journalists posit the idea of using data to achieve a more just world free from gender-related violence, and in the process they enact alternate imaginaries of what data science is, who does it, and who it benefits. This means that data activists and journalists are claiming data science for uses other than extraction and wealth hoarding, and they are demanding that it operate in the interests of people other than elite, white, settler men from the Global North. In her 2019 book *Intersectionality as Social Theory*, sociologist Patricia Hill Collins discusses the importance of theorizing from praxis—developing concepts and frameworks from the already existing practices of activists and changemakers.[66] Thus, instead of looking to academics, corporations, governments, and think tanks to inform policy and practice around data ethics, this book makes a case that grassroots data science led largely by Latin American feminist activists opens a door to teach those of us who seek to use data science in the service of restoration, transformation, and, ultimately, liberation.

While feminicide is one of the most extreme forms of physical violence, many other areas where data and AI are being deployed involve engagement with different forms of structural inequality, violence, and trauma: education, policing, health, housing, transportation, and more. To live in an unequal society is to be a witness to and, often, a survivor of historic and ongoing violence, whether that violence is physical, spiritual, cultural, and/or economic in nature. Before parachuting in automated systems to algorithmically allocate resources or make predictions that exacerbate trauma and violence, it may be worth asking first what the path is toward healing and transformation.

This is why I advance the idea that there is potential to generalize from the work of feminicide data activists to other domains where inequalities are historical, durable, and structural. What might a restorative/transformative data science around mortgage lending look like? What might a restorative/transformative data science around planning new bus and bike routes look like? What might a restorative/transformative data science around community safety look like? Indeed, there are grassroots data practices in these domains that are already doing this work (see chapter 8 for examples). The goal of this book is not only to describe activist practices, but also to show how they present lessons that are, potentially, generalizable to data science in domains beyond feminicide.

Restorative/transformative data science, then, means mounting an explicit, and usually collective, effort to systematize and circulate data in the service of addressing inequality, oppression, and violence. It seeks to restore life, living, vitality, rights, and dignity to the people and publics harmed by structural violence. And it places that work in service of social and political transformation to eradicate the conditions that produce structural violence in the first place. Restorative/transformative data science may be undertaken by activists, journalists, academics, nonprofits, community-based organizations, and others. It often involves the production of counterdata—systematic counting—as well as encompasses what happens with those records downstream—that is, all of the data science practices that groups use to secure, systematize, verify, analyze, publish, and circulate their data in the world. Finally, it is important to note that restorative/transformative data science may be undertaken—and mostly is undertaken—with minimal computing resources and basic data literacy skills.

In the case of feminicide, restorative/transformative data science is connected to a feminist politics of refusal, discussed further in chapter 6, which declines to accept a status quo that delivers violence disproportionately to women. This politics also refuses the system that makes such violence invisible, interpersonal, and normal. At the same time that restorative/transformative data science about feminicide appropriates the methods, tools, and evidentiary models of hegemonic data science, it bends and hacks these toward shifting power across multiple domains of the matrix of domination.

METHODS TO ARRIVE AT A RESTORATIVE/TRANSFORMATIVE DATA SCIENCE

The Data Against Feminicide team's approach to understanding activist data practices has been ground-up, inductive, and community-based. One of the first questions that our team had as we embarked on the project in 2019 was: *How widespread are practices of data production about feminicide and fatal gender-related violence in the Americas?* Helena Suárez Val, colead on the project and founder of Feminicidio Uruguay, had been maintaining a list of feminicide mapping projects on her own website and had interviewed several other *mapeadoras* (women mappers) for her research.[67] We initially started a spreadsheet with that list. As our team read research papers or hosted community events, each time we came across a mention of a group producing data about fatal gender-related violence, we did additional research about them and added them. Thanks to dedicated researchers on the team, especially Angeles Martinez Cuba and Alessandra Jungs de Almeida, this informal spreadsheet became more systematized and now includes more than 180 efforts from around the world. It includes individuals and groups who monitor feminicide, femicide, MMIWG2, LGBTQ+ killings, Black women killed in police violence, and other forms of fatal gender-related violence. We did not include efforts that register forms of nonfatal violence, such as sexual harassment, sexual assault, or nonfatal hate crimes. The focus is on those efforts that *produce data* about feminicide and fatal gender-related violence. There are many more projects that use data and statistics about feminicide in a secondary way (e.g., for social media campaigns, news stories, protests, artworks, drafting policy, supporting families), but these efforts were not included unless they specifically engage in data production themselves. And since our focus was on civil society—activism, journalism, academia—we did not include government-led efforts on this list.[68]

The infographic in figure 2.3 broadly characterizes this set of feminicide observatories, groups, and projects. The vast majority of works that we cataloged monitor femicide or feminicide in the Americas, with more than one hundred monitoring efforts based in Latin America specifically. We cataloged at least one feminicide monitoring project in twenty-six out of thirty-five total countries in the Americas. This prevalence reflects the fact that our area of interest in the Data Against Feminicide project has been the Americas. Because of this geographic focus, our catalog cannot be viewed as globally comprehensive. It would not be appropriate, for example, to use figure 2.3 to make geographic comparisons like "there is more feminicide data activism in Latin America versus Asia" as we did not equally scan all world regions. And yet it is striking that there are so many efforts globally—at minimum 180 (and there are certainly many more that are unknown to us)—that are producing grassroots data about feminicide and fatal gender-related violence.

As can be seen in figure 2.3, most counterdata production projects emerge from either feminist collectives, political collectives, or nonprofit organizations, but there are significant numbers of efforts undertaken by individuals as well as a smaller number of projects coming from academia and data journalism. Not all of the projects we cataloged are active and ongoing monitoring efforts. Some efforts—like the *Persistence of Indolence* report in Puerto Rico—had a specific time period and have ended. Others operate as *observatories*, meaning that they continuously monitor cases of fatal gender-related violence in a specific geography. Here the case of Puerto Rico is interesting: after the publication of the report, a group of eight feminist organizations came together to form an ongoing coalition called the Observatorio de Equidad de Género Puerto Rico. This group continues to work closely with Castelló, drawing from her data to monitor feminicides in an ongoing way, as well as collecting and publishing data about other gender equity concerns.

In regard to geographic scope, most efforts monitor feminicide at a national scale, like María Salguero, whose goal is to register feminicides across all of Mexico. At the same time, there are important examples of larger scale and smaller scale monitoring. For example, Sovereign Bodies Institute produces data about MMIWG2/MMIP across North America and is increasingly including cases from Central and South America and globally. In large countries like Brazil, Mexico, and the United States, where covering territory is challenging for a small group, there are more projects that produce data at the scale of the state or province. For example, the Fórum Cearense de Mulheres (Cearan Forum of Women) in Brazil registers feminicides for the state of Ceará, Brazil. And there are a handful of efforts to monitor feminicide at the scale of the city—for example, sociologist Julia Moñarrez Fragoso's pathbreaking work on Ciudad Juárez, or the group Néias–Observatório de Feminicídios Londrina, which tracks feminicide in the city of Londrina, Brazil.

The dimension of time is interesting in relation to counterdata production about feminicide and gender-related violence. As is evident from the timeline in figure 2.3, data activism precedes the #NiUnaMenos uprising in 2015, and yet many projects have taken shape in the wake of those transnational protests. For the most part, observatory-style projects start from the present and document cases moving forward in time. This matches the news cycle, which is often a primary source of information about cases. Yet there are some interesting departures from this majority pattern. For example, when Suárez Val started Feminicidio Uruguay in 2015 she incorporated data shared by a prior activist, Haydée Gallego, who had done that same work from 2001 to 2014, so her database stretches back in time and builds on activist labor and care that came before her. Dawn Wilcox, from Women Count USA, undertakes historical research to include femicides and cold cases stretching back to 1950, though she faces many hurdles to obtaining such information.

As the Data Against Feminicide team learned about the scope of transnational data activism, produced annual virtual events and taught public courses, we also had questions about activists' motivations and data practices that we sought to answer through in-depth interviews: Why do groups and individuals begin their monitoring projects? What sources do they use for information? How do they classify and categorize cases?

WHO ARE THE FEMINICIDE DATA ACTIVISTS?

In 2019, our Data Against Feminicide team asked, *How widespread are data activism practices about feminicide and fatal gender-related violence?* We began a spreadsheet of violence monitoring projects that now includes more than 180 efforts from around the world (and there are certainly many more that are unknown to us).

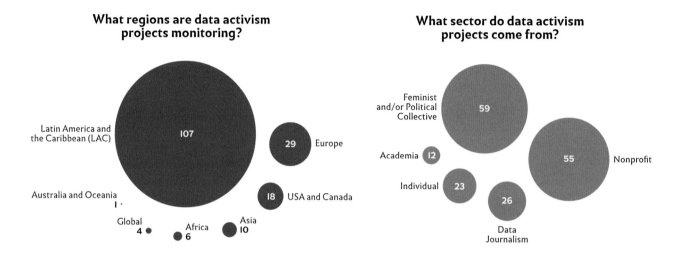

What regions are data activism projects monitoring?

Latin America and the Caribbean (LAC) 107

Europe 29

USA and Canada 18

Australia and Oceania 1

Global 4

Africa 6

Asia 10

What sector do data activism projects come from?

Feminist and/or Political Collective 59

Academia 12

Nonprofit 55

Individual 23

Data Journalism 26

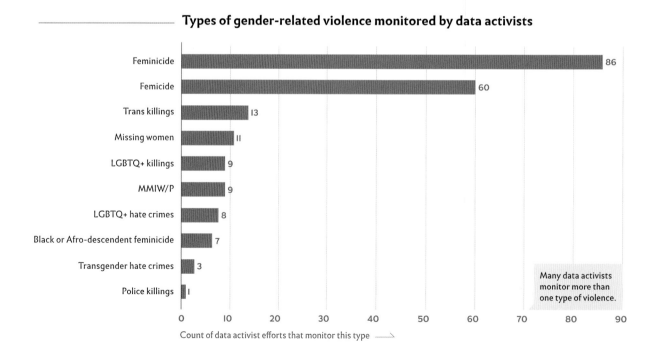

Types of gender-related violence monitored by data activists

Type	Count
Feminicide	86
Femicide	60
Trans killings	13
Missing women	11
LGBTQ+ killings	9
MMIW/P	9
LGBTQ+ hate crimes	8
Black or Afro-descendent feminicide	7
Transgender hate crimes	3
Police killings	1

Many data activists monitor more than one type of violence.

Count of data activist efforts that monitor this type ⟶

(x-axis: 0, 10, 20, 30, 40, 50, 60, 70, 80, 90)

What year did data activism projects start?

Of the 180 data activism projects the Data Against Feminicide team has counted so far, 14 had unknown start dates.

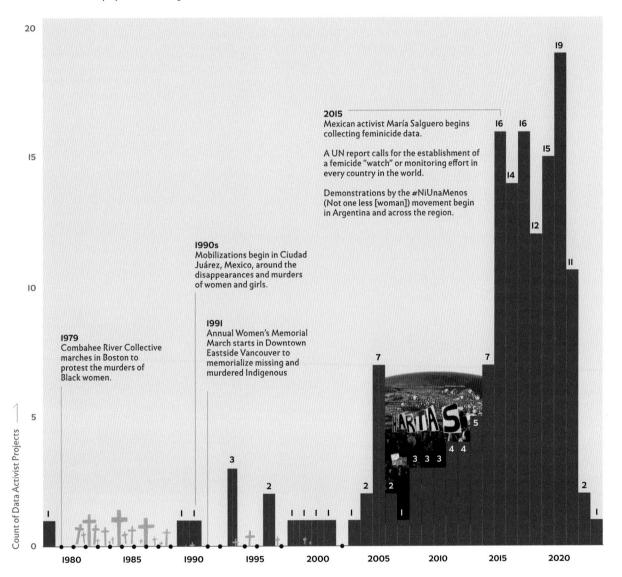

2015
Mexican activist María Salguero begins collecting feminicide data.

A UN report calls for the establishment of a femicide "watch" or monitoring effort in every country in the world.

Demonstrations by the #NiUnaMenos (Not one less [woman]) movement begin in Argentina and across the region.

1990s
Mobilizations begin in Ciudad Juárez, Mexico, around the disappearances and murders of women and girls.

1991
Annual Women's Memorial March starts in Downtown Eastside Vancouver to memorialize missing and murdered Indigenous

1979
Combahee River Collective marches in Boston to protest the murders of Black women.

Count of Data Activist Projects

FIGURE 2.3

This infographic characterizes the data activists that the Data Against Feminicide team has cataloged over the course of four years. Courtesy of the author. Analysis by Angeles Martinez Cuba. Design by Melissa Q. Teng.

The Process of
RESTORATIVE / TRANSFORMATIVE DATA SCIENCE

RESOLVING

Developing a theory of change

Background and motivation for why activists start a database of cases. Their theory of power and conceptual influences; their framing of the problem; how they encountered missing data; why they believe counting feminicides, MMIW/P, or gender-related killings may help to challenge the problem.

RESEARCHING

Finding + verifying information

Activists seek relevant information to add to their database. This can include sourcing existing datasets, mining media and other sources of information, and triangulating across sources to verify details. Such research either discovers new cases or adds information to existing cases in the database.

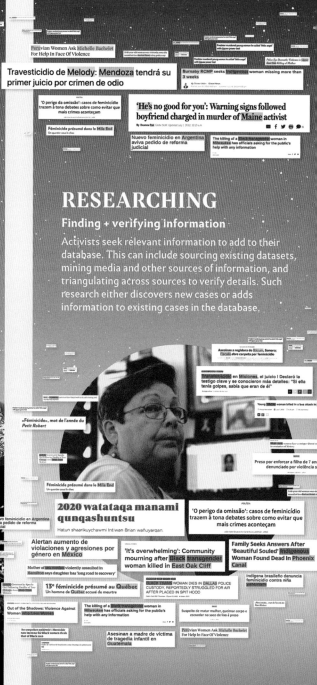

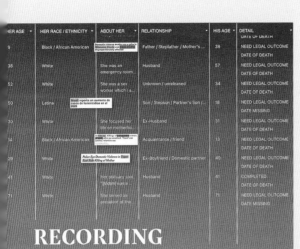

HER AGE	HER RACE / ETHNICITY	ABOUT HER	RELATIONSHIP	HIS AGE	DETAIL
					DATE OF DEATH
9	Black / African American	Domestic violence deaths can occur in families that are disproportionally affected	Father / Stepfather / Mother's ...	38	NEED LEGAL OUTCOME
					DATE OF DEATH
38	White	She was an emergency room...	Husband	57	NEED LEGAL OUTCOME
					DATE OF DEATH
52	White	She was a sex worker which i a...	Unknown / unreleased	34	NEED LEGAL OUTCOME
					DATE MISSING
50	Latina	Brasil reporta um aumento de casos de feminicídio no el 2020	Son / Stepson / Partner's Son /	18	NEED LEGAL OUTCOME
					DATE OF DEATH
30	White	She focused her life on motherho...	Ex-Husband	31	NEED LEGAL OUTCOME
					DATE OF DEATH
14	Black / African American	In Ethnic killings of women are anal...	Acquaintance / friend	13	NEED LEGAL OUTCOME
					DATE OF DEATH
29	White	Police Eye Domestic Violence in Upper East Side Killing of Mother	Ex-Boyfriend / Domestic partner	40	NEED LEGAL OUTCOME
					DATE OF DEATH
41	White	Her obituary said, "[Victim] was a ...	Husband	41	COMPLETED
					DATE OF DEATH
71	White	She served as president of the...	Husband	71	NEED LEGAL OUTCOME
					DATE MISSING

RECORDING

Information extraction + classification

Activists transform unstructured data from various sources into structured datasets located in databases, spreadsheets and/or text documents. They classify cases according to diverse typologies. They manage data, including ethics, access and governance of the database.

- Canadian Feminicide Observatory for Justice and Accountability
- Sovereign Bodies Institute, Women Count USA, African American Policy Forum, Black Femicide US, Jane Doe Inc
- Feminicidio.net
- Maria Salguero, Letra Ese
- Kilómetro Cero y Matria, Seguimiento de Casos, Observatorio de Equidad de Género
- Grupo Guastemalteco de Mujeres
- Recordar-LAS
- Utopix
- Red Feminista Antimilitarista, Colombia Diversa
- Alianza para el Monitoreo y Mapeo de los Feminicidios
- Counting Dead Women - Kenya
- Manuela Ramos
- data_labe, Uma por Uma, Observatório de Feminicídios de Londrina
- Cuantas mas
- Feminicidio Uruguay
- Observatorio Nacional MuMaLa, Mujeres de Negro, Rosario Mundosur, Observatorio Ahora Que Sí Nos Ven, Agencia Presentes, La Casa del Encuentro

REFUSING + USING

Where data go, who uses them

Reform / Working with state to formulate new laws, policies and practices

Remember / Memorializing killed people; Grieving in public

Revolt / Protesting, mobilizing; Performing resistance in public space

Reframe / Storytelling that challenges stigma; Reframing violence as structural

Repair / Supporting families and communities who have lost beloved members

FIGURE 2.4

The Process of Restorative/Transformative Data Science is a process model that describes the high-level workflow stages that data activists, journalists and organizations engage in to monitor feminicide. The graphics highlight individuals and groups of data activists and journalists, as well as the multidisciplinary restorative/transformative data science methods used in social movements across the Americas, including Say Her Name, Ni Una Menos, and MMIWG2. Collaged images courtesy of La Nación. Photo by Fabian Marelli; "Hartas" march in Neuquén, Argentina, June 3, 2017. Photographer unknown. María Salguero; Dawn Wilcox / Women Count USA: Femicide Accountability Project; Ana María Abruña Reyes / Todas / todaspr.com; Jaime Black and photo by J. Addington; #SayHerName figures from the African American Policy Forum's "Say Her Name (Hell You Talmbout)" song performed by Janelle Monáe featuring various artists. Video by 351 Studio and Wondaland; red shoes artwork by Elina Chauvet. Photographer unknown. Overall image courtesy of the author. Analysis and visualization by Angeles Martinez Cuba. Graphic design by Melissa Q. Teng.

How do they publish and circulate their data and with what impacts and effects? What challenges do they face in collecting, analyzing, and using the data they produce? What aspects of their work are they most proud of? To answer these questions, we interviewed data activists, academics, nonprofits, and journalists based mainly in the Americas starting in 2020. To date, our research team has conducted in-depth interviews with the forty-one grassroots data activism and journalism efforts listed in appendix 1, which also describes our qualitative data analysis process.

We went in seeking answers to our questions, yet as we analyzed these interviews, four stages of work emerged that were common across the varied contexts and geographies in which data activists and journalists were working. This is the descriptive model you see in figure 2.4, called the Process of Restorative/Transformative Data Science. You might immediately notice some significant differences between this process model and the typical diagrams that circulate about the process of doing data science. If you go out on the web and search images for "data science process", you will find lots of diagrams and charts with variations on linear processes like "Collect > Clean > Explore > Analyze > Visualize." These models, while they can be helpful for teaching (and I have certainly used them in technical classes and workshops) are too abstract and too apolitical to describe the work of data activists. The data science practiced by feminicide activists starts with an individual or group *resolving* to take action on the issue of feminicide and developing a theory of change that involves data. It involves a systematic process of *researching* various official and other information sources to find cases; then activists undertake *recording* of cases through an information extraction, verification, and categorization process. Finally, they employ multiple ways of *refusing and using data* to communicate their data and reach diverse audiences. Power and inequality permeate all stages of this process and feminicide data activists are creative, careful, and rigorous in navigating such ethical quandaries (which doesn't mean they always do things "right" or that there is no conflict). They also articulate many unresolved tensions. You can think about figure 2.4 as a wayfinding guide for the next four chapters of the book. In these, we will explore each of these workflow stages in detail and analyze how grassroots data activist practices depart from those of hegemonic data science. In doing so, we will draw out lessons for a restorative/transformative data science.

CONCLUSION

This chapter makes the case that official data, missing data, and counterdata are central concepts for understanding data activism about feminicide. Here I have reviewed literature on these concepts as well as formulated precise definitions that can serve us as we consider feminicide data activism. In particular, missing data as a concept is situated, relational, and political. Activists, statisticians, and theorists of ignorance would point out that there are structural patterns to missing data about feminicide; data and other forms of knowledge about feminicide and gender-related violence are actively neglected, silenced, and invisibilized.

Yet missing data do not exist outside the political demand that they should exist. In the case of feminicide, these demands come from families and social movements. When they are not fulfilled (which is most of the time), activists and journalists across the Americas are taking action to produce counterdata—resistant records that attempt to tally cases of feminicide and fatal gender-related violence. Yet missing data are not always a normatively "bad thing"; activists might deliberately hide or protect data that the state actively demands, for example. And it is certainly not a given that counterdata production will be emancipatory. Moreover, feminicide data activists do far more than count, as is evidenced by the case study of advocates in Puerto Rico. They verify and triangulate information; they produce analyses; and they circulate results in a variety of forms to challenge power across the matrix of domination. These practices constitute a restorative/transformative data science that seeks to restore life, living, vitality, dignity, and rights to wounded communities as well as to transform the structural conditions that make feminicide possible. As we will see when we go deeper into the data practices of feminicide activists in the following chapters, their restorative/transformative data science practices depart from those of hegemonic data science. They invite us to imagine a data science grounded in care, memory, rigor, and rage.

II THE PROCESS OF RESTORATIVE/ TRANSFORMATIVE DATA SCIENCE

3 RESOLVING

In 2012, members of the Red Feminista Antimilitarista (Feminist Antimilitarist Network) were seeking information about a spate of recent murders of women in Medellín, Colombia. The group had been around since the late 1990s and their name reflects their position against war, violence, and militarism in Colombia. Composed mainly of feminists, working-class women, and lesbians, they seek "other ways of exercising and building collective power," mainly through creative political actions in public space and popular education initiatives.[1] At the time, the group was leading a project to gather information about feminicide in Medellín. But local institutions had no answers. According to Estefanía Rivera Guzmán, a lead organizer with the group, this was the moment when "we realized that our institutions didn't have information systems that would account for the characteristics of the murdered women and the contexts of their deaths." So they decided to create their own registry of feminicides.

Monitoring primarily from news media articles and supplementing with information from social media and their own feminist networks, they started logging cases in their database and quickly came to an important insight: "We began to realize that feminicides happened not only because women were targeted for being women, but because they were women from specific social classes: impoverished women, women street vendors, trans women, women who practiced prostitution." This insight led the Red Feminista Antimilitarista down a path of reading and discussion for the next several years. Especially influential were works by theorists such as Rita Segato, Jules Falquet, and Verónica Gago, all of whom have emphasized the connections between gender-related violence and economic violence.[2] The group found that discussions centered only on intimate feminicide (where the aggressor is an intimate partner or ex-partner) were too limiting to explain violence against women in Colombia, where there is a high degree of neoliberalism and militarization, even in a time of supposed "peace."[3]

The Red Feminista Antimilitarista published their first report in 2014, in which they develop their concept of *neoliberal feminicidal violence*: "the extreme violence of capital on women who find themselves impoverished, stripped of power and significance in the modern, capitalist and patriarchal coloniality."[4] This included the map shown in figure 3.1, "Cartography of Neoliberal Feminicidal Violence in the Center of Medellín." The map not only shows where feminicides occurred in the district (the yellow icons) but conceptually and spatially links those killings to gendered exploitation of women's labor (the orange icons), including sex work, factory labor, secretarial work, and teaching. The top-left corner of the map describes how the group struggled to find nonmisogynist women icons.

With this report and map, the Red Feminista Antimilitarista is pointing to the economic roots of gender-related violence in Colombia. They articulate how the state, paramilitaries, and organized crime groups fight for control over urban spaces and terrorize poor and working-class women and communities in the process. Likewise, the Red Feminista Antimilitarista describes the extortionary economic practices called *pagodiarios* in which (disproportionately) women must borrow large sums of money from criminals at high interest rates to pay other criminals simply to avoid violence. For the group, even intimate feminicides are a manifestation of neoliberal feminicidal violence because the basis of intimate violence is so often connected to economic disparities, domination, and control. Colombia passed a law defining femicide in 2015 (see table 1.1), but in practice the law has been applied mainly to cases of intimate partner violence, ignoring many of the cases that the group had started documenting and passing over their intersectional economic analysis of violence.

In 2017, the Red Feminista Antimilitarista expanded their efforts to the whole country and renamed their project to the Observatorio Feminicidios Colombia (Colombia Feminicide Observatory). This is similar to the majority of our interviewees, who operate counterdata projects as ongoing monitoring efforts, often called *observatories*, that seek to document feminicide. For Red Feminista Antimilitarista, the observatory serves two major purposes. First is to comprehend a group-based, structural phenomenon at a national scale: "Beyond the data, it is the possibility of understanding what is happening to women in our country." That is to say, for Rivera Guzmán, one of the goals is to examine power using concepts and theories of power appropriate to the Colombian context and introducing novel concepts where necessary. Second, "the Observatory is a tool to build protection strategies." While the group monitors at the national scale, they continue to operate protection strategies for women through direct services such as what they call *protection circles*. This effort works with individual women facing violence to develop a community of people in their everyday lives who are briefed on the situation and who can step in to offer support to protect them from that violence. Leveraging their knowledge from the observatory, the group crafts these protection strategies as a way to prevent feminicide, another reminder that counting feminicide is less about counting the dead than it is about defending the right to life.

This chapter explores *resolving*—the initial stage of a restorative/transformative data science project and the first of four chapters that surface themes from our interviews with grassroots data activists (figure 3.2). Resolving surveys how and why grassroots groups begin counting feminicide—how they get started, how they conceive of the problem, how their understanding of the problem evolves, and why they determine that counting and registering data will be an effective method to address feminicide. As we will see, all activists mobilize an analysis of power. And all groups integrate data science into their working theory of change—their idea for what types of interventions are needed to challenge feminicide. While resolving is ostensibly the "first" stage of a restorative/transformative data science project (see figure 2.4), it is also an ongoing stage that persists and evolves. Groups are continually refining their analysis of power, their theory of change, and their data epistemology with time, experience, and learning.

Many monitoring efforts emerge from feminist activism, like the Red Feminista Antimilitarista, but other projects come from the nonprofit sector, from academia, or from data journalism. The latter is the case with Cuántas Más. The project was started by three journalists in 2014. That same year, Bolivian journalist Hanalí Huaycho was killed by her ex-partner Jorge Clavijo in front of their five-year-old son. She had reported him to the police fourteen times, but he was a lieutenant in the police force and no action had been taken. Clavijo escaped, the case was mishandled in a number of ways, and a new Bolivian government reopened the investigation in 2020. Huaycho's murder was the first in Bolivia to be classified as a feminicide under Law 348, passed in 2013.[5]

It was in this period of intense public attention on feminicide that journalists Raisa Valda Ampuero, Ida Peñaranda, and Marcelo Lazarte started Cuántas Más (How Many More [Women]). At first the team sought official data on murders and feminicide cases, but Bolivia had no open information laws, which made obtaining official records nearly impossible. So, like the majority of groups in this book, they turned to the media to monitor cases of feminicide. From 2014 to 2017, the team scoured articles from Bolivian news media and extracted details about each case of feminicide for their database. This was supplemented by information and case files contributed directly by families who were seeking—and not finding—justice. Peñaranda explained why families began to come to them: "One way to exert pressure is to make your case public. And not everyone has been able to do it."

ICONOGRAFÍA

Una simple búsqueda de iconografía y representaciones gráficas sobre la mujer evidencia la hegemonía casi completa de un cuerpo que alude a los sexual o una especie de "pose de modelo" para cualquier rol de la mujer.

Por éste motivo y desde la propuesta iconográfica crítica de los iconoclasistas utilizaremos otras representaciones de la mujer más acordes con sus propuestas alternativas.

ÁREAS, LUGARES Y LÍMITES

Río y/o quebrada
- - - Límite de comuna
—— Límite de barrio
Sistema de transporte masivo METRO
Vías destacadas

Espacialidad de la gentrificación.

Moda / Textíl / Confección
Servicios diversos y comercio
Administrativos
Aglomeraciones comerciales
Vivienda y especulación inmobiliaria
Degradación
Turismo / Ferias de negocios
Grandes proyectos urbanísticos

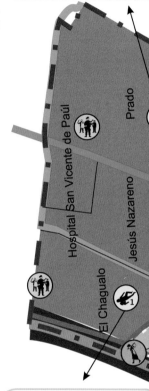

Prado centro:
Sandra Milena Pineda Otálvaro. 15 años edad.

VIOLENCIA NEOLIBERAL
Trabajos realizados por las mujeres.

Maquilas.
Mujeres que desempeñan actividades de manufactura de mercancías, predominantemente en el sector de la confección de prendas de vestir. Aunque también es amplia la versión del trabajo maquilado en la casa de la mujer, en el centro hay unos lugares que concentran en edificios estas actividades. Uno de esos espacios que se resaltan es el sector del Centro de la Moda, entre Colombia y San Juan, entre Carabobo y la Avenida Regional sentido sur norte.

Ventas y comercio informal
Mujeres que desempeñan actividades de venta de mercancías en espacios públicos, de forma estacionaria y/o ambulante, asociada al rebusque, a la venta de dulces, chazas, mercancía al menudeo. Principalmente en los encuentros de los flujos de mayor movilidad, alrededor de las calles más transitadas. Se destacan mujeres pobres, desplazadas, de la tercera edad, cabezas de familia, mujeres afrodescendientes e indígenas.

Trabajo sexual.
Mujeres que desempeñan su actividad económica en la prostitución, webcam, masajes y pornografía. Actividad ligada a los hoteles, moteles, licoreras y negocios nocturnos. Algunos lugares reconocidos socialmente donde se concentra ésta actividad económica son en el Barrio Colombia, el sector del raudal, Barbacoas, el sector entre Ferrocarril y Bolívar, entre Colombia y el Bazar de los puentes.

Oficios varios.
Mujeres que desempeñan actividades de aseo, terminados de mercancías, meseras, litografía, telas, publicidad y actividades variadas dentro de su labor. Prácticamente en todos los lugares donde hay comercio, industria y servicios. Principalmente mujeres pobres, desplazadas, cabezas de familia y afrodescendientes.

Administrativas y servicios.
Mujeres que desempeñan actividades administras y de gerencia en instituciones públicas y privadas, relacionadas con el gobierno o grandes empresas de servicios. Se reconocen lugares como las diferentes sedes del Estado, la alcaldía y la gobernación, hospitales, y grandes empresas como EPM y entidades financieras.

Estudios y servicios educativos.
Mujeres que desempeñan actividades de enseñanza y docencia en instituciones educativas públicas y privadas, y mujeres estudiantes de diversas edades en dichos centros de enseñanza. Hay una concentración de servicios educativos, en seguridad, tecnología, validación, salud y belleza, etc., entre la Av. Oriental y Boston, entre Prado Centro y Bombóna.

Ventas y comercio formal.
Mujeres que desempeñan actividades de venta de mercancías en diversidad de formas: alimentos, medicamentos, vestuario, hogar, ferreterías, etc. Prácticamente en todos los lugares donde hay comercio.

Habitante de calle y mendicidad.
Mujeres que sobre todo han sido excluidas de sus núcleos familiares y han buscado la calle como refugio.

Secretariado.
Mujeres que desempeñan actividades de atención al público, manejo y organización de información y manejo de procedimientos en las empresas. Prácticamente en todos los lugares donde hay comercio.

El Chagualo:
"Marcela" - Gabriel Mario Duque López. Entre 39 y 46 años de edad. Transexual.

MILITARIZACIÓN Y CONTROL

Actores armados masculinos.
El control militar en el centro como estrategia de dominación de clase se evidencia fuertemente. Actores y autoridades como la policía y el ejercito, las empresas de seguridad, y organizaciones del crimen y el narcotráfico, confluyen en una red donde existe en circulación muchos recursos que pueden definir cambios de poder dentro del control militar efectivo de los territorios. Principalmente por hombres de diversas edades.

Acoso sexual y laboral.
El neoliberalismo ha agudizado la explotación laboral y las condiciones de las trabajadoras. Así como en la calle, en los lugares de trabajo, la mujer es víctima de acoso sexual por parte de compañeros y jefes.

Feminicidios.
En el mapa se puede identificar los 17 asesinatos en el centro de Medellín durante el 2014.

MOVILIZACIÓN

Protesta y manifestación de las mujeres.
Diferentes colectivos y organizaciones de mujeres que tienen como lugar de movilización y protesta social espacios en el centro. Plazas, parques, edificios y rutas, han sido formas en que las mujeres organizadas manifiestan sus problemáticas y protestan denunciando los abusos y agresiones hacia ellas.

San Benito (Minorista):
Marlen Yesenia Diosa Cruz. 21 años de edad. Habitante de calle.
Leidy Orania Murillo Mosquera. 23 años de edad. Habitante de calle.
Tifany Alexandra Álvarez Moreno. 26 años de edad. Habitante de calle.

FIGURE 3.1
The Red Feminista Antimilitarista produced a map, "Cartography of Neoliberal Feminicidal Violence in the Center of Medellín," in 2014. Courtesy of the Red Feminista Antimilitarista.

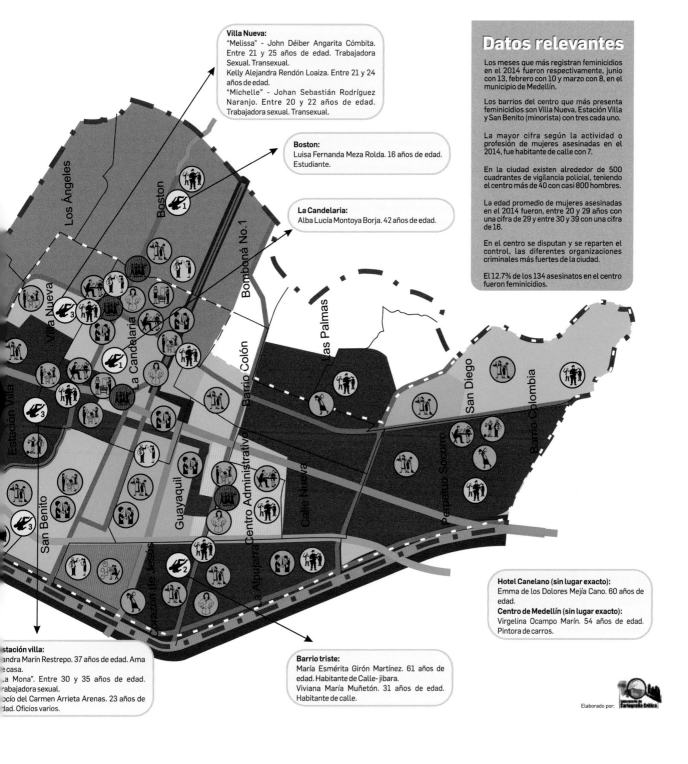

Villa Nueva:
"Melissa" - John Déiber Angarita Cómbita. Entre 21 y 25 años de edad. Trabajadora Sexual. Transexual.
Kelly Alejandra Rendón Loaiza. Entre 21 y 24 años de edad.
"Michelle" - Johan Sebastián Rodríguez Naranjo. Entre 20 y 22 años de edad. Trabajadora sexual. Transexual.

Boston:
Luisa Fernanda Meza Rolda. 16 años de edad. Estudiante.

La Candelaria:
Alba Lucía Montoya Borja. 42 años de edad.

Hotel Canelano (sin lugar exacto):
Emma de los Dolores Mejía Cano. 60 años de edad.

Centro de Medellín (sin lugar exacto):
Virgelina Ocampo Marín. 54 años de edad. Pintora de carros.

estación villa:
...andra Marín Restrepo. 37 años de edad. Ama ...e casa.
...La Mona". Entre 30 y 35 años de edad. ...rabajadora sexual.
...ocío del Carmen Arrieta Arenas. 23 años de ...dad. Oficios varios.

Barrio triste:
María Esmérita Girón Martínez. 61 años de edad. Habitante de Calle- jibara.
Viviana María Muñetón. 31 años de edad. Habitante de calle.

Elaborado por:

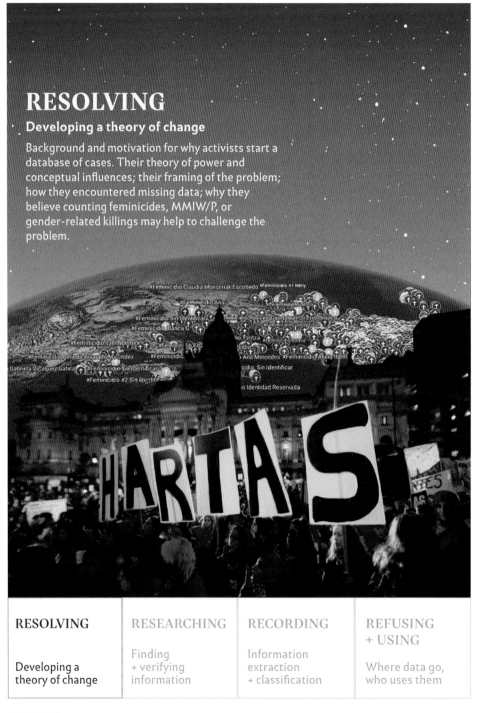

RESOLVING

Developing a theory of change

Background and motivation for why activists start a database of cases. Their theory of power and conceptual influences; their framing of the problem; how they encountered missing data; why they believe counting feminicides, MMIW/P, or gender-related killings may help to challenge the problem.

RESOLVING	RESEARCHING	RECORDING	REFUSING + USING
Developing a theory of change	Finding + verifying information	Information extraction + classification	Where data go, who uses them

FIGURE 3.2

Resolving is the first of four workflow stages in undertaking restorative/transformative data science about feminicide. Courtesy of the author. Graphic design by Melissa Q. Teng. Collaged images: Courtesy of La Nación. Photo by Fabian Marelli. Map courtesy of María Salguero.

While Cuántas Más aspired to collect every feminicide case they could find, the point of the project was not only to produce numbers. As Valda Ampuero framed it, "[We wanted] not only to fill in the missing data that are urgently needed, but also to be able to give a face, an image, to those data . . . here we are not talking about cold data but about people. Each of them is a person who has lost their life in terrible circumstances."[6] Cuántas Más published a map and and included details about each person from their life. For example, one young woman was a mariachi, and they included a clip of her favorite song. Their goal was to humanize and remember victims in life and also to challenge the misogynist, victim-blaming narrative framing of feminicide coming from police and media reports. For example, when the group went to the police for records about women murdered in domestic violence, the only data fields collected by the police were marriage status and level of intoxication, two fields that lend themselves easily to victim-blaming narratives (e.g. "she was a loose woman," "she was a public woman," "she was not in a heterosexual relationship," "she was drunk"). Likewise, Valda Ampuero found the media reports on feminicide from fellow journalists to be dehumanizing: "'100 deaths this year.' No, no, we're not going to do it like that." As they frame it, the goal of Cuántas Más's production of counterdata was to understand and explain the relationships of power behind the phenomenon of feminicide and to push back on the reductive and dehumanizing narratives that were widely circulating.

RESOLVING TO COUNT FEMINICIDE

Resolving to start a feminicide monitoring project can be both the hardest and the easiest part of a long and arduous process. Activists and journalists have various motivations for starting to register cases of gender-related violence. In the case of Cuántas Más, these factors included recent changes in legislation, a high-profile case of the feminicide of a journalist, and increased public attention to the issue. The Red Feminista Antimilitarista, on the other hand, felt that they needed more knowledge in order to understand and generate analyses of the realities of women in their country, influence political processes, and build strategies for grassroots violence prevention in Medellín.

For some activists, the motivation can be deeply personal. At least five interviewees from different groups disclosed to us that they themselves were survivors of gender-related violence or else had close family and community members who had been killed. For example, Annita Lucchesi, founder of Sovereign Bodies Institute, is a survivor of sexual and domestic violence, trafficking, and police violence. She comes from the Northern Cheyenne Tribe, whose reservation is located in Big Horn County, Montana, the county with the highest MMIP rate in the United States. Her own experiences, as well as those of her relatives and community, led her to seek data about MMIWG2 when working on her master's degree in geography.[7] What she found was both a lack of data and a lack of care for documenting the stories of women and families. This became one of the motivating factors to start first a database, and then a nonprofit organization, Sovereign Bodies Institute. In another example, the Néias–Observatório de Femi-nicídios Londrina, in Brazil, is named for the sister, Néia, of one of the founders. Néia was brutally assaulted by her husband when she tried to leave him in 2019, leaving her incapacitated, and she died two years later of a heart attack at age thirty-five. During the trial of her aggressor, Néia's sister and colleagues ran a successful campaign to attract public attention to the case. While campaigning, they were struck by the fact that they encountered many other cases of feminicide that had received no media attention and whose defendants got off with misogynist arguments about "strong emotions." The group decided to track and systematize these cases and they founded the observatory in order to "raise a voice for so many Néias."[8]

Many data activists start informally and come to understand the magnitude and complexity of the task at hand later in the process. For example, when Lucchesi initially started to compile MMIWG2 cases she thought it was a time-delimited task. As she characterizes it, "I was pretty naive in thinking that I could just kind of cross reference in a spreadsheet and come up with a number that way and that there would be a start and an end. And now, five years later, I'm still doing it and now it's my life's work." María Salguero had a similar start. She wanted to use her technical skills to help the women-led Mexican investigative journalism network Periodistas de a Pie (Journalists on the Ground), which was struggling to find data about feminicides.[9] What she thought was a one-time map has grown into a years-long mapping and data production project that has gone on to have a wide impact in policy and global news media.

When feminicide counterdata projects become highly visible, like Salguero's, they inspire and motivate others to follow their lead. Aimee Zambrano described how she had followed Salguero's work and the work of other observatories for some time before deciding to start an observatory for Utopix, an activist media group based in Venezuela. In other cases, the tipping point is a specific case of feminicide. This is how Letra Ese, a group that monitors LGBTI+ killings in Mexico, came to start their observatory. The group had been protesting and working toward justice in a number of cases of hate crimes and were frustrated by the lack of judicial response and the impunity that surrounded the cases. They started the Comisión Ciudadana de Crímenes de Odio por Homofobia (Citizen Commission Against Hate Crimes due to Homophobia) following the 1995 murder of María Elena Cruz, a trans woman. Their primary theory of change was that if they could systematize cases then they could influence the media to report on them. According to Alejandro Brito, one of the founders, the group felt that if they were successful at influencing the media's agenda, then "other agendas would be influenced: other legislative agendas, human rights, the women's movement, the LGBTI+ movement."

In all of these cases, when the groups got started they found the existing responses—from the state, the media, and the public—both inadequate and discriminatory. From the state and its institutions, they encountered narrowly framed legislation that left out many types of feminicide, families that were unable to find justice, as well as missing data—either the complete absence of official records or else data that were inaccessible, incomplete, or collected with biased categories (e.g., marriage status and intoxication). From the media, Cuántas Más describes either the complete lack of reporting or the sensationalism and dehumanization with which cases were reported. These produced a public culture in which individual acts of violence against women were normalized and tolerated. If a woman was killed, the event was more often than not written off as an isolated "crime of passion," or else the victim was blamed for her death due to some behavior that did not conform to norms of cisgender, heterosexual, white, middle-class notions of womanness.

In response to these structural barriers, groups develop their own analysis of the problem. While data activists use diverse ideas and theories appropriate to their context, in all cases they have developed and continue to refine an analysis of structural power that undergirds their work. This power analysis is central to the resolving stage of a restorative/transformative science project. The Red Feminista Antimilitarista did this through a deep and collective dive into feminist and decolonial theory, from which they then produced several theoretical innovations such as the concept of neoliberal feminicidal violence. Cuántas Más did this through their critique of existing media coverage and the victim-blaming culture that it produced. A group's power analysis directly informs their theory of change—the rationale and means by which they will challenge existing, biased responses to feminicide. Both groups—and all groups in this book—integrate counting into their theory of change. They resolve that counting and registering cases of gender-related killing will be a useful strategy to challenge power. Why? What purpose does counting serve across these diverse projects?

FROM THE PERSONAL TO THE POLITICAL

For grassroots data activists, it's not about collecting data for data's sake; none of them see data collection as a "solution" to gender-related violence. All of our interviewees were circumspect about the role of numbers and data. Rivera Guzmán asserts that the underlying analysis of inequality in Colombia is more important than counts: "Beyond the data and figures, it's the analysis of the context . . . [we have learned to] carry out analysis of the context on a daily basis, always critically." In a similar vein, Peñaranda from Cuántas Más states, "We have to understand that numbers are important, yes, but they are not the full representation of what we are seeing, rather it's the characteristics of the violence that we were seeing." As I will argue in the rest of this chapter, groups don't initiate data monitoring projects to "solve" a problem; rather they produce data to remake and reframe the problem space entirely—to transmute gender-related violence from the *personal* realm to the *political* realm.

There is a famous feminist saying, used by the Combahee River Collective and written about by activist Carole Hanisch: "The personal is political."[10] This came out of US activism in the 1960s, where women were told not to bring "personal" problems such as sex, domestic violence, forced sterilization, childcare, and abortion rights into "political" organizing work predominantly led by men. In response to the exclusion of these issues, women formed consciousness-raising groups: small-group dialogue sessions where they would share personal experiences. While consciousness-raising groups for white women focused mainly on sex and gender, the Combahee River Collective ran consciousness-raising groups that integrated discussions of race and class oppression along with sex and gender.[11] A key insight of consciousness-raising groups was that oppression works by stigmatizing and privatizing a structural problem—like domestic violence—so that an individual would face her struggle alone and ashamed, blaming herself. The goal of these groups, then, was to provide a space for processing stigmatized issues, to recognize those experiences as a structural pattern, and to channel that understanding toward collective political action and transformation. The consciousness raising group, therefore, is a space of scalar transmutation—a place where personal pain is heard and held and lifted into collective struggle.[12]

Counterdata groups do something similar by resolving to document each case of feminicide in painstaking detail. The Red Feminista Antimilitarista states it like this in their 2019 report: "Every violent death is a public death. Those murdered women not only deserve the grief of their relatives and loved ones, they also deserve a public grieving. Documenting feminicide is a way of doing this and, at the same time, a form of social reparation."[13] Groups see the production of counterdata as a way to move feminicide from a personal problem to a political problem; from a private issue to a public issue. It is the counterdata groups' analysis of power, combined with the case documentation and numbers that they produce—the aggregation of data—that allow them to attempt this scalar transmutation.

EXAMINING POWER

All feminicide data production projects begin by examining power. Here there is great resonance with the first principle of data feminism, which states that a feminist approach to data begins by analyzing how power operates in the world. Activists examine the underlying social inequality that produces feminicide. This includes consideration of the biased and unjust responses to feminicide from the state, the media, and the public. Examining power may be an explicit process that takes groups some time to develop—such as the several years of reading and discussion that the Red Feminista Antimilitarista undertook. Or a power analysis may be a shared understanding that is developed implicitly, through practice and dialogue, like the three collaborators in Cuántas Más who were determined to shift the way feminicide was narrated publicly in Bolivia and use their project to show that "[feminicide] is a problem of society, it is not a private problem." An examination of power may also emerge and develop more informally as activists start their work. For example, the activists from Néias— Observatório de Feminicídios Londrina became aware of the structural nature of the problem as they campaigned for justice for the specific case of Néias and saw how many cases did not have advocates like themselves.

Counterdata groups draw from a wide array of influences to examine power. These range from social movements (Ni Una Menos, Latin American feminist movements, Say Her Name, the MMIW movement) to individual people (María Salguero, Marcela Lagarde y de los Ríos, and Diana Russell) to international organizations (the UN, CLADEM) and also to more academic bodies of knowledge like Latin American decolonial feminist thought, queer theory, Black feminism, and Indigenous scholarship. While there is great variation in the guiding concepts that groups use to frame their analysis, all groups seek to move from a personal to a structural explanation for gender-related violence. Rather than explaining incidents of gender-related violence as isolated events, "crimes of passion," or the work of pathological deviants, data activists work to establish the deeper reality that violence is rooted in structural causes: from the feminization of poverty and deeply unequal gender structures to white supremacy and colonialism to neoliberal economic policy and territorial dispossession (see the glossary in chapter 8 for definitions of some of these systems).

But for grassroots data activists, this structural analysis is not enough. This is where their approach shows resonance with the second principle of data feminism: a commitment to using data to challenge unequal power structures and to work toward justice. Here is where the counting comes in. Just as data activists are using conceptual and analytic tissue to link diverse incidents of violence together, logging these cases in a spreadsheet or database operates as a way to empirically represent the power analysis of a given group. Feminicide data activists and journalists are not only *asserting* that intimate partner feminicides are related to feminicides of sex workers in public space are related to deaths from unsafe abortions, but are *evidencing* that fact by placing them together and by counting them together.

Counting and quantification have the power to render diverse things alike, each item neatly placed in its own row. Typically, in feminist theory this affordance has been viewed negatively. In the twentieth century, there was a long-running "quantitative/qualitative debate" stemming from feminist critiques of the patriarchal nature of knowledge production. Many feminists argued that the use of quantitative methods had the effect of silencing the standpoints of women and minoritized people, though it is worth stating that feminist researchers had also always used quantitative methods. While this discussion was never fully resolved, numerous authors have advanced the idea articulated by sociologist Liz Kelly and colleagues in 1992 that "what makes feminist research feminist is less the method used, and more how it is used and what it is used for."[14] Debates about quantification continue today. For example, Sally Merry demonstrates convincingly that global indicators about gender-based violence strip context and hide important cross-cultural variation.[15] Many critical data scholars have shown that quantification, especially arising from big data, holds the potential for harm and erasure, in both novel and painfully familiar forms.[16] In some cases where quantification is weaponized against minoritized groups, the path toward justice leads *not* through more counting but through refusals, bans, and moratoria. This is the case, for example, with facial recognition in the United States. Coalitions of nonprofit and community-based organizations led by people of color have called for bans on the technology due both to its technical inaccuracies and to law enforcement's use of it as a tool of mass surveillance and incarceration.[17]

We can hold it as true that quantification can cause harm. Yet at the same time, what about the harm of allowing a widespread, political phenomenon like gender-related violence to be endlessly pushed into the private and domestic realm, divided into individual cases, always treated separately? There is harm in refusing to see a structural pattern right in front of your eyes. To refuse to bring related things together or to give a problem a name participates in the willful ignorance and invisibilization discussed in chapter 2—the idea that what a society doesn't know or doesn't choose to know follows the lines of gendered and racialized power. In other words, while there may be harm in counting and aggregating, there may also be harm in *not* counting and *not* aggregating.

COUNTING, NAMING, AND FRAMING

This is what feminicide data activists mean when they speak of trying to *visibilizar* (visibilize) feminicidal violence. As Dawn Wilcox from Women Count USA states, "And it [femicide] just seemed like it was so ubiquitous that it was invisible . . . you get this kind of drip, drip, drip and it's so fragmented that it's hard to get a good picture of exactly what's happening—to whom and by whom and where and how." Counting and documenting and aggregating and naming is a way of transforming the phenomena that are being measured from discrete, individual events into stable and solid examples of a widespread pattern. Or, as Kathleen Pine and Max Liboiron state succinctly, "measurements make things."[18] Scholars have described how technical artifacts like databases have a performative power to shape the ways that we see the world, which Lauren Klein and I discuss in chapter 4 of *Data Feminism*. For example, journalist Jonathan Stray discusses how the US census brought into being the category of "Hispanic" in 1980.[19] This is not to say people who spoke Spanish or immigrated to the United States from Latin America did not exist prior to that point, but there had not previously been a state classification of that social phenomenon as an ethnic identity category.[20] Once social concepts are named and measured, they may scale and circulate, and, when recognized by the state, serve as evidence for policy or decision-making, for better or for worse. For example, Merry recounts how British colonial rule played a role in solidifying and scaling the Indian idea of caste, which had previously been more local, situated, and contingent: "By redefining castes in terms of categories that applied across the subcontinent, the British rendered caste a far more fixed and intractable social entity."[21] This is clearly an example of a social concept that is named, measured, and scaled in the service of domination, imperialism, and economic extraction. Yet it is precisely this power of "thingification" that feminicide data activists are seeking to reclaim: to name, to measure, to bring into being, and to scale the recognition of systemic gender-related violence. Helena Suárez Val writes about how feminist activists have long understood that "naming the violence is a way to understand the problem, to intervene in it and to contribute to its eradication."[22] Such a naming is in the service of defending life itself. As sociologist and feminicide data producer Julia Monárrez Fragoso explained to our team, "We feminists, by using the term feminicide, are saying that the life of women is a life that matters. It is a *bios*, a political life."

Thus, the counting of feminicide, the assembling of spreadsheets and databases under that name and frame, does more than simply "count the dead." Rather, counter-data actions also play a narrative, affective, and political role. This is to say that they constitute a form of necropolitical struggle—a discursive struggle over the widespread deaths of women and what they mean and who is responsible. Counting inserts into the public realm the groups' analysis of power and their uniting of disparate events under the concept of feminicide—or the concept of femicide, or MMIWG2, or Black women killed in police violence, as the case may be. One of the ideas that activists were most adamant about was how missing data about gender-related violence were related to the lack of a name for the phenomenon and the lack of a power analysis that would link disparate events together. For example, Fabiola Ortiz from the Grupo Guatemalteco de Mujeres (Guatemalan Group of Women) stated, "[In] that moment where there was no official data, there was no acceptance that there was this problem of violence." This naming and framing, backed by counting and systematizing, is a central part of the necropolitics of feminicide data activism.

Naming structural violence and assembling counts and registries does not immediately move the actors that data activists care about impacting the most: the state, the media, the families of victims, and the people in their everyday lives. Groups find that they must use creative and intentional strategies to circulate data to influence these specific actors. These methods—along with their necropolitical effects—are addressed in more detail in chapter 6. As Mariana Mangini from the Movimiento Manuela Ramos (Manuela Ramos Movement) in Peru described, "There are some media outlets that still don't consider feminicides as such. They don't write about them, they don't name them. They just put 'murdered woman.'" Audrey Mugeni from Femicide Count Kenya recounted how *femicide* is not a known word in her country and how she is constantly teaching people about its meaning. And yet, Peñaranda from Cuántas Más also sees social transformation happening in regards to solidifying feminicide as a concept: "Before 2013, in Bolivia, there was talk of 'crimes of passion' . . . It's a fairly short history of how the fact of using a word changes a country."

CHARISMATIC COUNTERDATA

The grassroots groups we interviewed often alluded to the fact that feminicide is the most visible tip of an iceberg (figure 3.3); or, another common metaphor, it was an extreme of a continuum of violence. Violence, despite its prevalence, is a notoriously difficult concept to define and measure because it is "multifaceted, socially constructed, and highly ambivalent."[23] The influential concept of a *continuum of violence*, proposed in 1988 by Liz Kelly following her survivor-centered research, describes the idea that acts of sexual violence are not necessarily discrete events, but shade and blend into each other, and almost inevitably reinforce each other.[24] The iceberg metaphor was

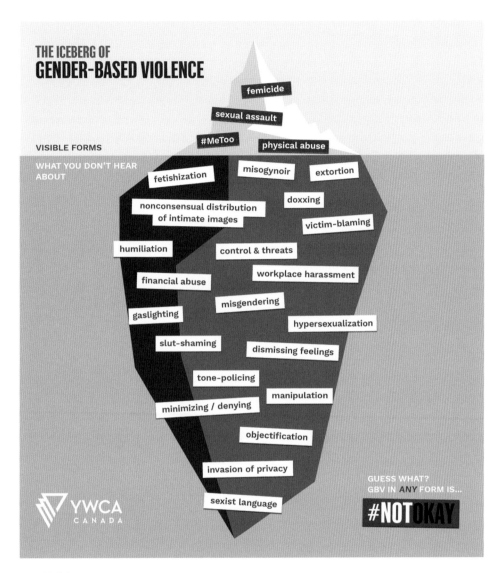

FIGURE 3.3

The iceberg is a common feminist metaphor for the depiction of gender-related violence, with some types of violence being more visible than others. Courtesy of YWCA Canada.

utilized by feminist activists to represent this idea and proposes that some parts of the continuum are more explicit and visible in public consciousness (murder, rape, insults, threats) while others remain below the waterline and are rarely seen as violence (sexist jokes, ads, microaggressions) either in law or in culture.

Myrna Dawson, scholar and founder of the Canadian Femicide Observatory for Justice and Accountability (CFOJA), puts it this way: "We all know that many of these cases start with violence, previous violence. They don't always become fatal right off the bat, whether it's intimate partner or non-intimate partner." This awareness—of the connectedness of fatalities to prior, less sensational forms of violence—was an explicit part of the analysis offered by more than half of the groups we interviewed. While they focus their attention and counting efforts on feminicide, they also try to use public attention on that concept to talk about less visible, less spectacular, and harder-to-count forms of gender-related violence: misogyny, sexist language, street harassment, harmful and degrading cultural stereotypes, economic violence, disrespect, dehumanization, racialization, erasure of trans and LGBTQ+ people, failure to hire and elect women leaders, microaggressions, exclusion of women and gender nonconforming people from masculinized professions, and so on.

This is where counterdata about feminicide becomes strategic and constitutes a form of *charismatic data*. This is a concept introduced by Max Liboiron and developed by Pine and Liboiron to denote measurements by activists to make a previously invisible form of harm visible, as well as to focus on metrics that promote dramatic or emotionally moving visions of that issue. Charismatic data functions simultaneously as "scientific evidence and proof of a moral imperative, and thus has the potential to launch action the way other data may not."[25] As the most extreme form of gender-related violence, feminicide is a (sort of) tractable measurement problem, but it is also a public-attention-gathering gateway to discussing the forms of violence that precede it and their structural roots. In this way activists can use the frame of feminicide to expand public dialogue to other forms of gender-related violence. This rationale is embodied in the way feminist journalist María Florencia Alcaraz explains the hashtag #NiUnaMenos: "It means 'no more femicides' but it is also a demand to stop all forms of oppression, from the most extreme violence to the invisible unpaid work of women; from unsafe abortions to daily harassment on the streets."[26] Each issue that Alcaraz mentions here is complex and multifaceted. Many of these issues—such as the invisible unpaid work of women—are less charismatic and harder to measure than feminicide. Yet connecting the charismatic issue to the less charismatic, but related, issues opens a gateway to a more comprehensive, structural public understanding of both the depth and the breadth of gender-related violence.

Some activist groups do in fact monitor other forms of violence. For example, Mumalá collects data about hate crimes against LGBTIQ+ people and has run surveys about street harassment in Argentina. In addition to documenting cases of MMIWG2, the Native Women's Association of Canada crowdsources stories of unsafe experiences. The Observatorio de Equidad de Género Puerto Rico has conducted studies about domestic violence during COVID. The Emergencia Comunitaria de Género run by CONAMI in Mexico logs all *violencias* submitted to them by Indigenous women or reported on social media in their area. Before cofounding the project Lupa Feminista Contra o Feminicídio, Télia Negrão had been keeping physical newspaper clippings of all cases of violence against women that she came across in the Brazilian media. Many individuals and groups expressed interest in collecting other types of data in order to reinforce their structural analysis of gender-related violence, but they are often constrained by time and resources. For example, the Argentinean observatory Ahora que sí nos ven (Now that they do see us) stated that although they work hard to produce data on feminicide, they "would like to have the time, tools and resources necessary to be able to work on other types of violence. On economic violence, on sexual violence, to be able to work on the issue of gender violence in a more structural way, and not just the tip of the iceberg." The Red Feminista Antimilitarista in Colombia aspires to expand its feminicide observatory to track cases as they proceed through the justice system so that the database doesn't only represent a registry of murders but a "registry of justice."

FEMINICIDE AS A CONTESTED CONCEPT

Grassroots activists produce data about feminicide in order to solidify and stabilize the concept of feminicide, and to bring a structural analysis of gender-related violence into law, policy, media and culture. But which feminicide? Who gets to define, describe, and typify the concept? Groups often mobilize their power analysis and their data to push back against narrow or exclusionary definitions of the concept coming from the state and the media, but also from other activists and advocates. For example, while the Red Feminista Antimilitarista wanted to connect their feminicide observatory to legal and international conceptions of feminicide, Rivera Guzmán describes how they found that the emphasis on intimate partner feminicide was limiting when so much gender-related violence was the result of armed conflict and militarization in Colombia. Given this imperative to understand feminicide in their region, the group created the aforementioned concept of neoliberal feminicidal violence. They also created categories of feminicide in their database related to other factors prevalent in the Colombian context such as organized crime, feminicide by paid assassins, and the Venezuelan refugee crisis. Rivera Guzmán sees these theoretical innovations as a way to localize to the particulars of Colombia as well as an offering back to the broader dialogue on the definition, categories, and causes of feminicide: "For us, in structural terms, we have to expand the analysis of feminicide, but it is not about expanding it for the sake of expanding it. It is that it is necessary to do so."

The desire to use data production to deepen and contextualize a geographic and regional analysis of violence is echoed by other groups. Sociologist Mariana Mora, who runs the project Cartografía de Femicidios en Costa Rica (Cartography of Femicides in Costa Rica), described how feminist scholars in the country, such as Montserrat Sagot Rodríguez, had been instrumental in formulating early concepts of femicide drawing from Central American experiences of gender-related violence. However, in Mora's estimation, there had been less scholarship since the passage of Costa Rica's legislation in 2007, and the circumstances of gender inequality had changed. This motivated her to initiate her mapping project: "[There are] new contexts of violence and inequality in which femicides are now happening in the country." This includes rising violence against women from organized crime and drug trafficking. Likewise, in Mexico, María Salguero explained how she uses the UN's Latin American model protocol—the international standard described in chapter 2—as a foundation for defining and categorizing feminicide in her database, and yet "the Mexican reality is different, it overflows any protocol." This is specifically due to the prevalence of feminicide due to organized crime and narcotrafficking in Mexico, so in recent years Salguero has focused her attention on analyzing how women's bodies are used as weapons of war, profit, and territorial domination.

Activists use their counterdata projects not only to contest, expand, and adapt definitions of feminicide to their unique regions and geographies, but also to center the importance of intersectional differences along the axes of race, class, sexual orientation, gender identity, and/or Indigeneity. For example, Rosalind Page is a Black nurse based in Arkansas. Over her decades of professional experience doing intake with patients in hospitals and clinics, she routinely asked questions about sexual abuse and domestic violence and was struck by how many of her Black women patients had experienced violence. And yet the issue was not discussed publicly: "I noticed that within the Black community, the focus is not on women and girls. The focus is on Black men and boys who are victims of police brutality, you know, any kind of law-enforcement-involved violence. And, not to take away from that issue, but I feel that within the Black community, there is a deeper conversation that we need to have." Page started Black Femicide US—a database and social media presence—to try to instigate such a public conversation about gender-related violence specifically within the Black community.

While Page uses the broad frame of femicide, and sees her work in dialogue with the work of activists in Latin America, other groups choose to be more specific in how they name the violence. For example, as part of the #SayHerName campaign, the African American Policy Forum compiles a database of Black women killed in police violence and is primarily focused on highlighting the ways in which Black women and gender nonconforming people are left out of the national US narrative around police brutality. The Sovereign Bodies Institute centered their data production efforts on missing and murdered Indigenous women, girls and two-spirit people (MMIWG2) across the Americas, and then expanded to include men and boys at the request of families, so they now monitor missing and murdered Indigenous people (MMIP). Lucchesi describes how they have in fact faced challenges collecting data in Latin America, where femicide is so central to the discussion of gender-related violence, because "Indigeneity or even race in general is just kind of left out of that narrative. There's lots of national and international discussions of femicide, but the racially specific portion of it is missing." This point of view was corroborated by the Coordinadora Nacional de Mujeres Indígenas (National Coalition of Indigenous Women, CONAMI), which leads the project Emergencia Comunitaria de Género (Community Gender Emergency) that catalogs feminicides and other violences against Indigenous women in Mexico. Stated Laura Hernández Pérez of the feminist and women's movements in Mexico, "Nobody was talking about feminicides of Indigenous women which, to this day, is our demand and something that we speak about. Feminicides of young women, of girls, and Indigenous women are the most invisibilized of the invisibilized." While data activists seek to stabilize and scale the concept of feminicide, groups also use data production and case documentation to contest the concept and draw attention to who, what, or where is disregarded and marginalized within that frame from an intersectional perspective.

Thus, as part of developing their analysis of power, data activists and journalists are enacting what Patricia Hill Collins calls the *heuristic use* of intersectionality: asking who and what and where is central to activist knowledge production about feminicide. For Collins, this kind of critical analysis is generative because it can help identify areas of overemphasis and underemphasis—to reveal, for example, that the specific violence that Colombian women face is not captured in dominant framings of feminicide or that Black women are disregarded in narratives about police killings or that Indigenous women are marginalized within activist discourses around feminicide.[27] Data activists' analysis of power—in their context, for their communities—serves as both a reinforcement of the importance of analyzing structural violence and also, crucially, a critique of which nations, communities, races, and genders are centered by reigning concepts such as *feminicide* or *police violence*. They are simultaneously critiquing the framing of these concepts (drawing the focus toward an intersectional analysis that includes gender *and* nation, gender *and* white supremacy, gender *and* settler colonialism) and expanding or contracting the frame to include people who have been excluded. What this action-tension enables is, potentially, what Collins calls "new forms of transversal politics" to confront violence.[28] *Transversal politics* include retaining groups' rootedness in their own contexts and an insistence on serving their communities while simultaneously building coalitions across differences. As we will see in chapter 5, these tensions play out around categorizing and classifying and standardizing data about feminicide and fatal gender-related violence. Fundamentally, they are questions of what—and who—counts as feminicide.

THE UNANTICIPATED BURDENS OF COUNTING

As grassroots activists resolve to monitor violence and develop their analysis of power, they do not always foresee the challenges involved in doing this work. Here the story of Cuántas Más is emblematic of these struggles. First and foremost, Valda Ampuero and Peñaranda described how their work required tremendous emotional labor, which is a theme that we will return to often in this book. "It is not easy to read ten cases of femicide and put them on a table, disaggregate them, to have to put a name, age, circumstances and all that detail, without it affecting you emotionally," explained Valda Ampuero. Caring for one's own mental health and for one's team as they encounter and document stories of brutal violence becomes paramount, a theme we will return to in the next chapter because the emotional labor required during the researching phase is intense. Peñaranda noted that, if they were to do it again, they would allocate resources differently to prioritize team care.

In addition, while the project was active, Cuántas Más gained the attention of the popular press, who often cited their figures; municipal governments, who sought them out for collaborations; and feminist journalists, who came to them seeking trainings and workshops. But this attention also proved to be challenging for a small team. Families reached out to the team with case files and requests, and the team got involved with policy advocacy. "We wanted to do everything," recounts Peñaranda. "We wanted to respond to the dizzying array of questions and proposals that came our way." They ended up spending a lot of time on capacity-building for individual journalists, but then those journalists would leave the news outlet. They realized that the people that really call the shots over cultural narratives are the owners of the media companies. The increased visibility of the project also led to media outlets using their data without crediting their work, which was extremely frustrating.[29]

A final and significant challenge was economic. Valda Ampuero asked, "How do we keep ourselves afloat—not only us, but the team of people who supported us at that time?" Cuántas Más was sustained by small grants from international development organizations, but these came with onerous reporting requirements. Rivera Guzmán affirms that economic security has also been one of the biggest challenges for the Red Feminista Antimilitarista. The observatory has been the least funded of their programs: "We don't know what is going to happen next year, if we are going to have resources or not, if we are going to continue or not. All of that is very emotionally distressing, beyond the emotional weight of seeing death all the time, murder all the time."

For these reasons and more, the labor of counterdata production is significant and difficult to sustain. This is true even (and especially) once projects acquire legitimacy and visibility in the eyes of media, policymakers, and families. Cuántas Más paused their feminicide monitoring effort in 2017 for all of these reasons. Nevertheless, some groups have found a way to persist across years and even decades (see appendix 1). Some of the longest-running efforts include the Red Chilena contra la Violencia hacia las Mujeres, which has documented femicides in Chile since 2001. La Casa del Encuentro, in Argentina, has been monitoring femicide since 2008 and has had the benefit of a secure funding stream from a national foundation. In Mexico, Letra Ese has been monitoring LGBTI+ fatal violence since 1996. Similarly, the Grupo Guatemalteco de Mujeres has been monitoring femicide for more than twenty-five years through a combination of small grants and tenacious volunteer labor.

LESSONS FOR A RESTORATIVE/TRANSFORMATIVE DATA SCIENCE

Paola Ricaurte has argued that feminicide data activism represents one approach to developing alternative data epistemologies "that are respectful of populations, cultural diversity, and environments."[30] This stands in contrast to what I have been calling *hegemonic data science*—mainstream data science that works to concentrate wealth and power; to accelerate racial capitalism, perpetuate patriarchy, sustain settler colonialism; and to exacerbate environmental excesses and social inequality.[31] How do these activist data practices differ from hegemonic data science? What might mainstream data scientists learn from grassroots feminist activists at each stage of their process? Starting in this chapter, and for the next three chapters, I will surface some of these differences and lessons in order to move toward a restorative/transformative data science.

First, feminicide data projects begin with a power analysis and the motivation to move a problem from the personal to the political realm. Counting and aggregating the phenomenon in spreadsheets and databases becomes a way to render diverse events alike, to give them a name (e.g., "feminicide" or "femicide" or "MMIWG2"), to place them under the same power analysis, and to assert them into the public realm as empirical evidence of a structural problem. In contrast to such intentional and political production of numbers, hegemonic data science has virtually nothing to say about the production of data. Data are assumed to exist as raw, apolitical observations and are either harvested through distant processes of extraction (e.g., scraping) or else produced by other people and institutions (i.e., not the data scientists themselves).

In hegemonic data science, the "science" part supposedly comes via the sophisticated methods for combining and analyzing data, not from a rigorous, scientific analysis of the conditions of its production nor of what and who is left out, erased, and misrepresented in available data, nor of how data and AI themselves may build on historical inequities to exacerbate harm. This is changing with work that investigates sources of harm in machine learning and datasets, as well as work that proposes more transparency about their conditions of production and their limitations.[32] But there is still little understanding in mainstream computer science that to produce a dataset is to bring a world into being—again, following Pine and Liboiron's description of how measurements make things.[33] This to say that data do not represent reality in some kind of one-to-one way. Rather, they play a strong role in producing, naming, and framing reality (and, likewise, in casting aside other possible realities). This makes them political, no matter the circumstances. Datasets, then, are more similar to other forms of representation and communication—such as photographs and maps—than technical disciplines would like to admit. For example, geographer John Pickles asserted that *maps make worlds* in a 2004 book on cartographic representation.[34] Practicing a restorative/transformative data science may mean to leverage this world-making affordance of data production more strategically. We see such intentional naming and framing happening with groups described in this chapter as they produce numbers to insert a feminist conception of structural violence into public discourse.

A second key lesson for a restorative/transformative data science has to do with the relationship of the numbers producers to the people they are counting. In his now classic 1995 book *Trust in Numbers*, Theodore Porter asserted that "quantification is a technology of distance," by which he meant that using numbers was a standardization technique, meant to overcome both geographic distance and distrust of doing business with strangers.[35] This holds true in many cases, particularly for knowledge produced by the state or corporations, where quantification is a management and disciplinary strategy to tame (and extract economic value from) vast territories, resources, and populations.[36] It often holds true in hegemonic data science, which values distance in knowledge production, prioritizes disinterested inquiry, and sees subjectivity and proximity to the object of study as a hazard. This stance is what Ruha Benjamin frames as "imagined objectivity" and is one of the many layers of "the ideology of white male supremacy," in the words of Adrienne Rich.[37]

However, quantification is not *always* a technology of distance. For feminicide data projects, their leaders and organizers are most often from the communities that they are quantifying. Some are themselves survivors of gender-related violence, and many, like the Red Feminista Antimilitarista, work directly in their communities on a variety of violence prevention, justice-seeking, and power-building projects. In these efforts, the positionality of the knower—as a person in the community and of the community—is crucial. The knower is proximate to what is being investigated. As Brandy Stanovich from the Native Women's Association of Canada told us, "I have very, very close friend-ships with these families." She told our team about the case of Chantel Moore, which had stayed with her because Moore was murdered by police during a wellness check in Edmonton and Moore's family "is like a second family to me." Project leaders' connec-tion to families and communities is valued, and they draw from this to build trust as well as to deepen their power analysis. They also have a clear sense of who the work is for, who they are building relationships with, and who is served by documenting cases. For example, Geraldina Guerra Garcés, from the Alianza Feminista para el Mapeo de los Femi(ni)cidios en Ecuador, insisted that their counts are not about producing statistics but rather to serve families impacted by violence to seek reparations. The Alianza has done this by providing their data to families and family-led activist groups, as well as by undertaking artistic collaborations with families to document the stories of their loved ones through multimedia maps.[38] Instead of quantification functioning as a technol-ogy of distance, then, it becomes a technology of proximation and pedagogy: a method for undertaking consciousness-raising in their community and working toward a col-lective understanding of structural injustice.

One final point on this matter, emphasized by both Cuántas Más and the Red Femi-nista Antimilitarista, is that the most technically challenging part of the work is not managing or analyzing the data nor producing data visualizations. Both groups felt that the hardest part of the work to transmit to new group members is the feminist examination of power at the heart of the work. "Sure, you can automate certain things," reflected Peñaranda, "but there are other things that you can't, and you have to invest in your team's understanding of this feminist conversation which is so important." For Rivera Guzmán, "the Excel sheet is very easy. You go along putting the person who killed her, it's very easy. But to understand which case is or isn't feminicide requires this conversation, this analysis of context, this theoretical comprehension." Here both groups articulate the importance of developing their own and their teams' collective understanding and analysis of power, along with their way of working with data using that analysis of power—meaning, their data epistemology.[39] This is a theoretical, interpretive, and pedagogical challenge—a consciousness-raising challenge—not a "can you use this API" or "can you make a predictive model" challenge. This is a third and final lesson for a restorative/transformative data science. Hegemonic data science tends to focus downstream in the data science process, developing innovative methods for analyzing data, but not for analyzing power, nor framing problems, nor challenging dominant epistemologies, nor producing data. The resolving stage of a reformative/transformative data science project invites mainstream data scientists to back up from a narrow technical focus on innovation in data operations and invest time in examining power upstream and building capacity on the technical team to navigate the informational inequalities (and community trauma) downstream.

In his 2022 book *Design as Democratic Inquiry*, Carl DiSalvo makes an eloquent case for exactly such problem-making (rather than problem-solving) as part of an experimental and democratic design practice.[40] This is to open the possibility that specific methods—such as design, computing, or data—may be used to reframe problems rather than to accept dominant problem-framings. Feminicide data activists are using data precisely in this way to refuse and reframe gender-related violence, remaking the problem itself rather than "solving" it with data.

CONCLUSION

Resolving is the first stage of a restorative/transformative data science project in which activists decide to address structural violence and determine how and why counting and registering data will be one effective method to do so. As they get started, all activist projects examine power: they develop a structural analysis of inequality. They also begin to challenge power with data: they develop a theory of change that involves systematically documenting cases. While activists have a variety of motivations for initiating their projects, this chapter makes the case that they use counts and registers of violence not to "solve" the problem of feminicide but to remake the problem, to give it a name, and to transmute it from the personal realm to the political realm. That said, feminicide itself remains a contested concept, and restorative/transformative data science projects not only use their analysis of power to challenge narrow state and media conceptions of the issue but also to contest dominant activist framings of the term.

While I have characterized resolving as a "first" stage, it is important to note that groups' analysis of power is continually evolving as they research and record cases and as they circulate their data to various audiences. Each of these stages feeds back into the resolving stage, leading groups to solidify or modify their analysis of feminicide, to refine their theory of change, to operationalize that in their data epistemology, to allocate resources differently, or to form new alliances and relationships. Resolving, then, describes a dynamic and ongoing process of problem-making, where activists are continually clarifying their analysis, seeking new strategies to acquire and record information and to influence the state, the media, and the broader public with counterdata.

4 RESEARCHING

It was 2018, and the group Mumalá in Argentina had been documenting feminicide for three years. Mumalá is the short name of the organization Mujeres de la Matria Latinoamericana (Women of the Latin American Motherland). The group was formed in 2001, during a period of economic crisis and great social need in Argentina. Mumalá characterizes itself as federal, feminist, popular, and dissident. The organization works across urban, rural, and remote parts of the country on gender rights, especially those issues that intersect with neoliberal economic violence. Mumalá's lead organizers started a femicide observatory in 2015, following the #NiUnaMenos uprising. At the time, the Argentine government had no central database of femicides (and it still does not have one that activists consider comprehensive).[1] Mumalá saw a need for independent monitoring that integrated a feminist analysis of the issue with specific political demands, and so it initiated its registry. Over three years the organization had amassed an extensive, centralized database of hundreds of cases, carefully researched and recorded by their members. But in early 2018, those same members walked away over political disagreements with the rest of the group—leaving the organization and taking the database with them.

Losing their data was a wake-up call for Mumalá. The organization used that painful moment to regroup and reformulate the way the group researched cases and stored data, as well as to expand its focus to monitor other forms of violence in addition to femicide. In late 2018, the organization relaunched what is now called the Mumalá Observatory: Women, *Disidencias*, Rights.[2] Members research and record cases of femicide, trans/travesticidios, femicide attempts, missing women, LGBTIQ+ hate crimes, and suicides of femicide perpetrators. As much as possible, they also try to record cases of indirect femicide, such as deaths from unsafe abortions.

Their research structure is now *federated*, meaning they have Mumalá affiliates based across Argentina's twenty-four provinces who monitor those specific locales for cases. This is better aligned with Mumalá's organizational structure, which is based on chapters that self-organize and coordinate local protests for gender rights around the country (see figure 4.1a). Thirty-two people in total produce data for the observatory. Like the majority of data activists we interviewed, their primary source of information is media reports. Mumalá members—whom they refer to as *compañeras* and *compañeres*—scan news media reports, conduct web searches, monitor Google alerts, and scan social media in order to discover new cases in the province they are responsible for.[3] They may also seek out official sources of information and work with their network of feminist journalists, who have more success than activists in speaking directly with prosecutors and judicial officials. Following the data loss in 2018, there is now not a single database but rather many multiple copies of the database. Activists work with their own local copies and then synchronize cases periodically. Mumalá publishes monthly and yearly reports to their social media accounts, often accompanied by analysis and infographics like the one in figure 4.1b.

The subject of this chapter is the work of *researching*—how data activists and journalists seek, find, and verify cases of feminicide and related information, especially in the absence of official data. Researching is the second stage of a restorative/transformative data science project about feminicide (see figure 2.4), and the researching process is at the heart of counterdata production work. Groups' analysis of power (from the *resolving* stage) is sharpened by encountering missing data from media and institutions during the researching stage. And *recording* cases is always a back-and-forth process with researching. As activists follow cases, sometimes over the course of months, years, or even decades, they continually seek and find new information and proceed to log that into more complete cases in their databases. As we will see, activists develop deep expertise in the information available in their contexts and engage in a vast array of creative informatic strategies to research missing data.

Compared to the groups we interviewed, Mumalá's research staff is quite large. Indeed, many feminicide data projects are conceived and run primarily by individuals. Such is the case with Dawn Wilcox, who runs Women Count USA from her home in Texas. She has a full-time job as a school nurse, but she spends many hours on nights and weekends scanning news media articles and following up on tips sent by strangers. Her goal is to document every case of femicide in the United States going back to 1950. A survivor of domestic violence herself, Wilcox became curious about femicide statistics in the United States in 2016. She started what she thought was a simple online information search and was shocked when she realized how difficult it was to

FIGURE 4.1

(a) Mumalá protesting on the seventh anniversary of #NiUnaMenos in June 2022. Courtesy of Mumalá. (b) Mumalá's observatory registered 221 femicides in 2021, including 6 transfemicides/travesticides. Courtesy of Mumalá.

find systematized information on women's murders. She found that the FBI's Uniform Crime Report was woefully incomplete, not only for women but for all homicides, as it relies on data voluntarily reported from law enforcement agencies. Other citizen-run archives, like the National Gun Violence Memorial, only register gun deaths. And domestic violence organizations tended to limit their scope to intimate partner relationships, which left out women killed by other family members, by neighbors, by strangers, in the context of sex work, and more. Wilcox felt that the fragmentation of the information contributed to the invisibility of the problem: "I felt like if I could bring all of this data into one place that, first of all, it could tell a story about what was really happening to women and where it was happening, how it was happening, who was doing it, who was killing women, what sort of relationships they had. And I felt like it would memorialize these women, which was very important. It would show that they were more than just statistics on a page."

Wilcox sources cases of femicide daily from digital news articles that she finds by using a search engine and typing a set list of queries such as "woman's body found" or "husband kills wife". Since Wilcox started the project, her work has become more widely known, and people will often email her news articles about femicide. Like many activists that we spoke with, she cross-references or "triangulates" multiple news media articles against each other and with other sources of data in order to verify information and arrive at the details she needs to log a case in her database. Reading a news article will reveal a victim's name, which she will then use to search for more details. She is adamant that every woman in her database needs a photo, and this is the piece of information that takes the longest amount of time to find. When the photos that she finds are low quality, she will retouch them. The work is immense: "I think even if I did this work full time, I would still need help. It's just . . . the sheer number of cases is just staggering." Wilcox is always working with a backlog of articles to review, cold cases to search up, and tips to follow up on.

All activists face challenges in the researching stage, but groups that monitor MMIWG2 and racialized feminicide face even more hurdles. News outlets cannot be their sole source of discovering new cases because they know that the media systematically neglect to report on this violence. Such is the case with the Sovereign Bodies Institute, introduced in the prior chapter, which monitors MMIWG2 and MMIP across the Americas and beyond. Official data are woefully inadequate, and Sovereign Bodies Institute has quantified exactly how inadequate they are. In a scan of official records in California, Sovereign Bodies Institute found that 91 percent of missing Native girls in California are also missing from at least one official database. Of those actually recorded in state databases, 56 percent of missing Native women and girls in California were classified incorrectly as a different race.[4] Government and law enforcement claim that part of the problem is jurisdiction confusion—what they refer to as *jurisdictional mazes*—meaning that it is unclear to officials themselves which county, state, federal, or tribal agency is responsible for investigating a crime. Sovereign Bodies Institute and other groups refute that claim by creating clear flow charts such as that in figure 4.3 and educating families about jurisdiction.[5] For Sovereign Bodies Institute, trying to detect cases from media articles is challenging because the media, first, don't report on MMIWG2 cases and, second, even when they do consider those cases newsworthy, according to Annita Lucchesi, founder and director of Research + Outreach, the "press don't really do a great job of acknowledging victim Indigeneity."

Instead of relying on official data or news media, then, Lucchesi describes how Sovereign Bodies Institute researchers "get creative with the data." They use Indigenous networks, both digital and analog, to discover and gather information about new cases. They review social media posts and do direct outreach. They have done public records requests through the Freedom of Information Act (FOIA), partnered with tribal enrollment offices, and used historical archives. As a family-centered, survivor-led, and survivor-centered organization, Sovereign Bodies Institute pairs its informational work with direct services to families and tries to respond to all family requests for help on specific cases. At the time of our interview in 2020, Lucchesi was working directly with a medical examiner for the state of Montana to carefully review state autopsies for a handful of cases that families wanted reopened, challenging the state ruling of "suicide" as a misclassification. Sovereign Bodies Institute's sources of information end up needing to be "multipronged" and "diverse" to counter the additional research hurdles faced by groups monitoring violence against Indigenous women.

RESEARCHING CASES

These three groups—Mumalá, Women Count USA, and Sovereign Bodies Institute—use diverse research strategies to find information about feminicide in their contexts. Researching is arguably the most time-consuming part of any counterdata production effort. This is where an individual or group spends the bulk of their time and effort: seeking, finding, and verifying cases of feminicide. The researching stage includes discovery of new cases as well as ongoing research to follow, add to, and verify information for other existing cases (see figure 4.2). During this stage of a restorative/transformative data science project, activists continually assess existing sources of information, including official databases, other counterdata and citizen data projects, news media, social media, and more.

RESEARCHING

Finding + verifying information

Activists seek relevant information to add to their database. This can include sourcing existing datasets, mining media and other sources of information, and triangulating across sources to verify details. Such research either discovers new cases or adds information to existing cases in the database.

RESOLVING	**RESEARCHING**	RECORDING	REFUSING + USING
Developing a theory of change	**Finding + verifying information**	Information extraction + classification	Where data go, who uses them

In their research process, all groups encounter many examples of missing data—data that are either wholly absent or else incomplete, incorrect, biased, or inaccessible. Missing data can function as both a motivating factor for beginning a project and as an ongoing challenge to navigate as activists research cases. For Mumalá, it was the failure of the government to create a federal registry of feminicide in Argentina, even after the #NiUnaMenos movement had made that a central demand and the government had committed to fulfilling it. For Wilcox, it was the incomplete and underreported federal homicide database combined with civil society data projects focused on subsets of femicide, but not femicide itself. And in Sovereign Bodies Institute's case, they encounter incomplete and misclassified official databases, law enforcement who use "jurisdictional mazes" as an excuse to not investigate, and heightened media bias, making cases of MMIWG2/MMIP especially challenging to document.

All activists and journalists doing this work are deeply concerned with getting the facts and details of a case straight. They learn to be suspicious of early media reports about a feminicide because these can often be filled with errors. They typically need to follow a case for some time until details are clarified. All groups seek to use multiple sources of information about a given case of feminicide to verify details. Once they obtain the name of the victim or some identifying details, activists can search for more media articles about that case as well as seek out institutional data such as police reports, court records, or death registries, a process they called *triangulation*. For certain groups—such as the groups described in chapter 2, the Fórum Cearense de Mulheres in Brazil, and the Grupo Guatemalteco de Mujeres—advocates start with the official data and then triangulate that with media articles in order to understand case details, context, and whether the state had properly classified the death. Activists learn which outlets cover the issue, how it is reported, and which terms and language and framing are used. For example, Feminicidio.net is a group that has been monitoring feminicide in Spain and producing reports since 2010. Given over a decade of experience navigating different sources of information, they have developed extensive expertise in the Spanish digital media ecosystem. Nerea Novo, a researcher with the group, says it like this: "The methodology [for searching for cases] can be taught quickly but then practice

FIGURE 4.2

Researching is the second stage of a restorative/transformative data science project. Courtesy of the author. Graphic design by Melissa Q. Teng. Collaged images: Courtesy of Ana María Abruña Reyes / Todas (https://todaspr.com).

gives you certain tricks to get to know which sources are more reliable, which sources less, which sources dedicate more time to research and which do not."

The work of researching cases—reading and scanning stories of violence, sometimes for hours every day—takes an emotional and mental toll on data activists. Betiana Cabrera Fasolis, from Mumalá, concedes, "Yes it's true . . . the work we do is quite pessimistic. The compañeras never cease to be surprised at the amount of violence in new cases and old." Researching feminicide involves the continual emotional labor of reading about brutal violence and the secondary witnessing of the trauma and loss of others, a question we will return to at the end of this chapter. For survivors of gender-related violence, it is work that hits close to home.

POWER AND RESISTANCE IN THE INFORMATION ECOSYSTEM

In undertaking research, activists operate in an *information ecosystem*: a dynamic constellation of actors that includes infrastructure, tools, technology, producers, consumers, curators, and sharers of information about feminicide. The metaphor of the ecosystem is designed to capture the dynamic nature of information; it moves and flows across scales and sites and actors as it is produced, curated, transformed, and used.[6] In the case of feminicide information, the sites and actors that surfaced most frequently in our interviews included the state and the media as key producers of information about cases (and, paradoxically, also key drivers of missing and biased information). Families themselves also provide information to activist groups.

As we interviewed activists and journalists, they provided analysis and insider perspective on their information ecosystems. They have expert answers to questions like the following: Who produces information about feminicide cases? How reliable is it? How can it be verified? How public is it? If it is not public, what are other ways to acquire it? When is information biased or unreliable? Why and how is information missing, biased or dubious? Indeed, we found that activists connect their analysis of power (discussed in the prior chapter on resolving) into concrete observations about how that power is made manifest in the information ecosystem in their country or locale, and they then use creative strategies to address those informatic gaps and biases.

Table 4.1 places activist analyses of missing data in counterpoint to the strategies they use to overcome that missing data through research.[7] Activist analyses of power are grouped into the domains that surfaced most frequently: the state (including both laws and the implementation of laws), media, and families. Each of these domains is important in the information ecosystem because each significantly affects the production of data and information about feminicide. These groupings also map to the

Table 4.1

Activist analyses of missing data and research tactics they use to overcome it

Domain	Reasons for missing data mentioned by activists	Research tactics activists use to overcome missing data
State—laws	• Absence of laws on feminicide • Narrowness of laws on feminicide • No public disclosure laws • Law doesn't recognize certain groups (trans people, Indigenous status of people) • Fragmentation of laws	• Include cases based on structural analysis of power • Count people that are excluded by present laws, including LGBTQ+, children, Indigenous people not recognized by state • Use nonstate information sources
State— implementation of laws	• Narrow interpretation of law • State ignores or avoids disclosure requirements • Lack of state resources to investigate cases and/or publish information • State employee turnover (police, prosecutor, judicial) • Lack of state expertise in gender, race, feminicide • Public information not disaggregated (by gender, race, sexuality) • Public information shows signs of political manipulation • State is absent in rural/remote areas • State consistently misclassifies cases (e.g. suicide cases, cases related to Indigenous women, trans people) • Information collection and reporting is fragmented across state agencies	• Manually review state websites, state social media feeds, state-run WhatsApp groups • Search court records • Follow up on individual misclassification/suicide cases, especially where family and community are contesting state ruling • Triangulate state records with media reports to see what's conflicting • Visit morgues and medical examiners • Work in networks and coalitions to discover cases and share information • Solicit/receive crowdsourced reports of cases • Use federated media monitoring structure to cover large territories • Use media sources—TV, radio, print, digital—to discover cases • Use other citizen-led databases * Partner with tribes and cross-reference their records * Mine state-run historical archives * File public records requests, like FOIA in the United States * Call friends who work for the state and discuss discrepancies

Table 4.1 (continued)

Domain	Reasons for missing data mentioned by activists	Research tactics activists use to overcome missing data
Media	• Media report sensational details but do not provide info on race, tribe, sexuality • Media report on some deaths and not others (e.g. often don't report on trans, Black, Indigenous, migrant, rural/remote, poor people) • Media reporting about feminicide is biased, toxic and sensational • Media draw from police reports and framings, replicating state biases (like misgendering trans people) and use victim-blaming frame • Media are absent in rural/remote areas • Media do not report on killings related to organized crime/paramilitary activity for fear of retribution and violence • Media do not follow cases through the justice system • Early media reports on killings are full of incorrect information	• Use stigmatizing language to search for cases in media, like "crime of passion," "man dressed as woman," but reject their framing • Seek cases from hyperlocal and regional news outlets and blogs • Use social media (esp. Facebook) and private groups (esp. WhatsApp) to find and validate cases outside mainstream media • Use networks and partnerships to confirm details/verify information • Collect humanizing information (e.g., photos, details about life) • Triangulate state records with media reports to see what's conflicting • Follow individual cases through judicial system * Use death certificates to corroborate race (with caution, activists know these are often wrong) * Follow prominent journalists reporting on feminicide
Families	• Family/community doesn't report because too much work, already traumatized • Stigmatizing for families to report feminicide (e.g., elite families try to keep cases out of the press) • Minoritized communities have good reasons not to trust the authorities so they don't report • Families may face violence for reporting	• Partner and share info with groups that provide services to families • Families/communities contact activists directly to request inclusion or to share more details * Get individual case files from families, friends, or state leaks * Contact families directly to verify information or offer support/services

* = Strategy mentioned by only 1–2 groups. Not a common pattern.

domains of oppression outlined in Patricia Hill Collins' matrix of domination, introduced in chapter 2. While that chapter provided a high-level analysis of different factors at play in producing missing data about feminicide in each domain, our interviews helped to elucidate activists' own analysis of factors that led to missing data in their locale.

MISSING DATA: STATE LAWS

Many interviewees mentioned either narrowly framed legislation or the complete absence of legislation leading to the lack of state-produced information (see table 4.1, first row). As Myrna Dawson from the Canadian Femicide Observatory for Justice and Accountability stated, "In Canada, nothing is officially seen as a femicide because we don't have any legislation or any official recognition of femicide." When laws do exist, they may be narrowly formulated. A handful of laws only include intimate partner violence, and many exclude or make no provisions for transgender women, so these cases will not be included in official counts, lists, or statistics (see table 1.1). Even when laws and official data do exist, activists mentioned that the absence of public records laws can inhibit the availability of information.

Activists counter these legislative hurdles by deliberately defining and counting feminicide in a way that matches their own structural definition of the violence, which typically exceeds the state's legal definitions. For example, Mumalá counts induced suicides—when a woman is driven to suicide by repeated domestic abuse—as femicides. This is a concept not currently outlined in Argentine law but it is recognized in El Salvador's feminicide law, which Mumalá has drawn inspiration from. They also count deaths from unsafe abortions as femicides. This information became a central part of the massive national movement for the legal right to abortion in Argentina in the years leading up to 2020, because it showed that unsafe abortions were a leading (and preventable) cause of maternal mortality. Mumalá also seeks information on transgender killings, *travesticidios*, and also on the premature deaths of trans/*travesti* people, which Mumalá attributes to gendered forms of social and economic violence that reduce trans life expectancy to half that of the cisgender population. By necessity, they must seek nonstate sources of information to register such violence. But any such expansion is only done with careful consideration. For example, in the absence of adequate legal definitions for phenomena like transfeminicide, Mumalá relies on discussion, debate and consensus-building about individual cases through their WhatsApp group, what Cabrera Fasolis calls "the soul of the observatory." Through such collective deliberation, the group determines whether a case corresponds to their definition of feminicide or trans/travesticidio, using their analysis of power, and therefore whether it should be counted.

MISSING DATA: STATE IMPLEMENTATION OF LAWS

Likewise, in terms of the implementation of laws, groups mentioned a variety of factors leading to state oversight, bias, and mismanagement of cases that affected downstream information systems (see table 4.1, second row). For example, activists and journalists highlighted the lack of state resources to implement laws, investigate cases, and/or publish information. For the Alianza Feminista para el Mapeo de los Femi(ni)cidios en Ecuador, among many others, they saw an acute absence of the state in rural areas of the country, leading activists to speculate that cases in those areas are much higher than official counts. Some groups mentioned the frequent turnover in law enforcement officials, leading to the loss of institutional information about cases. As Julliana de Melo from Uma por Uma noted, "Local police chiefs in the Pernambuco police change constantly which hinders the investigation process a lot. We saw the same case passed on to several local police chiefs who did not have access to the amount of data that we had. [. . .] So, this shuffling, this inconstancy on the part of the police, was also something that we realized was a failure in the system."

Other activists surfaced questions of bias and training, noting how the public sector employees are not trained to recognize or investigate gender-related violence, leading to disregarding such factors in cases or to misclassify cases. State misclassification, especially of race/ethnicity (for Indigenous people), gender (for trans and gender nonconforming people), and cause of death (for accidental deaths, induced suicides, and police violence) was one of the most frequently mentioned drawbacks of official data. Even when the state may produce information on fatal gender-related violence, activists stated that there is often institutional fragmentation over which agency collects which cases, leading to multiple "counting" arms of the state that count different phenomena with different criteria. This is an observation corroborated by ILDA's work on their data standard in Latin America, discussed in chapter 1, which found that states lack both technical capacity and shared methodology to share feminicide data across divisions internally. Finally, many activists noted that official information about feminicide is often not published in a timely and disaggregated way, and it occasionally shows signs of tampering, as was mentioned in the case of domestic violence data in Puerto Rico in chapter 2.

Despite challenges in acquiring official information, activists get resourceful to mine what they can from diverse state sources. For example, many manually review state websites, attorney general pages, court records, state-based social media feeds, or state-sponsored chat groups on WhatsApp or Telegram. These never yield "data" in the sense of systematized information in rows and columns, but activists can extract important details about individual cases for their databases. Activists based in diverse places—Guatemala, the United States, and Canada—have also resorted to visiting morgues to interview state employees and review specific cases together. For example, Sovereign Bodies Institute's aforementioned work with a medical examiner in Montana led to a case of accidental death being reexamined as a murder. As Lucchesi recounted the story to us, she described the impact of the collaboration: "It was very graphic and traumatic at first but it was also really empowering. And I was able to explain things to the family that I wouldn't have been able to explain otherwise."

While it can be acquired creatively, information from the state may still be biased, misclassified, or unreliable, so activists have various strategies to triangulate it with nonstate information. These include incorporating cases from other counterdata or citizen-led projects. For example, in the United States, both Wilcox and the African American Policy Forum periodically copy relevant cases from the Fatal Encounters database, a grassroots effort that documents police violence. Lucchesi said she tried to do the same for Indigenous women, but Fatal Encounters had placed Native cases into the "other" racial category, illustrating how counterdata projects can themselves replicate the same biases present in official data sources.

Activists develop novel and collaborative human networks of information-sharing in order to fill in gaps in official data. This is evidenced by Mumalá's federated monitoring structure where they are able to cover a vast geographic territory by having members responsible for discovering cases in each province. The Alianza Feminista para el Mapeo de los Femi(ni)cidios en Ecuador works in a similar way where their coalition includes regionally based groups that serve as a key source for information about cases in rural and remote areas. Many groups have developed relationships with feminist journalists and leverage those relationships to source information. And finally, once activists' work becomes well-known, like Wilcox's and María Salguero's, they begin to receive many crowdsourced tips about cases through email, messaging apps, and social media.

MISSING DATA: MEDIA

By and large, the most prevalent source of nonstate information for feminicide is the news media (see table 4.1, third row). Several times a week, members of Mumalá will type search queries such as "death of a woman Cordoba" into search engines to source recent news reports. They also plug those queries into Google Alerts so that they get notified when Google indexes new web pages that meet those criteria.[8] Because Mumalá distributes monitoring across individual provinces, the geographic modifier— "Cordoba"—is meant to limit the search to reports from the province of Cordoba. But search engines don't always index the crime tabloids, small regional news sites, and local blogs that activists find most helpful in their research. So compañeras and compañeres from Mumalá find they still have to go to the website of each news outlet in that province and read through different sections where feminicides may be reported. Most groups do this digitally, going site by site and section by section, but a small number of groups work in other formats. Members of the Grupo Guatemalteco de Mujeres read and clip physical newspapers, and Carmen Castelló watches the daily news on TV to discover new cases in Puerto Rico.

While all groups use news media as either a primary or secondary source for researching cases, all groups are also deeply critical of media reporting on feminicide and gender-related killing. It was variously called sensational, irresponsible, shameful, victim-blaming, misgendering, dehumanizing by design, stigmatizing, toxic, transphobic, lesbophobic, racist, xenophobic, demeaning, trauma porn, misery porn, and poverty porn. Activists described how media often draw directly from police reports and quote law enforcement as an authoritative source, which leads to the transmission of bias—a kind of collusion between the state and the media. A story recounted to us by Toni Troop illustrates this point. At the time of our interview, Troop was the executive director of Jane Doe, an organization that monitors fatal domestic violence in the US state of Massachusetts. She described how she was deeply frustrated that news articles always quoted police and never reached out to domestic violence prevention organizations. She invited three local Boston reporters to a meeting and asked them directly why that was. "And that was the one of the most enlightening conversations of my career," she emphasized, "because what they said was, 'We're going out to get the facts. And the facts change between our first story, versus the second story, until there's an arraignment. Some of the police officers and the DA's offices, they're not even calling it domestic violence yet. It's not up to us to do that.'" At that point, Troop had an aha moment in which she realized that the press felt unqualified to name the violence as *domestic violence* until a state official had uttered that term. While the press felt that they could not name the violence unless police named it, they did not consider the harm they might be doing by *not naming* the violence or by solely relying on police as the authoritative namers, framers, and definers of violence.

In many cases, activists have to adopt the stigmatizing language used by the media into their search queries in order to retrieve the articles they need (i.e., googling "crime of passion"). However, activist groups generally reject the framing of the news articles, which tend to report on killings in ways that sensationalize the violence, depict them as isolated incidents, and blame victims for their own deaths. This was why Cuántas Más ran so many workshops, as discussed in chapter 3, trying to improve journalists' understanding of gender-related violence. This is to say that media articles are useful for extracting specific fields needed for activist databases—such as victim age, method of death, or the relationship between perpetrator and victim—but they are toxic and harmful for the more important tasks of framing and analyzing the phenomenon. This is why groups' development and use of their own analysis of power, described in the prior chapter, is so important.

In response to harmful media narratives, activists reject dominant media framings of cases, use social media, and work with organizations on the ground, and often directly with families, to verify important details about cases. Like Wilcox, many activists seek out humanizing information such as photos or details about people's lives in order to give them some form of memory justice that the media has denied them. Where news media tend to overrepresent the point of view of the state, activists strive to believe, support, and follow the lead of victims' families and communities. For example, La Casa del Encuentro's observatory was founded in 2008 because the family of Adriana Marisel Zambrano, an Indigenous woman from a rural province in Argentina, could not get the attention of either the judicial system or media and reached out to the feminist organization for support.

Another built-in flaw in relying on media reports is that, in the words of Silvana Mariano from Néias—Observatório de Feminicídios Londrina in Brazil, "the press are selective." All groups recognize this as an inherent limitation of using news articles to systematically detect cases of feminicide, but some groups experience it more acutely. Activists like Mumalá who monitor large geographic territories discussed how media don't cover those areas or, in some rural and remote places, simply don't exist. There are other regions where media do not report on violence related to narcotrafficking for fear of retaliation from organized crime networks. Groups that monitor gendered violence at the intersection of race, Indigeneity, and sexuality, such as MMIWG2, LGBTQ+ killings, and Black women killed in police violence, must look beyond news articles to discover cases because they are systematically underreported in the media, as discussed in chapter 2. These groups develop extremely creative strategies, such as the Sovereign Bodies Institute mining historical archives, doing FOIA requests, and developing partnerships with tribal enrollment offices. Gregory Bernstein, who worked at the African American Policy Forum on their database of Black women killed in police violence in the United States, stated that such work requires "finding different, inventive ways of learning these stories." But at the same time, this work requires activists to acknowledge the incompleteness of their work; because the information is so challenging to find, Bernstein says, "we know that there are so many that we are missing."

MISSING DATA: FAMILIES

While mentions of families and communities connected to missing data were less frequent, activists did note some situations that led to families not reporting cases and thus to resulting gaps in information (see table 4.1, last row). First, while the pain of loss may lead some families to seek to make their cases public in order to secure justice, others handle their trauma by refusing to engage with either the state or media around the case. Numerous groups discussed that as *feminicide* has grown in usage as a term, a stigma has been associated with it. This has led to wealthier families trying to suppress information about cases, particularly when the perpetrator is another member of the family. In contexts of racialized state violence like the United States or widespread corruption such as El Salvador, families and communities have good reasons to not trust the authorities and may not report for fear of incurring retribution or further trauma. For example, Kimberlé Crenshaw, legal scholar and organizer of the #SayHerName campaign, and her colleagues wrote about the case of Kayla Moore, a transgender woman in Berkeley, California, who was undergoing a mental health crisis.[9] Her roommate called the police who, instead of providing support, attempted to arrest her and ended up suffocating her to death in her own bedroom. As their paper, and the broader #SayHerName campaign, demonstrates, these stories of state violence represent a widespread and systematic pattern, leading minoritized communities to not report and to not engage with institutions that harm and terrorize them.

Working with families can be an important source of information for activists to confirm details about a case, but activists are careful in how they engage with families. Some, like Sovereign Bodies Institute, do provide services and support to families, but Lucchesi emphasized the importance of waiting for families to contact them: "We don't reach out to families directly until they're ready. That trauma is so severe that there are huge unintended consequences that can come from directly soliciting family." The majority of data activists do not work directly with families, but rather offer various forms of *acompañamiento* (accompanying) to families and family-centered advocacy groups. This might mean providing space for their meetings, showing up for marches organized by family groups, providing data and information for family-led vigils, or other forms of support and solidarity. Through these relationships and partnerships, data activists and family groups often do end up sharing information with each other, sometimes even full case files. These can be important sources, especially when such information is not available from state or media sources.

INFORMATIC RESISTANCE IS CONTEXTUAL AND RELATIONAL

As a consequence of combining their power analysis with their researching practices, groups develop deep expertise in the flawed information ecosystems surrounding feminicide and gender-related killing. This is what Lauren Klein and I meant when we outlined the data feminism principle to *consider context*: "Data are not neutral or objective. They are the products of unequal social relations, and this context is essential for conducting accurate, ethical analysis."[10] Feminicide data can never be collected, compiled, nor used at face value, because the contextual conditions of their production so deeply affect their quality.

Activist expertise includes not only creativity and skill in locating individual data points to enter into their spreadsheets, but deep familiarity with the political, historical, legal, cultural, and geographic factors—the contextual factors—affecting the production, availability, and reliability of information about the issue. We can see this evidenced by Sovereign Bodies Institute's MMIWG2 Jurisdiction Flowchart in figure 4.3, which they use to teach families and advocates about the legal geographies of a particular case (as well as to refute law enforcement's claim that jurisdictional mazes are preventing them from launching an investigation). Sovereign Bodies Institute published the flowchart in their organizing toolkit in 2020, and it walks the viewer through a series of questions in order to determine legal jurisdiction of a particular case in the United States. Factors like whether the victim was taken over state lines, whether the crime happened on a highway, whether the crime happened on a reservation, and whether or not the perpetrator is a tribal citizen all shift which agency would be responsible for handling a case. Note that many of these details may not be known at the outset of a crime, so the responsible agency can shift while a case is under investigation, and information may be (often is) lost in the transfer. In-depth understanding of the legal landscape—as well as its shortcomings—has allowed Sovereign Bodies Institute to amass more complete and correct case data. In fact, their database has caught the attention of the federal government, which does not have comprehensive case information despite pressure from families and legal mandates to do so.[11] Federal agencies have asked Sovereign Bodies Institute multiple times for their data. Each time, the organization has consulted with families and the answer has been a resounding "no." Lucchesi explained the rationale behind their refusal:

> The federal government has never published any data or even just a number of how many cases have occurred in their jurisdiction. Members of the FBI and DOJ requested access to our database and we said we would be willing to sit at the table and discuss it if they (1) made data on cases in their jurisdiction available to us and to tribal governments and (2) increased the rate of prosecution of non-Indian sexual offenders in Indian Country (over 70% of the cases federal prosecutors decline in Indian Country are sexual assaults). The only response we ever received was 'We understand.' That tells us they are unwilling to provide data or hold perpetrators accountable—why would we trust our data with them if we can't trust them to keep us safe and document our deaths once we have been killed in their jurisdiction?

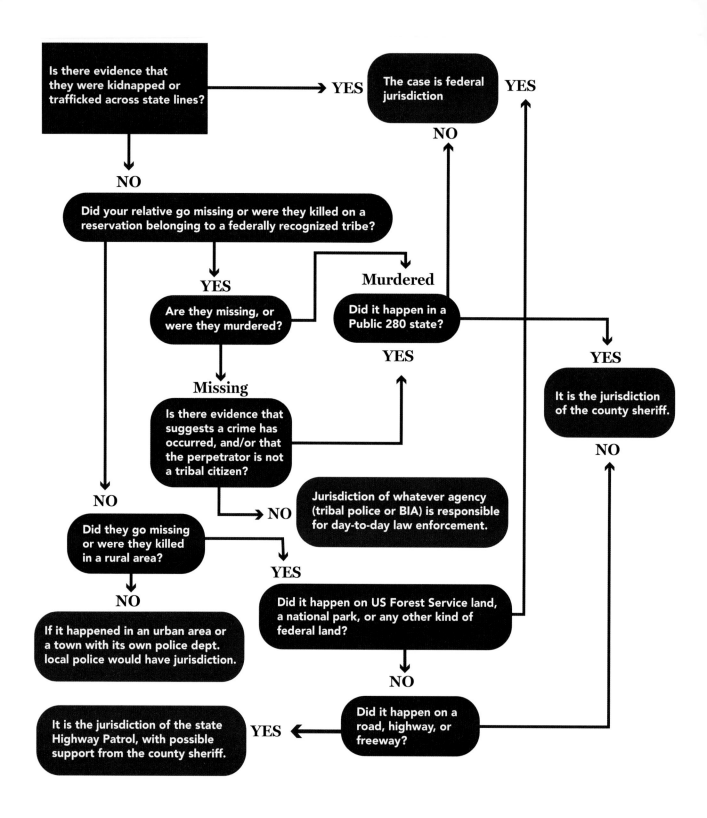

Sovereign Bodies Institute has a clear analysis of the context and circumstances leading to missing data, as well as what it would take to start to build trust and share data between government officials and civil society groups. As their work and table 4.1 show, while oppression may be organized to produce missing data and flawed information across multiple domains, creative informatic resistance is organized to respond in a multiscalar, transversal way. Some informatic tactics of activists are designed to respond to failures of the state and some to failures of the media and some to challenges of missing data in the familial domain.

This multiscalar resistance across domains of the information ecosystem demonstrates that feminicide data activists and journalists have a deep understanding of the legal, administrative, narrative, and interpersonal factors that produce missing data in their context and have also crafted creative responses to research information in the midst of those challenges. While prior research on data activism has tended to focus on how activist data circulate publicly and what political outcomes they achieve, examining the activist labor of data production demonstrates just how much expertise activists themselves must build about the unequal and deeply flawed conditions of information production about feminicide.

One important throughline across activist research tactics in all domains is not only their use of existing information sources but their creative production of coordinated, collective human relationships as a mechanism for case discovery, information sharing, and deliberative decision-making about cases. This resonates with data feminism's principle to *embrace pluralism*—to bring together multiple perspectives with priority given to local, collective, and Indigenous ways of knowing. This is embodied by Mumalá's federated media monitoring structure, where different individuals are responsible for monitoring different Argentine provinces. Or in their collaborative method for deliberating, through WhatsApp, about which cases should be included in the database. It is also manifest in their relations with feminist journalists whom they can call on to try to get information from prosecutor offices. We can also see this evidenced in Sovereign Bodies Institute's work to build relationships with families to help them access services, and with grassroots Indigenous groups across North America, who then notify each other about new cases. There is collective power in the development of these novel human-relational-informational configurations.

FIGURE 4.3

MMIWG2 Jurisdiction Flowchart from the Sovereign Bodies Institute's MMIWG2 & MMIP Organizing Toolkit. Courtesy of Sovereign Bodies Institute.

THE EMOTIONAL LABOR OF RESEARCHING VIOLENCE

As I mentioned at the beginning of this chapter, a persistent theme across the researching phase is the emotional labor required to research cases of violence as well as the mental and emotional toll of doing the work. Lara Andres, from Ahora que sí nos ven (Now that they do see us), an observatory in Argentina, says it like this: "The truth is that we can't spend all day, sitting down, reading about feminicides, because it does your head in. It makes you sick, so I think the most time I've spent registering feminicides, or reading news, would be two or three hours a day." Individuals talked about the psychological and emotional effects of the work on their well-being. Carmen Castelló found that watching the news to find cases early in the morning would leave her in a state of anxiety for the rest of the day, and Nerea Novo from Feminicidio.net discussed stories that she had read that she couldn't stop thinking about. It can be especially challenging for activists who work directly with families. As Brandy Stanovich from the Native Women's Association of Canada described: "When you're in certain settings, you have to be the strong one in that room because you're with the survivors of people who have gone missing or been murdered. But, later on, you might just want to lay down and cry because you can feel the emotions." Some people find that it is not only about the graphic details of some cases, but also that the counting and aggregating becomes too much to bear. Paola Maldonado Tobar, from the Alianza in Ecuador, reflected, "This work of marking how many women have been victims of feminicide and placing those on a map is painful, it's terrible." And for Audrey Mugeni, from Femicide Count Kenya, seeing the aggregated numbers is tremendously difficult: "When you go to the Excel sheet you're like, oh no, I need to close my eyes. So it's so much harder to look at the Excel sheet than it is to just read the stories."

Given the range of emotional and affective responses that activists experience as they research and record each case of feminicide, it is curious that the output of this labor—typically the neatly arranged spreadsheet or database—barely reflects this turbulence. In April 2022, I had a conversation with Helena Suárez Val, colead of the Data Against Feminicide project, about how challenging it is to *make labor visible* in relation to the research work that goes into feminicide data production. This is one of the principles of data feminism, drawing from feminist thinkers who describe how women's labor—including emotional labor and care work—is frequently invisibilized and devalued. An antidote, then, should be to show and credit that labor. Yet in the case of feminicide research, there are political and protective reasons to keep that labor hidden.

When you visit the publicly available spreadsheet that Helena produces for Feminicidio Uruguay, it is straightforward. Almost painfully simple to read. Each row

FIGURE 4.4
The open spreadsheet for Feminicidio Uruguay. Courtesy of Helena Suárez Val/Feminicidio Uruguay.

corresponds to a woman, to her violent death, and to related details surrounding the event: name, age, date, place, relation of the perpetrator to the woman. There is a matter-of-fact narrative in Column E that describes how her body was found, method of death, or whether she had been missing. Yet each row in this spreadsheet is the result of many hours of reading and research, triangulating details from multiple media articles, updating as new information surfaces about an investigation. Furthermore, because Helena produces a map based on her database, she often spends still more hours using photos, Google Street View, and media articles to find the most precise place to geolocate a particular case.

What is not visible in the spreadsheet is how Helena feels when she walks by one of the places of violence that she has geolocated on her map. One of them is about half a block from her apartment in Montevideo, and she remembers the blood on the side-walk every time she walks by. What is also not visible in the spreadsheet is her state of agitation when a case is initially reported in the press until the time that a perpetrator has been identified: "I would tell you that the day the alert arrives, I am involved in some way that whole day and probably a couple of days more to come. . . . You kind of stay a little tense until there's some sort of resolution." Still, for some cases, locating the perpetrator may take much longer or may never happen. What is also not visible in the spreadsheet are the many hours of reading stories of brutal violence; the careful collection of each painstaking detail of what is known about a case; the way it feels to leave so many fields blank.

In both her capacity as an activist and as a scholar, Helena has been thinking deeply about data production work as care work. We wrote a short essay together about this following Maria Puig de la Bellacasa's formulation of care work as an ethical-political commitment to "neglected things." For activists, painstakingly researching cases is one small way to recuperate and repair such public neglect. Matters of care, for Puig de la Bellacasa, involve not only revaluing the often invisibilized (and gendered) labor of maintenance and repair, but also disrupting the Western academic tendency to value distance, neutrality, objectivity, and impartiality between researcher and research subject. Instead, care work weaves researcher and researched together into affective relationships that produce new knowledge through proximity, emotion, and connection.[12] This resonates with the data feminism principle to *elevate emotion and embodiment*. Feminicide counterdata producers are deeply intimate with their data points—the people and places whom the data represent—because they have spent so much time researching and triangulating sources, rejecting stigmatizing media narratives, finding photos, and correcting the errors of the state and the media. As a result, many activists are deeply protective of their data, as Lucchesi and Sovereign Bodies Institute evidenced in their refusal to share it with federal authorities. For activists, paradoxically, these data are not data, never were data, and cannot be reduced to data. Yet their representation as data persists in the form of rows in databases or grids and columns of spreadsheets. The gridded orderliness hides the labor of its production.

But is this a feature or a bug? Do activists want their spreadsheets to be messy records of emotion, to be testimonies to their rage, or their attachment or their sadness? I will have to assume not, since nobody that we interviewed documents feminicide in such a way. Rather, the erasure of the material and emotional labor behind feminicide data production feels more strategic and calculated. It functions as a kind of *hack* to hegemonic data production. Presenting the results of a fraught and intimate and labor-intensive research process in the cool logic of the spreadsheet grid appropriates

Western, colonial, patriarchal penchants for distance and quantification and deploys them to elevate and amplify the feminist concerns and the feminist rage seething just under the surface of the lined, light-gray boxes.[13] But it elevates and amplifies rage precisely by visually obscuring it. This builds on and extends the data feminism principles *make labor visible* and *elevate emotion and embodiment*. It may not be desirable, in this case, to make labor visible for everyone, all the time. Sometimes, it is necessary to hide that labor behind the strategic deployment of the rhetoric of objectivity. Those within the feminicide data community share their labor and their grief with each other, but from the outside the work appears as neatly ordered rows and columns. A paper by communications scholars Yu Sun and Siyuan Yin on feminist data activism in China makes a related point, which is that under nondemocratic regimes, it is problematic and even dangerous to publicly credit activist labor because it makes activists targets for political oppression.[14] Likewise, it may not always be politically productive to elevate emotion when such emotion becomes the basis for discrimination—the familiar patriarchal approach of dismissing emotion as counter to reason or painting feminist concerns as too "personal." Moreover, some parties do not deserve our grief and our outrage, which are constantly leveraged in exploitative and voyeuristic ways by the media, leading to more than one activist characterizing media coverage of feminicide as "trauma porn." This leads me to ask the question: In the process of doing data science, when should one seek to make labor visible? In which contexts do we elevate emotion? And for whom?

LESSONS FOR A RESTORATIVE/TRANSFORMATIVE DATA SCIENCE

The information discovery tactics described in this chapter surface some key lessons for a restorative/transformative data science, especially if we look at differences between the researching labor of feminicide data activists in comparison with how researching and compiling data is treated in hegemonic data science projects. First, the labor of hegemonic data science is masculinized, overvalued, and overcompensated. The research group AI Now reported in 2018 that women comprise only 15 percent of AI research staff at Facebook and 10 percent at Google.[15] More broadly speaking, men dominate computer science occupations in the United States, comprising almost three-quarters of the workforce, and gender balance in computing has been declining since the mid-1980s.[16] In contrast, the work to research and monitor feminicide is almost the polar opposite. As we have seen, it is done almost exclusively by women and gender nonconforming people. Data activism that centers on racialized feminicide is almost exclusively undertaken by Black women, Indigenous women, and women of color. Most research is unwaged work. When projects originate from a nonprofit or from journalism, people may be paid for their time, but organizations have a hard time securing operational funds to sustain it. Thus, the labor of feminicide counterdata research is feminized, racialized, devalued, and undercompensated. This is consistent with the feminist concept of reproductive labor—the care work that sustains, maintains, and reproduces society.[17] In this case, instead of cleaning homes and raising children, activists are stewarding the memories of killed women, caring for wounded communities, and working toward repair and justice. It is women and queer people who are doing the reproductive labor to clean up, informatically speaking, after the structural excesses and negligence that led to the gender-related violence in the first place.

But this stark divergence is not only related to feminicide data research. Not all parts of the conventional data science pipeline are masculinized and highly compensated. Melanie Feinberg, an information scientist, describes how data collection and classification work is perceived as "unskilled and mechanical." When she runs data collection assignments in her classes, the students imagine (erroneously!) that "there is nothing to be learned from the process of generating data, because data collection is the mere recording of objects speaking for themselves."[18] Indeed, data collection, annotation, sorting, and labeling is often outsourced from the Global North to the Global South, where it is done for low wages, disproportionately by women and racialized people.[19] This goes even and especially for information tasks that resemble the work of feminicide data research in their emotional burden: content moderation on commercial platforms such as Facebook and YouTube. In these jobs, low-paid workers spend hours per day sifting through graphic, violent, racist, misogynist, exploitative content and labeling it according to platform policies.[20] This has led information scholar Julian Posada, among others, to assert that the AI industry profits from political instability and colonizes catastrophe to provide bad jobs to economically vulnerable people.[21] In contrast, the high-status part of data science supposedly comes *after* the data production activities of collection, classification, acquisition, and preprocessing. Drawing from these perceptions of status, Nithya Sambasivan and colleagues wrote a paper titled "'Everyone Wants to Do the Model Work, Not the Data Work,'" outlining the downstream harms of disinvesting in data quality.[22]

Feminicide counterdata activists provide a compelling model for what it looks like to refuse this (gendered, racialized, colonial) stratification of data labor. For better or for worse, they are intimately connected to their data due to the hours of time invested in researching each case. This labor engenders a deep expertise in the information ecosystem surrounding feminicide, as evidenced by activists' analysis of sources of missing data in table 4.1. Not only do activists understand the flaws and limitations of the information ecosystem, they understand how their own data also inherit those limitations. This is the essence of the *consider context* data feminism principle. In recent years, there has been more work that examines contextual knowledge as a unique form of expertise that has been undervalued by the mainstream data science community. For example, Annabel Rothschild and colleagues demonstrate how civic data workers, due to their proximity to the data collection process and their deeper domain knowledge, engage in information contextualization practices that mainstream data scientists stand to learn a great deal from.[23] In the case of feminicide data activists, knowledge of the limitations of their data lead to caution in how they use them and communicate them downstream. For example, all groups mentioned that they know their databases are not complete since not all cases are reported in the media or publicized on social media. They take care to communicate that any statistics remain undercounted, with the most intersectionally marginalized populations as the ones who remain most undercounted.

As Lucchesi reported, "The data requires a really intimate relationship in order to make it workable." She scoffed at some of the data requests that she has received from outsiders: "The requests that we get are based on assumptions that the data is just kind of like this divining tool that anybody can just jump in and use and that somehow all of the mysteries of this crisis will be solved from their armchair with casual exploration. And that's not the case. If it were, we would have fixed it already." The ignorance that Lucchesi is challenging here is a direct product of hegemonic data science's undervaluing of data research and collection work—leading data scientists to believe that they could just explore "the data" (Which data? Well, whichever are available and cheap.), treat them as ground truth, make a model, and then *voilà!* hidden wizardly insights about MMIWG2 are revealed.[24] At some level, this also points to a real failure in the education system: that educators in computer science, statistics, and mathematical domains have not been able to understand the urgency of integrating humanities and social sciences concepts, leaving their students impoverished—overconfident and underprepared—to design any kind of information system that relates to human life.

Finally, as I previously mentioned, many of the most creative tactics that counterdata groups develop to source information about feminicide come not through novel sensors or computational techniques but via the production of novel and collective forms of human relationships. These relationships tend to be nonextractive and authentic; they are not about acquiring data and then generating capital (financial or social) from it. This represents a contrast with hegemonic data science that, to the extent that it develops novel forms of human relationships, tends to be in the model of what Paola Ricaurte calls *data extractivism*, which operationalizes everything as a potential data source.[25] Driven by profit, modeled on colonial relations, hegemonic data science treats humans as interchangeable units of supply and demand (e.g., Uber drivers and their customers), as mechanical automatons (e.g., Amazon Mechanical Turk and microwork platforms), or as chauffeurs for sensors (e.g., Waze, Google Maps). This is not to say that feminicide groups' coalitions and human relationships are free of conflict (they are definitely not), but rather that the commitment to nonextractive human relations represents a key aspect of their data epistemology. It is a form of epistemic disobedience and a form of resistance to the data extractivist regime.

This epistemic challenge is resonant with non-Western calls for emphasizing relationality and collective responsibility in data and artificial intelligence. These include decolonial AI, Indigenous data sovereignty, and other data epistemologies described in chapter 8. Ultimately, a restorative/transformative data science requires such an alternative epistemology, and one that places care, context, and connection at the center of the data gathering process. This presents a profound challenge to the data acquisition process of hegemonic data science. As Feinberg states, "The real revolution in data labor will be in acknowledging that data collection should be celebrated for its skill and creativity."[26] It's not only about revaluing data research and production because it's the right thing to do, but also because it challenges harmful data extractivist regimes and it results in better data science. For example, feminicide data activists are profoundly aware of their data's limitations and quality issues, whereas hegemonic data scientists are often stunningly ignorant of or surprised by theirs.[27] This is especially true when the information ecosystem surrounding an issue is highly influenced by structural inequality, such as when there is rampant missing data, biased data, harmful and stereotypical information, or mis- and disinformation, all of which need to be assessed and evaluated for inclusion in a broader dataset. Indeed, a restorative/transformative data science involves dismantling the existing stratification of data labor, revaluing the proximity of the data producers to their subject matter, and doubling down on contextual knowledge as a unique and unautomatable form of expertise, something to teach and cultivate in community.

CONCLUSION

Researching is the second stage of a restorative/transformative data science project in which an individual or group seeks information about individual cases to add to their database. This can include sourcing existing datasets, mining state and media sources of information, verifying details, and triangulating across sources. In the researching stage, activists conjoin their analysis of power with their information-seeking practices, leading them towards a skillful, on-the-ground understanding of the sources of missing data and biased information that permeate the feminicide information ecosystem. Consequently, they navigate informatic gaps, biases, and errors by employing a variety of creative strategies to source information from the state, the media, and families. Activists often develop novel human-relational-informational configurations—using relationships of trust and solidarity to establish networks of information providers. Still, the work demands extensive emotional labor to sift through stories of violence and precisely document details of a human life and death.

Researching feminicide cases goes hand-in-hand with recording such details into spreadsheets and databases, the subject of the next chapter. Thus, while these stages are described separately, in practice they are tightly linked, with activists going back and forth between seeking information about a case and copying new details into structured fields and categories. The work is not only challenging because of the violence but also because there is no end in sight. Activists seek a world in which gender-related violence has been eliminated but women continue to be killed and cases continue to surface. In the face of state injustice and media stigmatization, researching cases is an informatic strategy to challenge such structural bias. Activist researching practices point toward a restorative/transformative data science that values proximate relations between data scientists, datasets and data subjects, as well as elevates the labor of data sourcing, labeling, and curation practices.

5   RECORDING

In 2017, a map and a manifesto went viral in Ecuador. Both were published by the Alianza Feminista para el Mapeo de los Femi(ni)cidios en Ecuador (Feminist Alliance for Mapping Femi(ni)cides in Ecuador), a coalition of feminist groups from around the country. The map depicts feminicides by province in Ecuador with both raw counts and shaded colors over a period of seven months (see figure 5.1). The accompanying manifesto spelled out the Alianza's demands, which included, first and foremost, that the state recognize that the violent death of a woman every fifty hours is a public emergency and prioritize the issue in public action and policy. Other demands included the legalization of abortion, a substantial shift in mining and extraction laws that had long put Indigenous and poor women at risk of violence, and the immediate creation of a national registry of information on violence against women and LGBTQ+ people. The manifesto even spelled out detailed requirements for such a national registry. The database, they insisted, must

(1) specify gender, ethnicity, age and other demographic characteristics about both the victim and the perpetrator;

(2) include geographic information so spatial patterns can be understood;

(3) show patterns of the various forms of violence; and

(4) be open and easily accessible to the general public.[1]

This chapter is about exactly these details that the Alianza was demanding. It is about the process of *recording* feminicide and why the variables, categories and classifications used to document cases of feminicide matter, as well as how activist data schemata are created, how they vary across counterdata groups, and how they are used, shared, and harmonized across groups. A couple notes on the terminology used throughout this chapter: All groups have a *data schema*, which refers to the set of *fields*

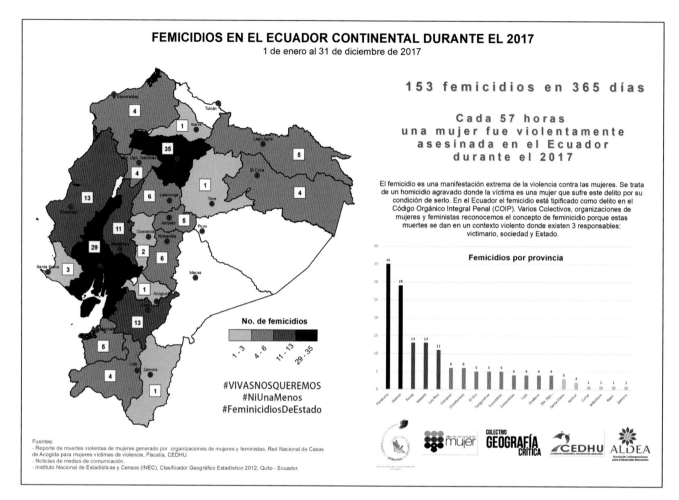

FIGURE 5.1
The first map published by the Alianza Feminista para el Mapeo de los Femi(ni)cidios en Ecuador
in 2017. Courtesy of the Alianza Feminista para el Mapeo de los Femi(ni)cidios en Ecuador.

or *variables* that they collect about each case of feminicide, such as name, age, date, and so on. If you are thinking in terms of spreadsheets, the data schema consists of the names of all the columns in the spreadsheet. One row of data corresponds to one case of feminicide. Activists enter data *values* or *points* into particular rows and particular columns. For example, if a victim were thirty-eight years old, then that would be the value for the age field in a particular row. Activists also have *categories* for cases—such as linked feminicide or transfeminicide—which they apply once a row of information is reasonably complete and they are able to classify a case with one or more categories. Such categories might be drawn from international standards, like the Latin American model protocol, or drawn from activists' own analysis, or a mix of both.

Recording, then, encompasses the transformation of unstructured data that activists find through research into structured data: cases organized into rows, values organized by columns, and lists of categories and classifications to apply to cases for analysis. Here there is a continuous back-and-forth relationship with the prior stage of a restorative/ transformative data science project—the stage of researching—and activists are constantly moving between researching cases, verifying information, and recording values.

The Alianza's 2017 map seemed to instantly be on everyone's mind. "It shocked the country," reports Paola Maldonado Tobar, president of Fundación ALDEA (ALDEA Foundation), one of the organizations in the Alianza. Behind the scenes, the spreadsheet that produced the map was simple—it consisted of just four columns: the date of the feminicide, the age of the victim, the name of the victim, and the province where it happened. The Alianza sourced these cases primarily from the media, but it also drew from official data from INEC (Ecuador's national statistics agency) and from women's and feminist organizations that had been keeping handwritten ledgers of feminicide in certain provinces since 2014, when Ecuador passed its first law codifying femicide (see table 1.1).

In fact, it was the possibility of amplifying the existing monitoring labor of women's and feminist groups that had inspired the creation of the Alianza. Maldonado Tobar felt that this was work that by necessity had to be done in community. As they started the work, she said, "I realized that we had to generate a system, and that for that we needed an alliance to be able to validate cases so that I wouldn't be the only one responsible for them. It seemed like a huge responsibility to me to issue a single figure, as if I were the only one in charge." Together, the coalition discusses, validates, and classifies cases through multiple WhatsApp groups—a pattern of pluralistic deliberation and harmonization that we saw across groups.

Like all activists we spoke with, the Alianza's data schema is dynamic. As a result of this systematizing, it expanded from four variables in 2017 to eighty in 2023, including nationality, gender and ethnic identity, sexual assault, pregnancy status, whether the woman had been missing, information about the perpetrator, including whether he was a member of the military or police force, and many more fields. The coalition added various intersectional categories of identity to their database as they came to understand how violence against certain subgroups was under-registered. Their most recent report has analysis and maps focused on Afroecuatorian women, Indigenous women, and trans women whose deaths are erased and invisibilized in the larger narrative about feminicide.[2] And while the Alianza seeks to align its data schema and categories with national law and international data standards, they deliberately exceed them in some cases. For example, Ecuador's law does not recognize femicides of trans women as femicides, while the Alianza does. The state has a separate law covering killings of children, while the Alianza counts all ages. The state does not recognize induced suicides as femicides, while the Alianza does. And if a perpetrator commits suicide, the case disappears from the judicial system and can never be named as a femicide, whereas the Alianza still considers it so. These divergences are intentional and strategic—designed to push the state to expand its conception of the problem.

The Alianza's more recent maps reflect the addition of these newer analytical variables. Like the map from 2017, the 2021 map of feminicides (figure 5.2) also plots spatial patterns, showing raw counts of feminicides per province in Ecuador along with using shaded colors. But this map is more complex—it depicts transfeminicide separately, also by province, using proportional circles. The infographics surrounding the map show a breakdown of cases by age range, by month, and by type of weapon used. And various parts of the graphic connect feminicide cases to patterns of sexual abuse, disappearances, suicides, prior registered complaints, and organized crime. While the purpose may seem purely analytic—connecting these factors to the incidence of feminicide—the bottom left of the graphic shows the group's larger demands for *memory*, *justice*, and *reparation*. In fact, this first demand—memory—surfaced as a central theme for many data activists in the recording stage. This connects the work of recording feminicide to the memorialization and collective memory of lives lost to structural violence, and it is also a call out to the multiple, ongoing movements for disappeared people across Latin America who use similar slogans. The appeal for reparation is an explicit link to the Alianza's connection of their work to the demand for comprehensive reparations for families harmed by feminicide. Ecuador's constitution guarantees comprehensive reparations to citizens whose constitutionally guaranteed rights have been violated.

Activists in Canada are familiar with the need to develop their own counts and classifications of femicide. There is no law defining or typifying femicide or feminicide

FIGURE 5.2

Feminicides in Ecuador 2021. Courtesy of the Alianza Feminista para el Mapeo de los Femi(ni)cidios en Ecuador.

in Canada, yet there have been activist efforts to register cases going back decades. In the 1990s, the Women's Memorial March started the annual practice of reading out names of missing and murdered Indigenous women in Vancouver's Downtown Eastside neighborhood (see chapter 1). Between 2006 and 2010, the Native Women's Association of Canada compiled a database called Sisters in Spirit and published an extensive report that linked violence against Indigenous women to poverty and dispossession.[3] The Canadian government defunded the effort in 2010, but the organization revived its monitoring efforts in 2019 under the name of "Safe Passage." The grassroots effort Woman Killing: Intimate Femicide in Ontario started its registry in the late 1980s.[4] As a young volunteer and graduate student, Myrna Dawson worked on the latter project with sociologist Rosemary Gartner and a group of workers at a local women's shelter, who came together around their dismay and grief at the murders of women in their shelters. While the effort eventually dissipated, Dawson continued producing data about femicide in her academic research. In 2015, the UN special rapporteur on violence against women and girls called for femicide watches in every country following the global uprising sparked by #NiUnaMenos. At that point, Dawson took action: "I looked at the data that we had and realized if we were going to do this work in Canada that this was the moment in time."

The Canadian Femicide Observatory for Justice and Accountability (CFOJA) is based at the University of Guelph and seeks to be a reliable source of primary information about justice and accountability for femicide victims in Canada. Like the Alianza, the CFOJA also publishes its research in reports and infographics (see figure 5.3). And like Maldonado Tobar, Dawson realized that she could not do the work alone as an academic: "I really wanted to have the voices of the people who were working in the sector—of people who had experiences with domestic violence and intimate partner violence." She assembled an expert leadership panel of over forty people, which seeks to represent the geographic and gender diversity of Canada, the perspectives of racialized women, and especially Indigenous women, disabled women, and survivors. This form of pluralistic governance has created some major debates and disagreements, but the observatory has managed to weather them to date.

The CFOJA is an interesting outlier in terms of the sheer number of fields and categories it uses to describe femicide in its database. While many counterdata groups, like

FIGURE 5.3
Infographic from the 2018 report on femicides in Canada produced by the Canadian Femicide Observatory for Justice and Accountability. Courtesy of the CFOJA.

Canadian Femicide Observatory for Justice and Accountability

Observatoire canadien du fémicide pour la justice et la responsabilisation

2018 Report

148 Women and Girls Killed in Canada in 2018

Rate of women and girls killed in Canada, 2018

5.31
21.85
0.45
0.95
1.04
1.94
0.38
2.08
0.62
0.90

*Rate of killing per 100,000 women/girls residing in province/territory, 2017

Overview

- **Nunavut, Yukon, New Brunswick, Manitoba** have the highest rates of women killings

- **34%** of victims were killed in **rural areas**, but only 16% of the population live there

- The most common method of killing was **shooting** followed by stabbing and beating

- **27%** of the perpetrators of intimate femicides committed suicide after the murder

- **91%** of the perpetrators identified were **male**

Who are the victims

Age distribution of the victims

Women and Girls Killed		General Population
10%	17 and under	19%
40%	18-34 years old	23%
23%	35-54 years old	27%
27%	55 and over	32%

Average age	Youngest	Oldest
40 years old	< 1 year old	94 years old

Gender-based motives/indicators

1. **Misogyny**
2. **Sexual violence**
3. **Coercive control**
4. **Separation/estrangement**
5. **Overkill**

In red: specific to intimate femicides

Race/Ethnicity of the victims   N=93*

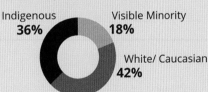

Indigenous **36%**
Visible Minority **18%**
White/ Caucasian **42%**

*This graph shows a total of 96% and not 100%. The additional 4% is constituted of victims believed to be Indigenous women, but this has not been confirmed.

Relationship with male accused   N=92

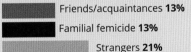

Friends/acquaintances **13%**
Familial femicide **13%**
Strangers **21%**
Current/former partners **53%**

Intimate Femicides: Relationship types

Legal spouse **36%**
Common-law spouse **38%**
Dating **27%**

22% were estranged

femicideincanada.ca
cfoja@uoguelph.ca
  CAN_Femicide

the Alianza, started their monitoring with a handful of fields, the CFOJA launched its effort with a data schema of almost two hundred variables (perhaps reflecting the project's academic origins). These include detailed information about the victim, the perpetrator, and their relationship, as well as many variables that attempt to capture the case's path through the justice system: charges laid, type of trial, type of plea, length of trial, defense's argument, type of sentence, length of sentence, date of deposition, name of the judge presiding over the case, and so on. This reflects the CFOJA's mission to not only focus on the violent murders of women but also monitor how and whether justice was served by the state. One of Dawson's long-term research interests, for example, is in studying the "intimacy discount"—the fact that perpetrators of intimate femicide nearly always receive lighter sentences.[5]

Despite the fact that Canada has no national legislation on femicide, the CFOJA still draws from international standards like the Latin American model protocol to determine both the variables that it tracks and the types and categories of femicide it uses to classify cases in its database. And, like many counterdata groups, there are cases where the CFOJA will intentionally and directly contradict the state's ruling on a case. Dawson says this happens most often in cases related to Indigenous women where the state does not classify a death as a homicide but the family and community insist that it is. In these cases, the CFOJA expert panel discusses and sources information from whichever members are closest to the geographic and cultural communities in the case. In most of these deliberations, the CFOJA tends to side with families and communities. This is precisely the point—to count and include those cases that the state is systematically ignoring and misclassifying.

RECORDING CASES

Recording is the stage of a restorative/transformative data science project that encompasses extracting and registering information, classifying cases, and managing data, which in turn includes issues of ethics, governance and access to the database (figure 5.4). For feminicide data projects the unit of analysis is a case of feminicide or gender-related killing and so one row in a dataset corresponds to information about one case

---

FIGURE 5.4

Recording is the third stage of a restorative/transformative data science project. Courtesy of the author. Graphic design by Melissa Q. Teng. Collaged images: Courtesy of Dawn Wilcox / Women Count USA: Femicide Accountability Project.

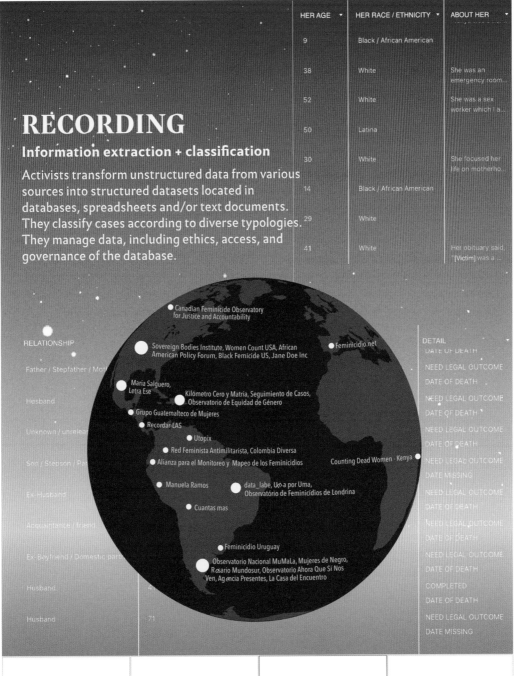

RECORDING

Information extraction + classification

Activists transform unstructured data from various sources into structured datasets located in databases, spreadsheets and/or text documents. They classify cases according to diverse typologies. They manage data, including ethics, access, and governance of the database.

HER AGE ▼	HER RACE / ETHNICITY ▼	ABOUT HER ▼
9	Black / African American	
38	White	She was an emergency room...
52	White	She was a sex worker which I a...
50	Latina	
30	White	She focused her life on motherho...
14	Black / African American	
29	White	
41	White	Her obituary said, "[Victim] was a ...

RELATIONSHIP

Father / Stepfather / Mot...

Husband

Unknown / unrelea...

Son / Stepson / Pa...

Ex-Husband

Acquaintance / friend

Ex-Boyfriend / Domestic part...

Husband

Husband

4

71

Canadian Feminicide Observatory for Justice and Accountability

Sovereign Bodies Institute, Women Count USA, African American Policy Forum, Black Femicide US, Jane Doe Inc

Feminicidio.net

Maria Salguero, Letra Ese

Kilómetro Cero y Matria, Seguimiento de Casos, Observatorio de Equidad de Género

Grupo Guatemalteco de Mujeres

Recordar-LAS

Utopix

Red Feminista Antimilitarista, Colombia Diversa

Alianza para el Monitoreo y Mapeo de los Feminicidios

Counting Dead Women - Kenya

Manuela Ramos

data_labe, Uma por Uma, Observatório de Feminicídios de Londrina

Cuantas mas

Feminicidio Uruguay

Observatorio Nacional MuMaLa, Mujeres de Negro, Rosario Mundosur, Observatorio Ahora Que Sí Nos Ven, Agencia Presentes, La Casa del Encuentro

DETAIL
DATE OF DEATH
NEED LEGAL OUTCOME
DATE OF DEATH
NEED LEGAL OUTCOME
DATE OF DEATH
NEED LEGAL OUTCOME
DATE OF DEATH
NEED LEGAL OUTCOME
DATE MISSING
NEED LEGAL OUTCOME
DATE OF DEATH
NEED LEGAL OUTCOME
DATE OF DEATH
NEED LEGAL OUTCOME
DATE OF DEATH
COMPLETED
DATE OF DEATH
NEED LEGAL OUTCOME
DATE MISSING

RESOLVING	RESEARCHING	**RECORDING**	REFUSING + USING
Developing a theory of change	Finding + verifying information	**Information extraction + classification**	Where data go, who uses them

of feminicide. In the process of recording cases, activists transform unstructured information into structured data, arranged into specific pairs of fields and values, and they apply categories to cases in order to detect patterns of violence.

With regard to the logistics of registering information, a large majority of the data activists and journalists we interviewed use spreadsheet software (mainly Google Sheets and Excel) to record data on individual cases. The Alianza, for example, started by recording cases using Google Sheets and Excel but later moved to Kobo Toolbox, which is open source and has more robust data security measures. Several projects, including the CFOJA, also keep individual case files in word processing documents where they can store free-form notes and the full text of media articles about the case. The CFOJA also uses SPSS, a statistical analysis software program, for recording variables and performing statistical analysis. Five of our forty-one interviewees use a database management system (DBMS) despite the added cost and technical knowledge required because they want more robust security, or to include multimedia files related to cases (like images), or to build more complex relational structures and queries with their data.

To record cases in their databases, activists tend to follow a manual process of copying and pasting individual data points (such as a victim's name or age) from news articles or other sources into the spreadsheet or database program. Some groups use a web form for this initial data entry and others enter the information directly into the spreadsheet. An individual case might be entered over the process of many recording sessions because only a certain amount of information will be available at the outset; the information will be corrected and updated; and new details will inevitably emerge as a case proceeds through the justice system. Thus, while researching and recording are ostensibly separate stages of a restorative/transformative data science project (and separate chapters in this book), they unfold as a continuous back-and-forth process over time. Especially for those groups that track cases through the justice system, a single case can take multiple years to fully research and record before activists consider the information complete, or at least *esclarecido* ("clarified" in Spanish). This is the word many activists used to describe cases once stable and verifiable information has come out and details are relatively certain.

STRUCTURING DATA

Before individual data values are registered in discrete rows and columns, activists have to decide which fields to record and how. Thus, recording also encompasses the process of determining which specific fields to use within activist spreadsheets and databases— that is, their *data schema* for how a feminicide will be registered. These range in complexity from the four fields that the Alianza started with to the CFOJA's 180 variables that it attempts to collect about every case. The median is around twenty-five fields. In the case of both the Alianza and the CFOJA, their fields and categories are produced from a collaborative governance process that tries to incorporate grassroots perspectives across large territories, that adheres to or intentionally diverges from legal definitions in their country, that coordinates with other observatories, and that aligns with emerging international standards such as the Latin American model protocol described in chapter 1. In this, groups are operating in a transcalar (community + local + national + regional + global) way to define, systematize, and analyze feminicide in their contexts, a point we return to later in this chapter.

What are the similarities and differences in activists' data schemata? How do they map onto emerging data standards? As a preliminary answer to these questions, the Data Against Feminicide project mapped activist data schemata in relation to the feminicide data standard developed by ILDA. The ILDA data standard was developed in collaboration with governments in Latin America (see chapter 1). Called the *Guide to Protocolize Processes of Femicide Identification for Later Registration*, it proposes the collection of sixty-five variables about each case, with groups of variables that relate to the victim, the accused, their relationship, the event and place of the crime, and the legal process. For Silvana Fumega, the director of research for ILDA, who led this process, "our aspiration is for there to be at least a minimum set of data that allows us to understand the situation of every country and for them to be comparable enough."[6] That is to say, the standard is designed as a base or minimum.

Figure 5.5 shows some of the commonalities and divergences between activist data schemata and the ILDA standard. To make this comparison, our team asked interviewees if they would be willing to share their data schemata with us (not their data, just their list of fields). Eighteen groups assented, with the majority of groups (thirteen of eighteen) coming from Latin America. It's important to note that this is a preliminary, descriptive mapping to investigate how activist variables and categories align (or do not) with international standards, but this cannot be considered representative of all feminicide data activists.

All data schemata we reviewed contain identifying information about the murdered person, which most frequently includes name and age (figure 5.5, top). Other frequently recorded fields included number of children, ethnicity, whether there had been prior complaints against the accused, and immigrant status of the victim. A number of fields suggested by the ILDA standard are very infrequent in activist data schemata, including place of birth, nationality, education, and protection measures (i.e., whether the woman had a restraining order or the equivalent), likely because these are extremely difficult to source from press reports and public information. And activists

Comparing the ILDA data standard to 18 activist data schemas

HOW DO ACTIVIST DATA SCHEMAS MAP ON TO EMERGING DATA STANDARDS?

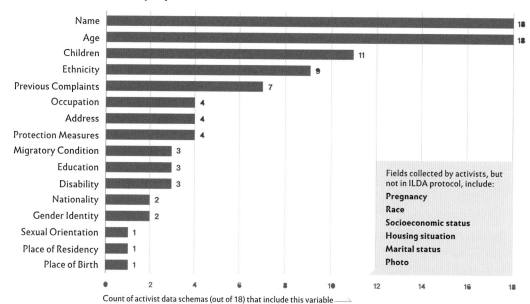

ILDA's proposed variables related to the victim

Variable	Count
Name	18
Age	18
Children	11
Ethnicity	9
Previous Complaints	7
Occupation	4
Address	4
Protection Measures	4
Migratory Condition	3
Education	3
Disability	3
Nationality	2
Gender Identity	2
Sexual Orientation	1
Place of Residency	1
Place of Birth	1

Fields collected by activists, but not in ILDA protocol, include:
Pregnancy
Race
Socioeconomic status
Housing situation
Marital status
Photo

Count of activist data schemas (out of 18) that include this variable ——→

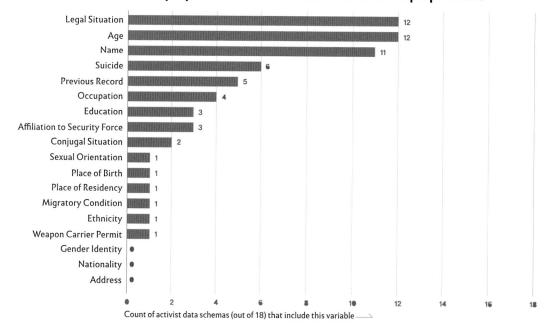

ILDA's proposed variables related to the accused perpetrators

Variable	Count
Legal Situation	12
Age	12
Name	11
Suicide	6
Previous Record	5
Occupation	4
Education	3
Affiliation to Security Force	3
Conjugal Situation	2
Sexual Orientation	1
Place of Birth	1
Place of Residency	1
Migratory Condition	1
Ethnicity	1
Weapon Carrier Permit	1
Gender Identity	0
Nationality	0
Address	0

Count of activist data schemas (out of 18) that include this variable ——→

frequently collect fields that are not mentioned in the standard. For example, around a third log whether a killed person was pregnant, and about half registered whether the feminicide involved suicide. This follows research that shows strong links between pregnancy and intimate partner violence.[7] Other fields in activist data schemata, but not in the ILDA standard, included race, housing situation, and socioeconomic status.

The majority of activist data schemata also include fields about the accused perpetrator (figure 5.5, bottom), but activists on the whole include comparatively less information about the accused than about the victim. Three groups log no information about the accused. This is the subject of ongoing conversations within activist and feminist circles around whether and how to *visibilizar* (visibilize) the alleged perpetrators of feminicides. These debates often surface more conceptual arguments around, for example, the importance of not only studying the effects of harmful systems like patriarchy (such as feminicide) but also studying the mechanisms that produce and reproduce those effects (such as toxic masculinity). Nevertheless, activists must navigate tricky ethical and legal territory in naming and potentially exposing the identities of individuals who have been accused but not convicted through the justice system. For example, Dawn Wilcox, who publishes the Women Count USA database openly online, has a disclaimer message prior to viewing the database and then also in every row of her dataset that reads:

ALL ACCUSED ARE ABSOLUTELY PRESUMED INNOCENT UNTIL CONVICTED IN A COURT OF LAW. INFORMATION ABOUT SUSPECTS OR PERPETRATORS IS OBTAINED FROM PUBLISHED NEWS SOURCES OR LAW ENFORCEMENT BULLETINS & CASE OUTCOMES WILL BE UPDATED AS TIME ALLOWS. PLEASE EMAIL CORRECTIONS TO womencountusa@gmail.com.

When activists do log information about the accused perpetrator, they most frequently include their name, their age, and the legal situation (status) of the case as it proceeds through the justice system. Six of eighteen groups logged whether the accused perpetrator committed suicide, which can often affect whether a case is prosecuted by the state as a feminicide or not. Less than a third of groups log variables such as perpetrator

FIGURE 5.5

Out of a group of eighteen activist data schemata, this chart (top) depicts how many activist groups collect the variables about the victim recommended by the ILDA standard and (bottom) shows how many activists collect the variables about the accused perpetrator recommended by the ILDA standard. Courtesy of the author. Analysis by Angeles Martinez Cuba. Visualizations by Wonyoung So and Melissa Q. Teng.

occupation, education level, and whether they were a member of law enforcement or other security force. Very few groups log more detailed information about the accused perpetrator recommended by ILDA, such as their address, ethnicity, and place of birth, likely because this information is extremely hard to find from public information and media reports. Interestingly, no groups logged the gender identity of the accused perpetrator. While the vast majority of perpetrators of feminicide are men, people of other genders also commit feminicide and other acts of gender-related violence.

Beyond the ILDA standard, many groups record fields that are not named in the standard and that no other activist group records, reflecting their different missions to interrogate economic, racialized, and/or homophobic violence. The Alianza, seeing how invisibilized the murders of Afroecuatorian, Indigenous, and trans women were, evolved community-based strategies to collect more information around race/ethnicity, gender identity, and location of the murder so that it could bring an intersectional focus to its analysis.[8] In another example, the Sovereign Bodies Institute has a variable to indicate whether an Indigenous woman was murdered within fifty miles of extractive industries because Indigenous communities and researchers have documented how these industries lead to a rise in sexual violence and trafficking.[9] Other fields—among many—collected by only one or two groups include history of substance abuse, mental illness, and whether the crime was perpetrated by police.

It is important to note that all activist data schemata are dynamic. Activists add or modify fields as their understanding, dialogue with others, theory of change, and political objectives evolve; this is a way in which the resolving stage and the recording stage influence each other. This means that they continually revisit and update existing cases. For example, Sovereign Bodies Institute has a policy to add any field that a community member requests to their data schema. This is an important example of the transcalar approach mentioned earlier, where community input can guide and shape the data recording, even when it means increasing activist labor. Lucchesi described this dynamism: "We're always adding new data points which unfortunately means we always have to go back and add those points for the 5,000 cases already in the system."

CLASSIFYING CASES

Counterdata producers also develop their own categorization for cases—assigning types and subtypes to cases of feminicide and gender-related killing. Both the CFOJA and the Alianza draw from the Latin American protocol, which outlines two broad categories for femicide—direct and indirect—as well as a typology that includes intimate femicide, nonintimate femicide, child femicide, racist femicide, and the nine other types pictured in figure 5.6 in taupe. As with the data schema mapping in the prior section,

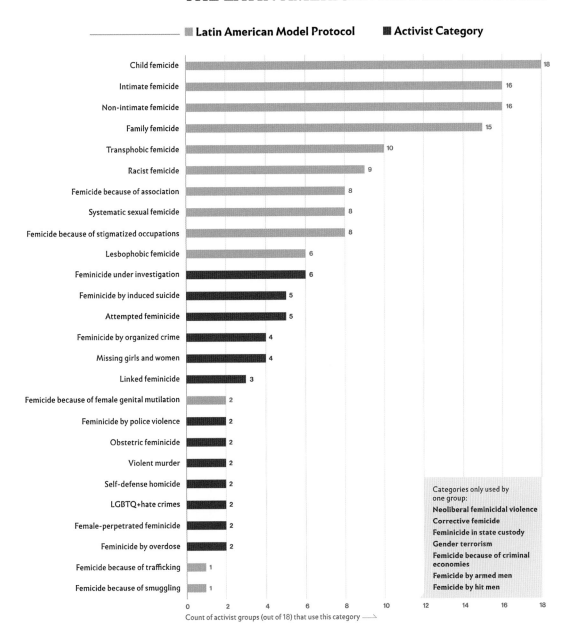

Comparing data categories of the Latin American Model Protocol to Activist Categories

HOW DO ACTIVIST CATEGORIES MAP ON TO THE LATIN AMERICAN MODEL PROTOCOL?

Latin American Model Protocol ■ Activist Category

Category	Count
Child femicide	18
Intimate femicide	16
Non-intimate femicide	16
Family femicide	15
Transphobic femicide	10
Racist femicide	9
Femicide because of association	8
Systematic sexual femicide	8
Femicide because of stigmatized occupations	8
Lesbophobic femicide	6
Feminicide under investigation	6
Feminicide by induced suicide	5
Attempted feminicide	5
Feminicide by organized crime	4
Missing girls and women	4
Linked feminicide	3
Femicide because of female genital mutilation	2
Feminicide by police violence	2
Obstetric feminicide	2
Violent murder	2
Self-defense homicide	2
LGBTQ+hate crimes	2
Female-perpetrated feminicide	2
Feminicide by overdose	2
Femicide because of trafficking	1
Femicide because of smuggling	1

Categories only used by one group:
Neoliberal feminicidal violence
Corrective femicide
Feminicide in state custody
Gender terrorism
Femicide because of criminal economies
Femicide by armed men
Femicide by hit men

Count of activist groups (out of 18) that use this category ⟶

FIGURE 5.6

Out of a group of eighteen activist data schemata, this chart shows how many activist groups use femicide categories proposed by the Latin American model protocol and also shows categories developed by activists themselves that derive from observing patterns in their countries and contexts. Courtesy of the author. Analysis by Angeles Martinez Cuba. Visualizations by Wonyoung So and Melissa Q. Teng.

we (the Data Against Feminicide research team) were interested in understanding how activist categories for feminicide aligned with or diverged from the typology proposed by the Latin American model protocol.

While the majority of groups that we interviewed had heard of the Latin American model protocol and many had drawn some concepts and categories from it, only one group—the CFOJA—directly imported its typology into their database for classifying cases. Some data activists' classification schemes remained relatively simple, and they simply flag whether the case is a feminicide or not. Other activists have evolved more complex categorizations with subtypes of feminicide. Some of these align with the Latin American model protocol typology. For example, the majority of groups have either a category or variable to track intimate versus non-intimate feminicide as well as to distinguish feminicides of cis women from feminicides of trans women. These are called *transphobic femicides* in the Latin American model protocol and *transfeminicidios* (transfeminicides) by most activists.

Around half of activist projects track a victim's racial, ethnic, or Indigenous identity in order to be able to name a killing as a *racist femicide*, and one-third have some indicator of sexual orientation in order to be able to classify a *lesbophobic femicide*. The relative dearth of these categories in activist typologies may be because of the difficulty of obtaining such information from media reports. Activists expressed discomfort with making guesses about race, ethnicity, and sexual orientation without it being explicitly mentioned in the press or confirmed through another source. (Even when it is mentioned, activists also know that the state and the media tend to misclassify race and ethnicity and misgender trans people, as discussed in the prior chapter.) The CFOJA, for example, has a field called *suspected racialized*, where activists mark whether they have reason to believe that a woman is Indigenous or a person of color, and they can then follow up with their expert panel or reach out to local partner organizations to try to ascertain the victim's race or ethnicity. Yet they are not always able to verify such information, as is evident in figure 5.3, where next to the pie chart about race, they have a statement that they believe an additional 4 percent of victims were Indigenous but could not verify that information. Finally, certain categories that appear in the Latin American model protocol are used very infrequently or not at all by the activists we worked with: *femicide because of trafficking, femicide because of smuggling,* and *femicide because of female genital mutilation* (FGM).[10]

On the flipside, there are a number of common activist categories not included in the Latin American model protocol that are used by at least a few of the groups that we interviewed (see the purple bars in figure 5.6). The most common are *induced suicide feminicide*—where a victim is driven to kill herself because of verbal or physical abuse—and *linked feminicide*—in which a person the woman loves is killed in order to hurt her. A number of groups have categories or separate databases for classifying cases related to feminicide: *missing girls and women, attempted feminicide*, and *feminicide under investigation* (denoting that it's unclear yet whether it is a feminicide). And there are a host of diverse activist categories that attempt to capture the relationship of a feminicide with organized crime and paramilitary activity: *gender terrorism, femicide because of criminal economies, femicide by organized crime/gangs, femicide by armed men, femicide by hit men*. These are different—and broader—than the Latin American model protocol categories, but all are trying to get at emerging patterns between organized crime and gender-related violence. Finally, there are a number of categories formulated and used only by one or two groups. For example, *neoliberal feminicidal violence* is a category from the Red Feminista Antimilitarista in Colombia (discussed previously in chapter 3) that links neoliberal economic policy with gender-related violence.

What are these categories for? Why do groups classify cases? Most of the activists elaborated something similar to Estefanía Rivera Guzmán's sentiment from chapter 3: "It is the possibility of understanding what is happening to women in our country." Categorizing cases is intimately bound up with the resolving stage of groups' work: both their analysis of power—understanding the workings of structural inequality—and their theory of change—developing strategies to intervene in and challenge those dynamics using data. Categories enable special focuses for analysis, advocacy, and communication. They permit downstream analysis of types of feminicide that may be less frequent quantitatively but take a different qualitative form than the majority of cases (e.g., *transfeminicide*, the murder of a transgender woman, or *ecofeminicide*, the murder of an Indigenous woman leader in the context of defending her community's land). Groups may then highlight those numbers or those cases differently in data visualizations, infographics, and reports—for example, the Alianza's map in figure 5.2, which notes the number of transfeminicides (8) and feminicides by organized crime (67), along with total feminicides (197). In addition, activists may see patterns of violence emerging in their region and then seek to formalize them through a category so that they might advocate specifically around such cases. This was the case with categories like *gender terrorism* and *feminicide by overdose*, both of which were categories under discussion (not fully formalized and incorporated yet) by a couple of groups at the time of our interviews.

One notable distinction between activist and legal classifications of cases is that while the judicial system attempts to classify cases based on intent—the motives of the individual perpetrator—most activists either implicitly or explicitly use the idea of *scenarios of feminicide*, proposed by Montserrat Sagot Rodríguez and Ana Carcedo Cabañas, in order to classify cases in their databases. They define scenarios as "socioeconomic, political and cultural contexts that produce or support particularly unequal power relations between men and women that generate dynamics of control and overt violence that can lead to femicide."[11] The modus operandi is thus ascribed not to the killer or killers but to the societal context in which the murder takes place. Thus, rather than trying to know the perpetrator's motives (not information that would be readily available outside case files), activists look for common identifying factors and then apply the appropriate category.

Neither the UN's Latin American model protocol nor activists provide many categories for classifying indirect feminicides—those that are characterized as "passive" murders and include deaths from unsafe abortions, maternal deaths, and other forms of deaths due to neglect and lack of access to adequate health care, housing, and social services. For example, the Black maternal mortality crisis in the United States is an example of widespread indirect feminicide, where racialized access to food, housing, services, and healthcare leads to the disproportionate and preventable deaths of Black mothers during or following childbirth.[12] Only two activist groups attempt to classify *obstetric feminicides*—deaths caused by malpractice or violence in reproductive care. Mumalá, in Argentina, has a category for *social trans-travesticides*—here the group tries to log the premature deaths of trans and travesti people due to social exclusion and lack of access to services and employment. An important conceptual innovation that is emerging from work by the Observatorio de Equidad de Género in Puerto Rico, and elaborated in conversation with the Alianza in Ecuador, is the aforementioned *feminicide by overdose*. This places responsibility on the state for the proliferation of women's deaths by drug overdoses due to, among other things, factors such as the ready availability of opioids and the lack of prevention and treatment services that incorporate a gender perspective.

But this is not necessarily to say that there *should be* more Latin American model protocol categories or activist categories for indirect feminicides. Should a death from an unsafe abortion be considered part of the same phenomenon as a death from an intimate partner or as a death from a travesti person unable to find gainful employment or as an abduction that ends in murder? These questions are not resolved in either the scholarly literature nor in activist practices surrounding feminicide. One of the ways that they are increasingly playing out is through global and grassroots typologies of gender-related violence—where some groups insist that a broad frame for feminicide is important and strive to count many different types of violence within that frame, while other groups focus more on counting what they can (which tend to be those cases of intimate feminicide which make it to news reports). The challenge for groups that seek to document cases with the more expansive framing of feminicide is that many types of indirect feminicide are never made public so they must work with families and communities as well as activist networks to try to source those cases. From an informatic standpoint, for example, it will be nearly impossible to obtain information about unsafe abortions or maternal deaths or premature deaths of trans people, leading to an unknown amount of missing data.[13] This calls us back to several of the "counterdata caveats" outlined in chapter 1—compiling poor quality, severely undercounted counterdata may work at cross-purposes to the goal of visibilizing a widespread problem of structural inequality.

THE PRACTICAL POLITICS OF COLUMNS AND CATEGORIES

While there is significant variation in activist practices of recording, some interesting patterns emerge across them. Counterdata activists and journalists are intimately attuned to the politics of variables, categories, and classifications within their spreadsheets and databases—both what they afford and what they foreclose. Activists recognize that they are engaged in a process of, as scholars Bowker and Star put it, "deciding what will be visible within the system (and of course what will thus then be invisible)."[14] While activist typologies of gender-related violence are often informed by legal definitions in their countries, as well as the Latin American model protocol as we saw in figure 5.6, they often intentionally exceed these definitions and standards. In the recording stage, activists use counting, classifications, and categories as a way to *challenge power*—to contest the state's narrow or absent legal definitions of gender-related violence, to uplift what communities and families are naming as feminicide and structural violence, and to gain insight into emerging patterns of violence that are not named and not captured in laws or international standards.

The idea that columns and categories are political is old hat. Literature on the sociology of classification has long described how official counting, undertaken by the state, encodes a particular institutional agenda. Sociologists Aryn Martin and Michael Lynch name the concept of *numeropolitics* and outline the ways that counting is both enumeration and also, at the same time, classification, because each time you count something, you are also judging that thing to be a member of the class of things that you are counting.[15] To record a death as a feminicide, then, is to count and to classify at the same time. But it is one thing to describe and explain, in academic literature, the fact that all counting is political, and another to explicitly deploy that knowledge as a practical political tool. Feminicide data activists *operationalize* numeropolitics in the service of amplifying subjugated knowledge—forms of knowledge that have been excluded from mainstream institutions. Families, communities, and women's and feminist groups know what is happening and are witness to the structural patterns and interpersonal effects of feminicide, but often the media, the law, and the judicial system continue to treat cases as individual, isolated events. Says Guerra Garcés from the Alianza about the group's mission: "One of the most important things of this alliance is to give voice to the women who say that it is feminicide. We fight. I fight here at the national level . . . I have fought with the prosecutor's office many times because of the issue of the cases, because for them they are not feminicides." Thus the work of recording and classifying cases becomes, in the words of Patricia Hill Collins, a *resistant knowledge project*—an endeavor to understand and challenge the unequal systems of power that result in subordination and exclusion.[16] And while sociologists may have long known that counting and categorizing is political, the institutions that typically produce and use numbers have a tendency to forget this fact.[17] Thus, activists wield institutions' penchant for naturalizing and reifying and depoliticizing numbers against them.

This politics of counting and classifying unfolds through the rows and columns of the spreadsheet, what Helena Suárez Val calls a *data frame*. She asserts that it is important to understand how "different entities and relations are put in the frame through specific arrangements of data."[18] Helena's point is that activists are deliberate in their choices of fields and categories as a way to contest media and state framings of feminicide. Data schemata, in other words, are playing both a rhetorical and representational role, directing attention toward some variables (say, pregnancy status or induced suicide or connection to narcotrafficking) and foreclosing others (say, marital status or what the victim was wearing, both of which support a victim-blaming frame). We see this in the way that some counterdata projects use variables in their data schema in an aspirational way, which is to say they include columns in their spreadsheets that are nearly impossible to obtain from media reports about the event. These include, for example, variables such as education level, socioeconomic status, mental health history, and sexual orientation. Activists know that incorporating these fields into a data schema will produce many rows with missing values or hard to verify data. Says Betiana Cabrera Fasolis from Mumalá, "We do not consider the data on the socioeconomic and educational situation of both victims and perpetrators to be solid data for publication." Thus they do not include these fields in a downstream analysis for data quality reasons, yet formalizing these variables into their schema becomes a way to assert—using the data frame—that it matters to track the connection between socioeconomic status and feminicidal violence and to demand that the state monitor such factors as well. Counterdata activists seek, through columns and categories, to not only fill in gaps where official data are absent, but also to assert what, how, and which factors "count" in a case of feminicide.

But deliberately exceeding state and international standards for *what counts* as a case of feminicide and *how* each case should be elaborated and systematized does not mean that anything goes and everything counts. For example, the Grupo Guatemalteco de Mujeres (Guatemalan Group of Women), discussed further in chapter 6, triangulates information from press reports and federal death records to make a judgment on whether the violent death of a woman constitutes a femicide. As can be seen in figure 6.8, over a period of six months, they considered somewhere between 40 and 60 percent of violent deaths of women as femicides, which means that there are many violent deaths of women in Guatemala that did not meet their criteria for gender-related killing.[19] Groups also have to be careful to verify information they receive. For example, the data journalism group Agencia Presentes saw reports circulating on Facebook about a trans woman who died by being set on fire in a particular neighborhood in Buenos Aires, and they were preparing to include the case in their map of LGBT+ hate crimes. But after following up with local shopkeepers, police, and sex workers, the report turned out to be a rumor posted to social media by a well-meaning but wrong activist. Groups try to draw firm lines around their definition of feminicide, their standards for determining that from available information, and their process of verification.

To define and classify feminicide, data activists construct a data infrastructure—collective, dynamic, social, and technical—around their work. This resonates with the *embrace pluralism* data feminism principle, which states that the most complete knowledge comes from synthesizing multiple perspectives. Activists build on legal and international standards with participatory, dialogue-based processes that are transcalar. This includes dialogue and reflection about individual cases among members of the project and any advisors: for example, the CFOJA taps its advisory board for case knowledge, and the Alianza's member organizations deliberate on cases through WhatsApp. It also includes dialogue with other activists in which they share their typologies and influence each other. For example, Carmen Castelló shifted her classification from *asesinato* (murder) to feminicidio based on conversations with the Observatorio de Equidad de Género in Puerto Rico. Dawn Wilcox began recording information on Indigenous identity based on her friendship with Annita Lucchesi from Sovereign Bodies Institute. Friendships and dialogues across projects help activists cope with the emotional labor and, in some cases, the loneliness of doing the work as an individual. Several people shared the sentiment of Ivonne Ramirez, who runs the project Ellas Tienen Nombre about Ciudad Juárez: "I think that what has helped me to cope with this work is to collaborate with others who do similar work."

Thus the recording stage of a counterdata project is not only about the production of columns and categories and counts, but also represents a form of *infrastructuring*—the production of sociotechnical relations around activist data.[20] Recording cases in community lends credence and meaning to the spreadsheets, makes space for deliberating on categories, and surfaces answers—through dialogue and deliberation—to fundamental questions around what is and should be counted. For example, a number of the organizations we interviewed are in regular dialogue with each other because they participate together in events like the ones we have produced through the Data Against Feminicide project or networks like the Comité de América Latina y el Caribe para la Defensa de los Derechos de las Mujeres (Latin American and Caribbean Committee for the Defense of Women's Rights, CLADEM), or the Red Interamericana Anti-Femicidio (Interamerican Anti-Femicide Network, RIAF), or the Red Latinoamericana contra la Violencia de Género (Latin American Network Against Gender Violence).

This last group is a network of thirty-five-plus organizations that monitor feminicide by producing data and includes many of the organizations we interviewed. Facilitated by MundoSur, a nonprofit based in Argentina, the members participate in technical capacity-building sessions as well as group discussions of definitions and methods. This transnational coordination work accelerated during the COVID-19 pandemic, when video calls increasingly became the norm. Ignacio Piana, who leads data analysis for MundoSur, used the term *harmonize* when describing their work. He explains, "We thought that if we came with a methodology or a form of standardization already preconceived by us and tried in some way to impose it on the organizations—imagine! Thirty-five organizations with their particular struggles and some with decades of experience in this—it was very likely that it would fail right from the start. So, we said, we are going to co-construct it with the organizations, we are going to co-construct a methodology, we are going to co-construct definitions as a first step." While groups in the network share their data monthly to produce the regional map in figure 5.7 with aggregated statistics, the goal is not to impose a "standard" definition of feminicide nor a fixed set of categories, but rather to encourage alignment and articulation in order to create a space of ongoing dialogue.

This method is fundamentally different from the typical process of producing global indicators of gender-related violence. Feminicide data activists scale via *harmonization* over *standardization*. This process of harmonization unfolds over time, nurtured by community, dialogue, and relationships that provide the infrastructuring tissue. Activists are not only producing their own data about feminicide, they are also producing their own processes of data sharing and data integration, but on their own terms, respecting differences, local context, and multiplicity. This infrastructuring work contributes

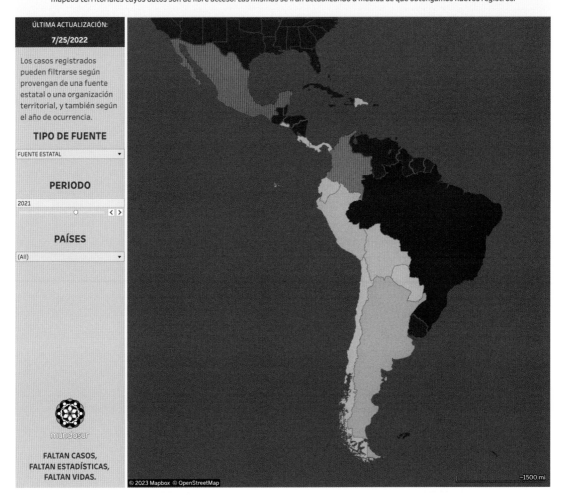

CASOS REGISTRADOS A NIVEL NACIONAL

2019	2020	2021	2022
5,474	**4,432**	**4,498**	**408**
FEMINICIDIOS	FEMINICIDIOS	FEMINICIDIOS	FEMINICIDIOS
en 16 países desde el 1/01/2019 al 31/12/2019	en 16 países desde el 1/01/2020 al 31/12/2020	en 12 países desde el 1/01/2021 al 31/12/2021	en 9 países desde el 1/01/2022 al 29/06/2022

Estas cifras fueron construidas con información proporcionada a MundoSur por organizaciones territoriales de América Latina u obtenida de mapeos territoriales cuyos datos son de libre acceso. Las mismas se irán actualizando a medida de que obtengamos nuevos registros.

ÚLTIMA ACTUALIZACIÓN:
7/25/2022

Los casos registrados pueden filtrarse según provengan de una fuente estatal o una organización territorial, y también según el año de ocurrencia.

TIPO DE FUENTE

FUENTE ESTATAL

PERIODO

2021

PAÍSES

(All)

FALTAN CASOS,
FALTAN ESTADÍSTICAS,
FALTAN VIDAS.

© 2023 Mapbox © OpenStreetMap

~1500 mi

FIGURE 5.7
Screenshot of a web map published by MundoSur drawing from the data of the thirty-five-plus organization members of the Red Latinoamericana contra la Violencia de Género (Latin American Network Against Gender Violence). At the bottom left, it says "Missing cases, missing statistics, missing lives." Courtesy of MundoSur.

to the dynamism of activist data schemata; we saw how the Alianza in Ecuador began by collecting four variables about each case and now collects more than eighty. This process of dynamic learning, sharing, and diffusion of fields and categories resonates with *rethinking binaries and hierarchies*: this is the data feminism principle that asks us to challenge the gender binary, along with other systems of counting and classification that perpetuate oppression. Far from seeing columns and categories as fixed or natural, activists are continually engaging in reflexive practice to add, remove, or align their variables and categories with those of others.

Yet pluralistic dialogue about definitions and categories is not free of conflict. Dawson described how, as they got started, the expert panel of the CFOJA was divided on two major issues. One was whether the term *prostitution* or *sex work* should be used to discuss femicide that occurs in this context. This relates to long-standing tensions in feminism around whether prostitution should be "abolished" (the abolitionist argument) or whether it should be legalized and made safer (the sex worker argument).[21] In the end, the category for this violence was named "Femicide in the context of what is referred to as sex work or prostitution," and it drew from an approach developed by the UN that both recognizes and promotes the rights of sex workers while also recognizing the context of trafficking and sexual exploitation that can occur with prostitution.[22] The second issue for the CFOJA's expert panel related to another long-standing tension in feminism—the presence of a vocal minority that has sought to exclude trans voices, and trans women, in particular, from feminist spaces, activism and scholarship.[23] Ultimately, says Myrna, "We decided that we were going to include transfemicides because there are evolving definitions of femicide. Some definitions focus on females whereas others focus on women, including trans women." Like a handful of other groups, the observatory decided to document *all* cases of trans killing, including both trans women and trans men, to better understand the contexts and circumstances surrounding violence experienced by transgender communities. That said, they do not include the murders of trans men in aggregate counts of femicide like those seen in figure 5.3, since that would be misgendering them. While the group moved through these tensions, their process highlights the challenges of pluralistic governance and the ways in which acts of classification can surface long-standing political debates.

THE DATABASE IS A PLACE; THE SPREADSHEET IS A MEMORIAL

While categories and spreadsheets permit downstream analysis, there is another function that the process of recording opens up. In our interview with Julia E. Monárrez Fragoso, one of the first academics to assemble her own database of feminicides in

Ciudad Juárez, we asked her thoughts on the meaning and impact of widespread feminicide data activism. She reflected, "It seems to me a very important task in the sense
that it is ultimately a counter-hegemonic memory against a state that does not deliver
justice and imposes its own memory. These will be the central memories, because while
people do take the official statistics into account, they also turn to the organizations
that keep the list, the count."

Monárrez Fragoso's focus on counterhegemonic memory as one of the primary contributions of these activist databases was striking to me and resonates with the Alianza's infographic in figure 5.2, in which their first demand was *memory*. A significant
proportion of data activists and journalists consider the database itself as a place of
remembering, witnessing, and caring for people, even in their death. This shouldn't be
confused with the use of data to undertake public memory work, such as vigils and art
installations that use data to represent absent bodies (discussed further in chapter 6).
Rather, for a number of feminicide data producers, the spreadsheet itself functions as a
memorial and the work of recording cases as memory work.

This surfaced in several ways in our conversations with activists. First, just as activists are attuned to the practical politics of categories and columns, they are also attuned
to the power of such rhetorical subtleties as the naming and ordering of fields in a
database or spreadsheet. It matters, for example, how women are described by recording their names in a field titled *Victim* versus a field titled *Name*. Helena uses *name of
the woman*. Dawn Wilcox of Women Count USA genders her database fields by using a
possessive pronoun: *her name, her age, her occupation*, and, for the alleged perpetrator,
his name, his age, his occupation. Why? A victim is defined by the event of her death,
whereas a person has a name and a community and a life that precede the event that
ended it. Likewise, an alleged perpetrator is also defined by the violent event of vic-
timizing another. In her in-depth analysis of feminicide databases as data frames—as
objects with representational, communicative, and narrative properties—Helena asks,
"Ontologically (and politically), this characterisation potentially has essentializing
effects: were the persons in the frame always already victims and victimizers?"[24] She
argues that the victim-victimizer binary works to displace attention from the structural
dimensions of pervasive gender inequality. Figure 5.8 illustrates this binary as mani-
fested in the dataset about domestic homicides of women published by the Uruguayan
government. While most government data schemata and many of the activist data
schemata set up this victim-victimizer binary, Helena invites us to *rethink binaries and
hierarchies* for representational and political reasons. In other words, how we name
single columns in a spreadsheet can communicate a whole worldview about the nature
of feminicide and the lives of the people involved.

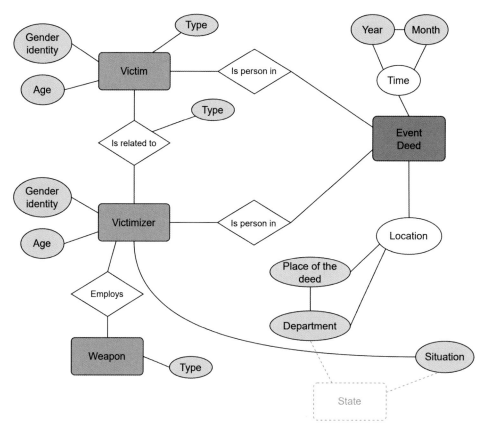

FIGURE 5.8
Helena Suárez Val analyzes the binary ontology of victim-victimizer that emerges from the official "Domestic Homicides of Women" dataset published by the Uruguayan government. Courtesy of Helena Suárez Val.

Likewise, activists also make key design decisions about which variables to record in their spreadsheets not only for downstream analysis but also for reasons of narrative and representational justice. Wilcox described how she began Women Count USA logging how many children a woman had, but later removed that category: "I think it's important to remember that these women mattered not because they were mothers, not because they're somebody's sister, daughter, but because they had value simply because of their own worth." This message is a core part of Wilcox's narrative around her project and resonates with the *elevate emotion and embodiment* data feminism principle—a way to foreground the fact that rows in a dataset about feminicide are people.

For example, Wilcox uses an image with large text front and center on the project's website, as well as for her business card. It reads:

> She is someone~~'s wife daughter sister mother~~.

The relations are crossed out so that it simply reads "She is someone." In this case, in contrast to the prior section about activists tracking variables like socioeconomic status, it is not so much that Wilcox is demanding that the state *not track* orphaned children, but rather that linking a woman's value to her status as a mother is misogynistic, even when it is "only" in a spreadsheet. In contrast, it is important to note that a number of other groups *do* track the number of children left behind by a feminicide. For example, in figure 5.3, the Alianza shows that 197 children were orphaned in 2021 in Ecuador. But the Alianza is using this field in the sense of reparative justice—that is to say, to track and demand reparations from the state to these families. In contrast, Dawn is using the category as representation and memory—that is to say, to think about what ideas about a woman's value are communicated within the narrative space of the spreadsheet itself. Both of these uses are intentional, both are political, but the locus of the political demand is placed differently.

The work of recording cases can constitute ritualistic work for some activists. Helena noted how there was a monthly rhythm to logging cases, and if there seemed to be fewer cases, she became alarmed and started scouring the media for cases she may have missed. Lucchesi from Sovereign Bodies Institute described the work of recording cases as "spiritual" and stressed the importance of culturally grounded data production and care practices, drawing inspiration from Indigenous data epistemologies. Accordingly, activists seek to protect their databases and spreadsheets from being manipulated and used in ways that disrespect the lives represented therein. For example, in 2022, when Helena and I were coteaching the online Data and Feminicide course, I proposed using the spreadsheet from Feminicidio Uruguay for an activity where learners would make a pivot table and a data visualization. Helena pushed back, feeling strongly that the rows of data in the spreadsheet—really the lives described by the spreadsheet—were not meant as playthings for experimentation, and so we chose a different dataset for the exercise.

Activist databases, then, are not abstract spaces of latent "raw" data waiting for activation through statistical analysis. Instead they are a form of memorial, a sacred space suffused with grounded connection to the embodied lives and people who are partially represented there. In this sense, feminicide data activists are also archivists and memorialists, engaged in acts of remembering that take place in databases and spreadsheets, where they link the recording of individual human lives and deaths with collective social memory and responsibility. Perhaps the sharpest example of this approach is

DATE	HER NAME	HER PHOTO	HER AGE	HER RACE / ...	ABOUT HER	CITY	STATE	RELATIONSHIP
1/31/2018			36	Latina/x		Whitewater	WI	Husband
1/31/2018			31	Unknown / U...		Pasadena	CA	Husband
1/31/2018			87	Unknown / U...		Dallas County	TX	Stranger
1/31/2018			31	Native Ameri...	She was a me...	Fort McDermitt I...	NV	Unknown/Unrele...
1/31/2018			40	Native Ameri...	She was a me...	Fort McDermitt I...	NV	Unknown/Unrele...
2/1/2018			77	Unknown / U...		Phoenix	AZ	Husband
2/1/2018			51	White		Oklahoma City	OK	Husband
2/1/2018			58	White		Highland Park	TX	Stranger
2/1/2018			33	Black / Africa...		Baton Rouge	LA	Boyfriend
2/1/2018			24	Latina/x		Corpus Christi	TX	Boyfriend
2/2/2018			47	White		Apollo Beach	FL	Husband
2/2/2018			26	Black / Africa...		Norcross	GA	Boyfriend
2/2/2018			43	Black / Africa...		Las Vegas	NV	Ex-Boyfriend
2/2/2018			43	Unknown / U...		Her obituary said	AZ	Husband
2/2/2018			56	Black / Africa...		Chicago	IL	Fiancé / Ex-Fiancé
2/2/2018			56	Latina/x		Chula Vista	CA	Uncle / Nephew ...
2/2/2018			15	Latina/x		Tucson	AZ	Father / Stepfath...
2/2/2018			38	Latina/x		Chula Vista	CA	Boyfriend

FIGURE 5.9
Dawn Wilcox from Women Count USA spends many hours seeking photos of killed women for her database. Courtesy of Dawn Wilcox / Women Count USA: Femicide Accountability Project.

Wilcox's. She can spend hours searching for a photo for each killed woman in her database, specifically looking for photos that are not stigmatizing, not mugshots, and not dehumanizing (see figure 5.9). When a photo that meets her criteria is low-quality, she will spend time, sometimes hours, retouching it before publishing it in her database. This embodies what Michelle Caswell and Marika Cifor have named, in archival practices, as a feminist ethics of care.[25] Posited as a form of radical empathy, the authors argue that such an ethics of care stands in contrast to archives of human rights violations that robustly document harms solely in order to seek legal redress. For Dawn, her database is a place for honoring individual lives and asserting the vast scope of the crisis. The spreadsheet, then, has value in its own right as a memorial, not only when it is activated in some future data analysis process or in some future case for legal redress. Through these small acts of care and witnessing and crafting, activists cultivate their databases into sites of active remembering, disobedient archives that refuse to submit to the normalization of gender-related violence.

LESSONS FOR A RESTORATIVE/TRANSFORMATIVE DATA SCIENCE

The work to record cases—to transform unstructured information into structured data, to document feminicides into rows and columns, to categorize cases—surfaces some key lessons for restorative/transformative data science projects. First, unlike hegemonic data scientists, activists have no pretense that their work to log cases and categories and variables about feminicide is neutral or objective (key values for hegemonic data science). Rather than thinking about data as "information collected, stored and presented without interest," activists are leveraging a key feature of data capture: "To collect, store, retrieve, analyze, and present data through various methods *means to bring those objects and subjects that data speaks of into being*" (emphasis mine).[26] This is what sociologist Evelyn Ruppert and coauthors speak of as *data politics*. Hegemonic data science, however, remains largely ignorant of such data politics because refusing the political dimension of data has largely worked, thus far, to mask the fact that it is undergirding and exacerbating deeply unequal systems of domination.

This kind of ignorance is blisteringly explicated by philosopher Charles W. Mills, who proposed that white ignorance is essential to the maintenance of white supremacy. Mills writes, "The Racial Contract prescribes for its signatories an inverted epistemology, an epistemology of ignorance, a particular pattern of localized and global cognitive dysfunctions (which are psychologically and socially functional), producing the ironic outcome that whites will in general be unable to understand the world they themselves have created."[27] White people, in other words, are ignorant about the methods and impacts of white supremacy, and this epistemological gap is, in fact, central to the maintenance of white supremacy itself. It constitutes a kind of plausible deniability about the harm caused by such a system. Palawa sociologist Maggie Walter has extended Mills's argument to the settler colonial state and the discipline of sociology specifically, making the case that the lack of engagement with Indigenous lifeworlds and processes of colonization "requires the application of a group epistemology of ignorance on an industrial scale."[28]

In the case of hegemonic data science, the ignorance—and the plausible deniability that ensues—comes from the inability or the unwillingness to admit that data have politics, that data have political effects, and that these mostly have to do with concentrating wealth and power in the hands of those who already have wealth and power. Historian Dan Bouk and colleagues frame this collective ignorance around *forgetting*: "We forget that official numbers have to be made even when things are going well."[29] In other words, even though institutional data are actively produced, those same institutions tend to forget the design and measurement decisions they made and take their numbers at face value, as one-to-one descriptors of the world. Likewise, Sally Merry discusses the process of naturalization that happens when creating indicators about gender-based violence: "Once decisions about categorization and aggregation are made the categories may come to seem objective and natural, while the power exercised in creating them disappears."[30]

But is ignorance always a "bad thing"? In recording cases of feminicide into spreadsheets and databases, activists exploit this hegemonic ignorance of data politics and bend it to their advantage. If numbers are neutral and data in spreadsheets are one-to-one representations of reality, then activists are simply producing feminicide facts—and who can deny the objective facts? This is a use of ignorance that feminist philosopher Alison Bailey names as *strategic ignorance*: "a way of expediently working with a dominant group's tendency to see wrongly. It is a form of knowing that uses dominant misconceptions as a basis for active creative responses to oppression."[31] Feminicide data activists thus appropriate some of the same formal language as hegemonic data science: numbers, systematic collection, categorization and classification, spreadsheets, databases, and statistics. They are not naïve about the contradictions inherent in this approach. Télia Negrão, from Lupa Feminista contra o Feminicídio in Brazil, described how the group was grounded in a feminist critique of the neutrality of science on the one hand: "Our knowledge, our speech, is a situated speech. We speak from a place. From a place of silenced voices." Yet Negrão went on to say that this meant that they themselves were responsible for being "the bearers of our own narratives. We had to build narratives, and these narratives had to be built from reliable instruments, testable instruments, and scientific methodologies." Following Gayatri Chakravorty Spivak's notion of *strategic essentialism*, Helena Suárez Val has called this a practice of *strategic datafication*—activists mobilize data for representation, evidence, and legitimation, despite their (acute) awareness that data and quantification have been historically tied to imperialism, colonialism, capitalism, and patriarchy.[32] In pursuing strategic datafication, activists borrow the (outsized) credibility and epistemological authority of the quantitative disciplines to birth the worlds that they purport to measure.

This is not to say that activists in any way distort the truth—indeed, all groups are earnest and rigorous in their quest to record accurate, verifiable information about every individual case in their databases. Rather, the point is that hegemonic data science prizes neutrality and (wrongly) sees quantitative measures as the surest route to expunging politics and bias from knowledge. So if data are simply facts without politics, a door is open to leverage them to boost activists' structural (political!) framings of feminicide to travel further and faster, influencing the actors that many groups target: the state and the media.

Another key lesson for restorative/transformative data science in relation to the recording stage has to do with the underlying conception of the spreadsheet or database that is being mobilized in each domain. For hegemonic data science, the dataset is a disembodied space of abstraction; it is "inert representation";[33] "truth divining";[34] the neutral record of facts; "raw" or latent information awaiting to be brought to life through sophisticated analysis, which then yields "business intelligence" downstream in the pipeline. In contrast, for a restorative/transformative data science, the spreadsheet or database is an embodied space of rich and relational connections to the world; an aggregation of prayers; a product of collective deliberation; a refusal to forget individuals; an insistence on visibilizing systemic violence. Feminicide data activists strategically leverage the epistemic authority that accrues to them for using "data-driven" methods, but rather than participating in the surveillance capitalism regime of "data as property"[35] they are proposing and modeling an alternative vision of *data as countermemory* and *data as megaphone*. Following Monárrez Fragoso's emphasis on activist data production as memory work, the countermemory function serves activist struggles to "keep the memory of hidden, everyday, and private violence fresh, public, and continuous."[36] The megaphone function amplifies the voices of grassroots activists and women leaders, creating a conduit for what Negrão framed as "silenced voices." Maldonado Tobar from the Alianza describes this function: "[High-quality data] allows us to give voice and enhance the voice of the women leading these organizations." Both uses fulfill the function outlined by Ruppert and colleagues in which "data is generative of new forms of power relations and politics at different and interconnected scales."[37] Both uses reclaim data and quantitative methods from epistemologies of violence and extraction and place them in the service of epistemologies that center life, living, vitality, rights, and dignity. This shift is fundamental to a restorative/transformative data science.

Through this data as megaphone function, feminicide data activism is concerned with scale. But it is in a fundamentally different way than hegemonic data science thinks about scale. Scalability is the unquestioned value that underlies virtually all of tech development (and venture capital) today. In their paper "Against Scale," social scientists Alex Hanna and Tina Park critique the concept and the value of scalability—the idea that a system can expand without having to change itself in substantive ways or to rethink its constitutive elements.[38] They link scale to standardization, classification, and colonization. What is scalability if not a modern-day reboot of the centralizing, extractive colonial violence that led to the massive transfer of global wealth into the hands of white European individuals, companies, and institutions? In her work on theorizing nonscalability, anthropologist Anna Lowenhaupt Tsing asserts that we need to pay attention to "the mounting pile of ruins that scalability leaves behind."[39] Ironically, or maybe not ironically, it is scale thinking that is directly responsible for creating the Americas that tolerates feminicide in the first place; scalability's pile of ruins normalizes gender and racial violence in the name of wealth accumulation.

Feminicide data activists have a different starting point for scale that orients us toward what scale might mean for a restorative/transformative data science. What if the purpose of scale is not to concentrate wealth in the hands of our billionaire overlords? What if the purpose of scale is also not to generate and circulate expert technocratic judgment, as Merry details in her account of how global gender-related violence indicators get produced? What if the purpose of scale is to amplify the voices, power, and knowledge of the people closest to the harms enacted by colonial, venture-capital scale?[40] A restorative/transformative data science scales on its own terms and in its own way. Feminicide data activists scale through building infrastructure—technical and relational and dialogical—through geographically-dispersed expert panels, through membership networks such as the Red Latinoamericana contra la Violencia de Género, and also through WhatsApp groups and *encuentros*. At the same time, they remain rooted in their local contexts and see their role as a megaphone—helping to propagate grassroots voices across these broader networks. A restorative/transformative data science may scale through harmonization—aligning definitions and categories and counting practices through dialogue and relational infrastructures. Such a system fosters divergence and multiplicity (hence the proliferation of the many activist categories that depart from the Latin American model protocol). In other words, for restorative/transformative data science, multiplicity is not a problem and standardization is not the answer.

CONCLUSION

Recording is the stage of a restorative/transformative data science project in which activists transform unstructured information into structured data comprised of fields and values, cases, and categories. Recording is the stage that most highlights how data—all data, and especially data about feminicide—are intentionally *produced*. Data are not facts waiting to be found, they are actively crafted into rows and columns, following processes of careful deliberation and consultation. Likewise, data schemata and categories—the systems by which activists collect and label information—are dynamic. They shift and change as activists learn from each other, observe patterns of violence in their region, harmonize with international standards, and respond to family needs.

Being at the heart of a counterdata production project, recording has close ties to the other stages of work. In their process of crafting data, activists move fluidly between researching and recording cases—reading news articles or searching social media, and then logging small details into specific fields, and then returning to research again for missing values. The categories activists develop and apply to their cases relate directly to their analysis of power developed in the resolving stage of a restorative/transformative data science project. Those categories then enable the downstream analysis and disaggregation of subtypes of feminicide. While activists themselves know better than anyone else that data are produced—not found—they leverage the legitimacy and apparent objectivity of numbers and classifications to amplify the voices of grassroots organizations and to shift public memory. These strategic functions of data—as megaphone and as countermemory—surface especially in the next chapter on data communication and circulation, in which activists simultaneously use data and refuse data.

6   REFUSING AND USING DATA

It was March 8, 2022, International Women's Day, also known as 8M, and I was navigating a jubilant Callao Street in Buenos Aires, the place that sparked the Ni Una Menos uprising in 2015. It was the first mass mobilization after the exhaustion of pandemic lockdowns and tens of thousands of Argentines, especially women and girls, *disidencias, travestis*, sex workers, babies, and families had turned out en masse to protest.[1] A document assembled and read by dozens of feminist groups, women's groups, workers' groups, and LGBTQ+ groups spelled out their main messages: "No to the government's trade agreement with the IMF! No to violence against women and *disidencias*! Stop the repression of the women, men and non-binary people leading this fight!"

I made my way to the National Congress, through enormous banners and drumming circles, people dressed in colorful smocks to match their group, vendors selling snacks and underwear (so much underwear!), and pins and scarves. Finally, I arrived at my destination: the banners, puppets, infographics, and signs organized by Mumalá. As described in chapter 4, Mumalá works across the country and focuses on the intersection of neoliberal economic violence and gender rights. At the march, one large banner read "Ni Una Menos / Vivas y libres nos queremos"—Not one (woman) less / We (women) want to stay alive and free. The other banner read "La deuda es con nosotres"—The [national] debt is to us (figure 6.1). These are slogans that link physical violence, economic violence, and the labor of women and nonbinary people. The government had recently made a new deal with the International Monetary Fund (IMF) to restructure Argentina's national debt and impose austerity measures. The IMF and the Argentine people have a decades-long and deeply contentious relationship, with both scholars and politicians asserting that the neoliberal reforms instituted by deals with the IMF represent a form of international imperialism that has exacerbated the

FIGURE 6.1

(a) The group Mumalá's mobilization for the International Women's Day March on March 8, 2022. "La deuda es con nosotres" translates to "Your debt is to us." Mumalá is using *nosotres* instead of *nosotros* (us men) or *nosotras* (us women). *Nosotres* represents the nonbinary way to say "us." Courtesy of the author. (b) An example of a report from Mumalá's Observatory. The group publishes their reports as extended social media posts with aggregated statistics and graphics. This is a Facebook post from March 22, 2022, and represents a report about five years of monitoring femicide in Argentina.

very economic conditions they are supposed to ameliorate.[2] Instead of owing the IMF, Mumalá asserts that the government owes women and gender minorities.

    The Mumalá group included around fifty people in glitter and costumes, some carrying smaller individual signs, and several hoisting a huge Uncle Sam puppet representing US imperialism. One of the most fascinating aspects of the display was how Mumalá incorporated data from its observatory, called Women, *Disidencias*, Rights. As shown in figure 6.1, members of Mumalá held up printed placards with infographics about femicides. A honeycomb infographic provided multicolored counts of different types of violence. The observatory had counted thirty-two direct femicides, two *trans/travesticidios*, and sixty-two attempted femicides so far in the year 2022. A bar chart

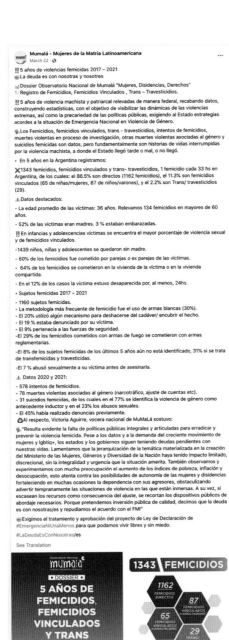

Mumalá - Mujeres de la Matria Latinoamericana
March 22 · 🌐

‼️ 5 años de violencias femicidas 2017 – 2021
🌎 La deuda es con nosotras y nosotres

📊 Dossier Observatorio Nacional de Mumalá "Mujeres, Disidencias, Derechos"
⚖️ Registro de Femicidios, Femicidios Vinculados , Trans – Travesticidios.

‼️ 5 años de violencia machista y patriarcal relevadas de manera federal, recabando datos, construyendo estadísticas, con el objetivo de visibilizar las dinámicas de las violencias extremas, así como la precariedad de las políticas públicas, exigiendo al Estado estrategias acordes a la situación de Emergencia Nacional en Violencia de Género.

🩸 Los Femicidios, femicidios vinculados, trans - travesticidios, intentos de femicidios, muertes violentas en proceso de investigación, otras muertes violentas asociadas al género y suicidios femicidas son datos, pero fundamentalmente son historias de vidas interrumpidas por la violencia machista, a donde el Estado llegó tarde o mal, o no llegó.

· En 5 años en la Argentina registramos:

✖️ 1343 femicidios, femicidios vinculados y trans- travesticidios, 1 femicidio cada 33 hs en Argentina, de los cuales: el 86.5% son directos (1162 femicidios), el 11.3% son femicidios vinculados (65 de niñas/mujeres, 87 de niños/varones), y el 2.2% son Trans/ travesticidios (29).

📊 Datos destacados:

- La edad promedio de las víctimas: 36 años. Relevamos 134 femicidios en mayores de 60 años.

- 52% de las víctimas eran madres. 3 % estaban embarazadas.

‼️ En infancias y adolescencias víctimas se encuentra el mayor porcentaje de violencia sexual y de femicidios vinculados.

-1439 niños, niñas y adolescentes se quedaron sin madre.

- 60% de los femicidios fue cometido por parejas o ex parejas de las víctimas.

- 64% de los femicidios se cometieron en la vivienda de la víctima o en la vivienda compartida.

- En el 12% de los casos la víctima estuvo desaparecida por, al menos, 24hs.

· Sujetos femicidas 2017 – 2021

- 1160 sujetos femicidas.
- La metodología más frecuente de femicidio fue el uso de armas blancas (30%).
- El 20% utilizó algún mecanismo para deshacerse del cadáver/ encubrir el hecho.
- El 19 % estaba denunciado por su víctima.
- El 9% pertenecía a las fuerzas de seguridad.
- El 29% de los femicidios cometidos con armas de fuego se cometieron con armas reglamentarias.

-El 8% de los sujetos femicidas de los últimos 5 años aún no está identificado, 31% si se trata de transfemicidas y travesticidas.

-El 7 % abusó sexualmente a su víctima antes de asesinarla.

📊 Datos 2020 y 2021:

- 578 intentos de femicidios.
- 78 muertes violentas asociadas al género (narcotráfico, ajuste de cuentas etc).
- 31 suicidios femicidas, de los cuales en el 77% se identifica la violencia de género como antecedente inductor y en el 23% los abusos sexuales.
- El 45% había realizado denuncias previamente.

🔊 Al respecto, Victoria Aguirre, vocera nacional de MuMaLá sostuvo:

🩸 "Resulta evidente la falta de políticas públicas integrales y articuladas para erradicar y prevenir la violencia femicida. Pese a los datos y a la demanda del creciente movimiento de mujeres y lgbtiq+, los estados y los gobiernos siguen teniendo deudas pendientes con nuestras vidas. Lamentamos que la jerarquización de la temática materializada en la creación del Ministerio de las Mujeres, Géneros y Diversidad de la Nación haya tenido impacto limitado, discrecional, sin la integralidad y urgencia que la situación amerita. También observamos y experimentamos con mucha preocupación el aumento de los índices de pobreza, inflación y desocupación, esto atenta contra las posibilidades de autonomía de las mujeres y disidencias fortaleciendo en muchas ocasiones la dependencia con sus agresores, obstaculizando advertir tempranamente las situaciones de violencia en las que están inmersas. A su vez, si escasean los recursos como consecuencia del ajuste, se recortan los dispositivos públicos de abordaje necesarios. Porque pretendemos inversión pública de calidad, decimos que la deuda es con nosotras/es y repudiamos el acuerdo con el FMI"

🌎 Exigimos el tratamiento y aprobación del proyecto de Ley de Declaración de #EmergenciaNiUnaMenos para que podamos vivir libres y sin miedo.

#LaDeudaEsConNosotras/es

See Translation

👍😮💙 79   67 Shares

👍 Like   💬 Comment   ↪ Share   🖤 ▾

Most relevant ▾

⚫ Write a comment...   🖼 🌅 😊 ⚙

Most Relevant is selected, so some comments may have been filtered out.

FIGURE 6.1 (continued)

placard showed a breakdown of the relationships between victims and perpetrators, with the majority—60 percent—of femicides due to intimate partner violence. Other placards depicted aggregate statistics about children left behind, the number of victims known to have filed prior complaints against their aggressors (20 percent), and the weapons used.

As artifacts of communication, these printed infographics were altogether curious. In the midst of the raucous dissent, drumming, and chanting, here were some charts and statistics that one might normally expect to find in a staid policy Powerpoint. What was the purpose of including infographics in a public protest? What other ways does Mumalá circulate their data, to whom, and with what effects?

This chapter surveys the fourth and last stage of a restorative/transformative data science project—how grassroots groups simultaneously use data and refuse data (figure 6.2). It describes how activists circulate their feminicide data in the world, what forms their data artifacts take, and how activists try to mobilize diverse audiences and shift power in multiple domains using data-driven communication. This is only possible after the accumulated efforts that go into producing data—the resolving, researching, and recording described in the prior three chapters.

As our research team interviewed activists about where their data go and why, it became clear that all groups use data to refuse feminicide. Within that larger refusal, five themes emerged that describe the goals and impacts of feminicide data.[3] Each represents a way that counterdata are used to refuse feminicide: to *repair* and heal families and communities; to *remember* the individual and collective lives lost to structural violence; to *reframe* feminicide as a structural issue, rather than an interpersonal problem; to *reform* the laws, policies, and practices of the state around feminicide; and finally, to *revolt* by using data in large-scale mobilizations in public spaces. This last category describes Mumalá's use of infographics in the 8M march. While these uses of activist data take different aesthetic forms and target different audiences, I will argue in this chapter that they all constitute ways of enacting feminist refusal of gender-related violence through data. In the process, activists also end up refusing (while using) the data themselves.

FIGURE 6.2

Refusing and using data is the final stage of a restorative/transformative data science project. Courtesy of the author. Design by Melissa Q. Teng. Collaged images: Courtesy of Jaime Black. Photo by J. Addington. #SayHerName figures from the African American Policy Forum's "Say Her Name (Hell You Talmbout)" video by 351 Studio and Wondaland. Red shoes artwork by Elina Chauvet. Photographer unknown.

REFUSING + USING

Where data go, who uses them

Reform / Working with state to formulate new laws, policies and practices

Remember / Memorializing killed people; Grieving in public

Revolt / Protesting, mobilizing; Performing resistance in public space

Reframe / Storytelling that challenges stigma; Reframing violence as structural

Repair / Supporting families and communities who have lost beloved members

RESOLVING	RESEARCHING	RECORDING	**REFUSING + USING**
Developing a theory of change	Finding + verifying information	Information extraction + classification	**Where data go, who uses them**

REFUSING AND USING DATA

While prior chapters focused on how activists produce counterdata, this chapter centers how they communicate their data. As we will see, activists use data to enact refusal and they circulate their data in a wide variety of forms ranging from reports and graphics to large-scale artworks to vigils and ceremonies. Activists circulate data about feminicide first and foremost to refuse a status quo that disproportionately delivers fatal violence to women. Here the idea of necropolitics from Achille Mbembe is useful to pick up again from chapter 2. *Necropolitics* relates to the fact that death is not distributed equally; specific groups are targeted for both direct and indirect violence. Necropolitics also relates to the performative and discursive element of interpretation around what violent deaths mean—this is to say: In a particular society, whose deaths matter?

While Mbembe himself had little to say about gender, multiple feminist scholars have drawn connections between feminicide and necropolitics. In a 2011 article, geographer Melissa Wright describes the necropolitical struggles between activist and government narratives around femicide on the Mexico-US border.[4] Government narratives blamed women themselves (or families, who should be ostensibly monitoring and controlling women's use of public spaces to avoid them getting killed) and reinforced the misogynist idea that, by being "inappropriately public," women victims were responsible for their own deaths. Activists countered with campaigns around government impunity, to focus the public's attention on the state's systemic deflection of its responsibility for ensuring women's safety and human rights. Media scholar Francesca M. Romeo advances the idea of *necroresistance* to describe work "that targets the authorities that perpetrate necropolitical violence and mobilizes the sight of the corpse as a means to sustain and ensure security for the living." In this way, activists are able to "translate corporeal suffering into social and political change."[5] For example, political theorist Shatema Threadcraft lauds the way Black Lives Matter has used necropolitical struggle to shift public discourse in the United States, though she expresses doubts about whether such strategies work to make visible the femicides of Black women.[6] And Bronwyn Carlson, writing about Indigenous femicide in Australia, describes how Indigenous communities document femicides of relatives on social media in defiance of the data silences produced by the settler state.[7] These too are examples of necroresistance—refusing to accept violent death, refusing to normalize it, and using imperfect but available and effective tools in the service of that refusal.

Data activists' circulation of feminicide data constitutes exactly such necroresistance. While their production of data in the prior stages—resolving, researching, and recording—might also be considered necropolitical, it is in the communication and circulation of data about feminicide that activists mobilize their data in the service of particular narratives and interpretations; they give their data particular forms in order to reach specific audiences, deploying data for necropolitical effects.

Activists communicate their data to refuse feminicide and to refuse the social, political, and informatic silences that surround it. Moreover, activists recognize the complex and diffuse structures of power that reinforce the necropolitics of feminicide across multiple domains: the state, media and culture, families and communities. While many feminist critiques responsibilize and try to reform the state, none of its neglect would be possible without the oppression that is actively circulated in other domains. In turn, activists resist across those multiple domains and their highly diverse and creative use of data communication strategies reflects that. Table 6.1 shows the five themes for activist data actions that emerged from our interviews that describe activist goals: to *repair*, to *remember*, to *reframe*, to *reform*, and to *revolt*. These are not hard and fast categories, but rather serve to explore the diversity of settings in which feminicide data circulate and the impacts that they have (or aspire to have).

Data communication undertaken by most activist groups falls into one or two of these categories. As we saw with Mumalá, they use their data to support large-scale protests (*revolt*), but they also produce detailed reports and infographics that they circulate on social media with the goal of attracting media attention (*reframe*). Other groups we will meet in this chapter use their data to interact with the state around laws and policies (*reform*) or serve impacted families and communities (*repair*). And activist data can circulate without the activists themselves in the driver's seat. Observatory projects often operate as data intermediaries that support other groups that may focus more deeply on one of these goals, such as providing information to survivor-led advocacy groups so that they may hold a vigil to *remember* their family members together. In some cases, official institutions draw from activist data for their research and reports, and the lines between counterdata and official data start to blur. For the remainder of this chapter, we will examine these five themes in turn.

But before we go there, I want to raise an ethical consideration that emerges from examining activism that engages with necropolitics. In their paper on trans necropolitics, scholars C. Riley Snorton and Jin Haritaworn posit the idea that the deaths of trans people of color serve as a kind of raw material for trans rights movements that are dominated by white trans activists from the Global North. Through two case studies, they show how trans people of color are disregarded in life but become "used" in their afterlife by social movements: "Immobilized in life, and barred from spaces designated as white (the good life, the Global North, the gentrifying inner city, the university, the trans community), it is in their death that poor and sex working trans people of

Table 6.1
Goals and audiences for activist data communication

Theme	Audience	Forms
Repair	Survivors; families and communities	Direct services to families; accompaniment; rallies for justice for specific cases;
Remember	Families; activists; public	public vigils with lists of names; artworks about victims; social media posts about individuals
Reframe	Public; media and culture	Journalism and storytelling, supported by maps, statistics and multimedia; activists give interviews to press about their data; workshops and trainings for journalists
Reform	State	Reports with accompanying graphs, maps and charts; policy recommendations backed by statistics; activist dialogue with state about cases and data
Revolt	State; media and culture; public	Data used in protests; large public art interventions; coordinated campaigns on social media

color are invited."[8] Rather than a site of recuperation or memory justice, then, necropolitical narratives can themselves be an extractive method—a way of using the afterlife to reinscribe some of the same forces of domination that activists claim to be working against.

Circulating data publicly for necroresistance engages exactly such fraught ethical territory. In the following explorations of the diverse ways that feminicide data circulate, we will continue to examine how data are *used to refuse*, at multiple levels, taking multiple forms, across multiple domains of power. At the same time, each theme necessitates ethical questions about value extraction, revictimization, privacy, and the production of deficit narratives about marginalized groups, and we will also engage these questions.

REPAIR

Uma por Uma, which translates to One by One, was a data journalism project based in Recife, Brazil. During 2018, a large team of more than thirty women journalists working for the digital news outlet NE10 compiled a database of every woman or girl killed in the state of Pernambuco. The first goal was to systematically collect information about each case. The next was also to do original reporting on each, including

FIGURE 6.3
Uma por Uma compiled a database and in-depth reporting about every feminicide in the Brazilian state of Pernambuco during 2018. Courtesy of NE10.

interviewing families and law enforcement and following cases through the justice system. The published investigation, shown in figure 6.3, leads with the words "There is a story to tell behind each murder of a woman in Pernambuco. One by one, we will count them all."[9] The multimedia site includes maps and infographics, a page narrating the life of every woman in their archive, as well as the impact that her death had on her relatives and community. Julliana de Melo, one of the reporters, explained their intentions: "We always tried to show that when a woman dies, it's not just the woman who dies—that woman's family also dies a little."

Around half of the groups that our research team interviewed engage directly with family members. Those who do develop intimate relationships with impacted relatives and survivors. They help source information about cases when families can't get any from law enforcement and they may support families to access state resources. For example, Uma por Uma secured protection from the State Department of Women for a family facing threats of violence for having reported their case, and they fought with the judicial system to demand that they not drop specific cases. Says Ciara Carvalho, executive editor of the project, "We ended up establishing close relationships with some families. And we were a support, too . . . The mother of one of the girls told me, 'If it wasn't for you, my daughter would have been forgotten.' The mother recognized that it wasn't only me, it was the project, it was all of us."

Repair is the term our team uses to refer to the ways that feminicide data circulate to address the grief, loss, and acute needs of families, relatives, and communities following a feminicide. This formulation of repair follows Joan Tronto and Bernice Fischer's definition of care as "a species of activity that includes everything we do to maintain, continue, and repair our world so that we can live in it as well as possible. That world includes our bodies, ourselves, and our environment."[10] Repair and reparative thinking have a long trajectory in feminist theorizations of violence and exclusion.[11] Perhaps the closest metaphor for how we are using this term comes from the work of sociologist Saide Mobayed, who parallels the work of feminicide data activists to the Japanese method of *kintsugi*, which involves mending broken pottery pieces with colored glue so that the cracks and fissures are visible.[12] The object is restored, it is still beautiful, and yet it wears the record of its trauma.

We saw repair surface in the impact on families of the inclusion of their loved ones in an activist database; in the humanizing, careful storytelling for families that data activists and journalists do; in the direct services and support that activists end up providing to families; and in activists providing data to family-led groups for their advocacy work. All of these are ways that activists use their data for repair and healing. Moreover, as we have seen, data activists are often themselves survivors of gender-related violence or have been touched personally by feminicide, so in some cases the repair and healing is also for themselves.

Annita Lucchesi from Sovereign Bodies Institute talked about how the simple act of logging a person's case can be a small but meaningful step: "Entering them in the database doesn't necessarily bring their case to justice and it isn't justice for them or their family on its own, but it does feel like we're giving them a little bit of peace knowing that they're being counted and they're being honored and they're being prayed for." Many data activists told us stories about family members who reached out to thank them for including their loved one in their database.

In the case of Uma por Uma, the repair function of their work included the impact, for families, of having carefully reported stories about their relative in a high-profile public forum. As de Melo insisted, the team wanted to "look more deeply at these stories and give these women the dignity of having their stories well told. . . . So there's a well-treated text, well-written, with dignity, so that the memory of this woman is kept properly and justice is done." Storytelling for families is central to the work of the

African American Policy Forum, originators of the #SayHerName campaign. The organization compiles a database of Black women killed in police violence and also runs a network for mothers. They worked with musicians to produce a song about the victims, and even commissioned a play for families that reimagined their daughters back into the world and depicted what they would have gone on to do. Stories, narratives, and "well-treated texts" created explicitly for families serve a reparative function to insist, whatever treatment they may be experiencing from the state or the media, that their relative was beloved and deserves dignity in death and in life.

Like many groups, Uma por Uma found themselves fielding requests for information and services from families. Data activists and journalists occupy a unique informatic position in relation to feminicide: they often have more information about cases than authorities (or are more forthcoming with such information) and they often have expert knowledge of the media, the judicial system, and their failings. Because of this, some activists and journalists end up doing direct service provision or case-by-case advocacy. For example, Sovereign Bodies Institute told us about buying groceries for families, and they have an extensive list of direct services for families on their website. Others, like Carmen Castelló in Puerto Rico, act as first-line information providers, telling families who contact her where to start to access different services. Most commonly, this work takes the form of *acompañamiento*. This Spanish word translates to *accompaniment*, and it can mean this literally, such as the Grupo Guatemalteco de Mujeres (Guatemalan Group of Women) who accompany families to appointments with state officials. In their words, this is a way for activists to "fight for access to justice" for individual families. But acompañamiento can also mean other forms of support, such as the Movimiento Manuela Ramos (Manuela Ramos Movement) organization in Perú, which provides space for family groups to meet at no cost. Or Mumalá members showing up for family-led rallies and protests around specific cases. Or Utopix in Venezuela joining in digital protests produced by family groups. Sometimes this accompaniment can be in direct support of preventing further immediate trauma in the community. As Mak Miller-Tanner from Justice for Native People stated: "If something bad happens, we'll see a number of suicides that follow. And so we've tried to do stuff in our community, when that happens, to gather around and be like, 'Hey, we're here. We're all here. Let's all get together and mourn and support each other,' so that people don't get too desperate and we start losing more people." For Justice for Native People, as with other Indigenous-led groups, these represent approaches to community safety as well as to sovereignty, forging autonomous relations of care and kinship apart from the settler state. In all these cases, data activists build relationships and ongoing links of support with families, impacted communities, and family-led activist groups.

Accompaniment—being in relationship with families—also changes the work of data production for activists themselves. Brandy Stanovich, a staff member at the Native Women's Association of Canada, described how she personally knows and has sat with so many of families whose loved ones are in their database: "It's not just the data collection. It's bigger than that." Alejandra Bengalia, a journalist who works with the Casa del Encuentro in Argentina, stated that "an organization that counts is not the same as an organization that counts and accompanies [families]. . . . Because when you go to read the data, it is not the same, because you know the face of the people you accompanied. It is indescribable." The work of accompaniment helps affirm activists' political commitments. For Geraldina Guerra Garcés, data production is always first and foremost about serving families. For her and the Alianza Feminista para el Mapeo de los Femi(ni)cidios en Ecuador (Feminist Alliance for Mapping Femi(ni)cide in Ecuador), this has led to a project called Flores en el Aire (Flowers in the Air), where the group worked closely with eight families to coproduce multimedia maps—with sounds, stories, videos, and images—that tell a spatialized story of their daughter's life and their fight for justice.[13]

A handful of data activists also described how they provide their data to social movements, family-led and survivor-led organizations that are producing reports, protests, vigils or other acts of data communication. In this, data activists operate as *data intermediaries*: individuals, groups, or organizations that "serve as a mediator between those who wish to make their data available, and those who seek to leverage that data."[14] This works particularly once a group is well-known for their data production. For example, data produced by the group Lupa Feminista Contra o Feminicídio is often sought out by leaders of social movements and organizers of protests across Brazil. For Télia Negrão, this was the real importance of their data, because "it gives women a guarantee that they have a source of information to be able to fight." Activists also provide key informational infrastructure to support the leadership of families in the antifeminicide movement. Sociologist Mariana Mora in Costa Rica told us how she dropped everything to fulfill a data request from a man who had lost his sister and young niece to feminicide and wanted to write an article about child feminicide for the local paper. Data requests from family groups were happening so frequently that, at the time we interviewed them, the Alianza in Ecuador was working on building an interactive dashboard so that families could run custom reports and access the data they needed in a more automated way. In general, all groups were eager for their information to be used by and useful to social movements and families.

There are a number of ethical challenges faced by activists in using their data to repair and heal communities and directly serve impacted families. Many groups talked about the importance of not revictimizing families in the process of producing data about feminicide. And while many had received thanks from families for including a loved one in their databases, all groups who published their data openly online had also received requests to remove names by family members. In all but one case, they removed the case or took down social media posts immediately upon request, but it served as a poignant point of reflection for activists that inclusion in a database is not always going to be healing nor a welcome gesture for all relatives.[15] For this reason, the majority of feminicide data projects—thirty-six of forty-one groups we interviewed—do *not* openly publish their databases with raw, disaggregated information so as to avoid exposing victims' names and identifying information. In the case of Jane Doe, a US-based organization, the executive director renamed their database from Domestic Violence Homicide List to Domestic Violence Homicides based on feedback from a family member that calling it a *list* was dismissive and reductive. Groups that deal with MMIWG2 and racialized feminicide also discussed the trust and care that was required to work directly with families due to the intergenerational nature of the trauma.

Data activists leverage their data and positionality in a variety of ways to support families and communities impacted by feminicide. These actions challenge power primarily by supporting families who are left navigating a confusing maze of institutions amid extreme pain. They are often shown that their relative did not matter to the state or to the media or to the public. Using data for repair helps survivors in multiple ways: to navigate their grief; to seek services; to validate that their relative mattered and deserves dignity; to fight for justice; and to build their voice and their power. For Lucchesi, reflecting on the impact of Sovereign Bodies Institute's work, the most important impact has been "to see how it has led other survivors to find their voices and help families find their voices. I think that's definitely what I'm most proud of."

REMEMBER

As I described in the introduction, Helena Suárez Val, Silvana Fumega, and I are the three coorganizers of the Data Against Feminicide project. In March 2022, we experienced a wholly different way of encountering feminicide data. We met in a cafe in the center of Montevideo and talked with the two artists who had coordinated a project called *Cortar el hilo*, which translates to "Cutting the thread." Alejandra García and Marby Blanco are textile artists based in Uruguay, and, in 2020, they created an exercise in participatory embroidery and collective memory. García and Blanco put out an open call on Instagram to women and *disidencias*. Each would embroider a single name with red thread onto a piece of linen fabric, and the individual pieces would be joined at the end to represent the women killed in feminicide in Uruguay in 2019 (figure 6.4a). For the two coordinators, this was a way to establish a relationship between the embroiderist and the woman killed: "We wanted each woman and *disidencia* that was part of the call to get to know a little more about that other woman (or sometimes a girl) who was no longer alive, and that she not remain only in name . . . that this woman, in some way, would continue to be present."[16]

Each of the twenty-seven participating artists was sent the linen, thread, a name, and accompanying information, sourced from *Feminicidio Uruguay*, the data activism project that Helena leads. The artists all met in person to deliver their finished works, and it was a moment of tremendous emotion "when a large number of new stories, feelings, sensations and connections emerged. Everything that happened while the piece was being embroidered, what was left on the canvas, could reach us and in some way give greater meaning to what we were doing."[17] Later, García and Blanco sat together and stitched the individual pieces into a single long scroll, paying careful attention to connecting the red thread from the end of one piece to the beginning of the next.[18]

Almost exactly two years after its initial creation, Helena, Silvana, and I had the opportunity to feel some of these connections because Alejandra and Marby brought the finished piece to our meeting in Montevideo. We walked into the Plaza de la Constitución, an expansive and historic public space with tree-lined diagonal walkways. Helena stood, holding one side of the artwork, and Marby unrolled it slowly across the wide stone walkway. At almost thirty feet long, it took up the urban space, blocking passage through the plaza (figure 6.4b). Unfurled, we could see the work up close, and appreciate the variation in size, in lettering, in the precision versus the drama of the stitches, in the small flourishes introduced by the hands of the artists, in the juncture points where individual fabrics and names were pieced together to make the whole.

In this case, the data from Feminicidio Uruguay were arranged explicitly to *remember* those lives lost to violence; to establish an intimate, material relationship between individual artists and individual women killed; and to stitch those relations together into a collective refusal to forget. Using activist data to remember draws collective deaths into the public sphere for mourning, memorial, and observance. It represents what urban studies and peace studies scholar Delia Duong Ba Wendel has called an *ethics of nonerasure*, representations of violence that make history known in spatial form to tell otherwise ignored or hidden truths.[19] While *repairing* is about using data to support impacted families and communities, *remembering* actions target a broader public and aim to answer the question posed by Ida Peñaranda from Cuántas Más: "How do we explain what we have lost?"

FIGURE 6.4

Cortar el hilo (Cutting the thread), a participatory embroidery project coordinated by Alejandra García and Marby Blanco. (a) *Cortar el hilo*, detail. (b) *Cortar el hilo*, installation. Courtesy of Alejandra García and Marby Blanco. Photos by the author. Stitched photograph by Melissa Q. Teng.

Using activist data for remembering can take many forms. There are many creative artworks that draw from feminicide data—like *Cortar el hilo*. For example, the Red Chilena contra la Violencia hacia las Mujeres created a 100-meter-long memorial with silhouettes and names of victims of feminicide that they place on the ground in public spaces.[20] Helena, Silvana and I, together with curator Jimena Acosta Romero and artist Melissa Q. Teng, have built off work about *artivism* (art + activism) and *craftivism* (craft + activism) to frame these works as *data artivism*—techniques that use data and a variety of artistic methods to engage in collective resistance and memory justice, and that have long histories in the Latin American context.[21] Another common aesthetic form is the vigil. For example, on the first Tuesday of every month, Movimiento Manuela Ramos in Peru hosts a public vigil and brings along a large-scale physical calendar to the event (see figure 6.5). It represents one month, and the names of women are

FIGURE 6.5
Calendar produced by Movimiento Manuela Ramos for their monthly vigils. Red hands mark the day that a woman was murdered. Courtesy of Movimiento Manuela Ramos.

placed on the dates when they were killed. These names are drawn from the group's ongoing counterdata production about feminicides in Peru. Next to each name is a red handprint, pressed into the calendar by one of its activist creators. Participants in the vigils touch the calendar as part of the ceremony of remembrance. As Mabel Barreto describes, "Each one of us would reach out and put our hand on that little hand as if we were trying to make a connection. More than a connection, it's like—each one of those victims is an inspiration of life for us. We consider these murdered women to be our sisters."

Data activists have developed other creative techniques for memorialization. Many activist projects remember individual women by creating social media posts about each woman as they enter their information into their databases. For some, such as Rosa Page, who runs Black Femicide US, the stream of posts with women's faces and stories on social media is the primary way the data are publicly circulated. An illustration collective out of Mexico called No estamos todas (We [women] are not all here) creates artistic renderings of victims of feminicide and publishes them to Instagram.[22] Other activists have produced digital videos set to music, using photos from their databases.[23] La Casa del Encuentro in Argentina raised money to produce a book with a story about every woman's life from over ten years of collecting data. Called *For the Women: Ten Years of Reporting Femicide in Argentina*, it is, sadly, very heavy because there are so many pages and so many stories.[24]

The ethical challenges that face activists crafting public remembrances relate to using the names and stories of victims. Virtually all activists who engage in remembering actions noted the tension around representing victims as full human beings while not revictimizing them, objectifying them, or hurting their families. Who has the right (and the consent and the blessing) to use whose names and stories? In *Remembering Vancouver's Disappeared Women*, Amber Dean narrates the story of artist Pamela Masik's planned and then canceled exhibition of large portraits of missing and murdered Indigenous women. Activist groups led by First Nations women challenged Masik's claim that she, a white settler woman, could act as a "witness" to the violence that the community had suffered. Gloria Laroque, one of the organizers of the Women's Memorial March, gave this quote to local media: "The show would have made Masik the 'spokesperson' for Aboriginal women's issues, denying the efforts and voice of Aboriginal and Downtown Eastside women, as well as causing pain to family members of the murdered and disappeared."[25] This resonates with the concerns raised by Snorton and Haritaworn around how trans rights movements extract value from the stories of trans women of color—a form of exploitation that centers trans women of color in death but excludes them in life.

To navigate these tensions, some activists take steps to anonymize victims—by only publishing their first names in artworks, for example. Others create works of memorialization that address the phenomenon of feminicide in the aggregate rather than the particular, such as the widely exhibited REDress Project by Anishinaabe and Finnish artist Jaime Black, which commemorates missing and murdered Aboriginal women in Canada.[26] Other activists coordinate directly with family-led groups and integrate families into their remembering actions. For example, Movimiento Manuela Ramos provides meeting space for family-led activism and strives to maintain ongoing "relationships of friendship and feminist camaraderie" with families. And still other activists continue to use full names in their remembering work but remove them at the request of families.

As we stood there in the plaza observing *Cortar el hilo* and the long linen line of names, in some minutes of silence, I thought about how the artwork in front of me was not only refusing feminicide, but also enacting a refusal that names and people should ever become data in the first place. The process of *Cortar el hilo*—connecting individual artists and individual women—serves a recuperative purpose: to try to recover names and data from their spreadsheeted abstraction and back into embodiment, albeit embodied through another set of hands and another body that cares for the one that was lost. This constitutes a deeply political action, but not one focused on the state. As Mvskoke urban planning scholar Laura Harjo has stated, remembering demonstrates a turning away from the colonial state and a refusal to reduce killed women to their status as subjects of the state, choosing instead to remember them as loved and missed people.[27]

Indeed, as described in chapter 5, many data activists see themselves as either directly doing or else supporting memory justice work—the insertion, contestation, and recuperation of public memory around feminicide. In her introduction to the Red Feminista Antimilitarista's 2019 report, *Stop the War against Women*, Graciela Atencio states it like this: "When we document a femicide, we make a montage. We put together the pieces of those stories, we weave into it the resonance left to us by the daily tragedy of patriarchal violence. And when we collect all the possible information that circulates in the media about the murdered women, it is because we resist accepting that the victims are nobodies."[28] In this formulation, circulating activist data is directed toward the ethics of nonerasure and the production of public countermemory. Activists collectively refuse to forget individual lives. They attempt to recuperate some amount of their dignity and fullness and to stitch those personal stories into the structural story of pervasive gender-related violence.

REFRAME

Journalists María Eugenia Luduena and Ana Fornaro cover LGBTI+ issues across Latin America and remember 2019 as a particularly terrible year for hate crimes, lesbicides, and transfeminicides. Recounted Fornaro, "It was a year in which, for some reason, there was a lot of violence at the beginning of the year. We were, like, we couldn't cope." Added Luduena, "We were uploading stories all the time." The two journalists cofounded and run a news outlet called Agencia Presentes that seeks to "visibilize human rights violations against LGBTI+ people, land defenders, Indigenous women and migrants using an intersectional lens."[29] Fornaro and Luduena started to keep a count of the violent episodes because they couldn't publish stories fast enough to keep up with their occurrence. Whereas most groups start out by counting and documenting feminicide and later seek out news outlets for media attention and amplification, Agencia Presentes *is* the media. Securing a small grant, they convened a group of five LGBTI+ activists and worked together to define the data fields, categories of violence, and visual design approach. In late 2019, they published the Mapa Periodístico Crímenes de Odio Contra LGBT+ (Journalistic Map of Hate Crimes Against LGBT+ People), which includes an open registry of cases and a map of where they have occurred across Argentina (figure 6.6). It has been widely shared and cited, and the journalists continue to log cases there even when they don't have the bandwidth to undertake full coverage of every hate crime.

The primary goal of the map, as with their overall reporting mission, is "to influence the political agenda and public conversation using communication."[30] They explained that LGBTI+ issues are rarely covered by mainstream media in Latin America, and when they are, it is nearly always with bias, stereotypes, and stigmatization. The map is one of many approaches Agencia Presentes has used to reach a broader audience about gender rights. Another artifact that garnered viral attention was their comic-book-style explainer of the legal case of the murder of the travesti activist Diana Sacayán (figure 6.7). The six-frame comic tells Sacayán's story through vibrant illustrations.

Sacayán was of Indigenous descent—from the Dieguita people, whose lands remain in the Chilean Norte Chico region and northwestern Argentina. She was nationally known for her leadership on LGBTI+ issues, instrumental in advocating for trans access to healthcare and state recognition of gender identity. In 2012, she became the first person in Argentina to have her national identity card updated to reflect her gender identity.[31] Three years later, Sacayán was brutally murdered in her home. The crime was investigated with explicit attention to her gender, a first for the country, and the

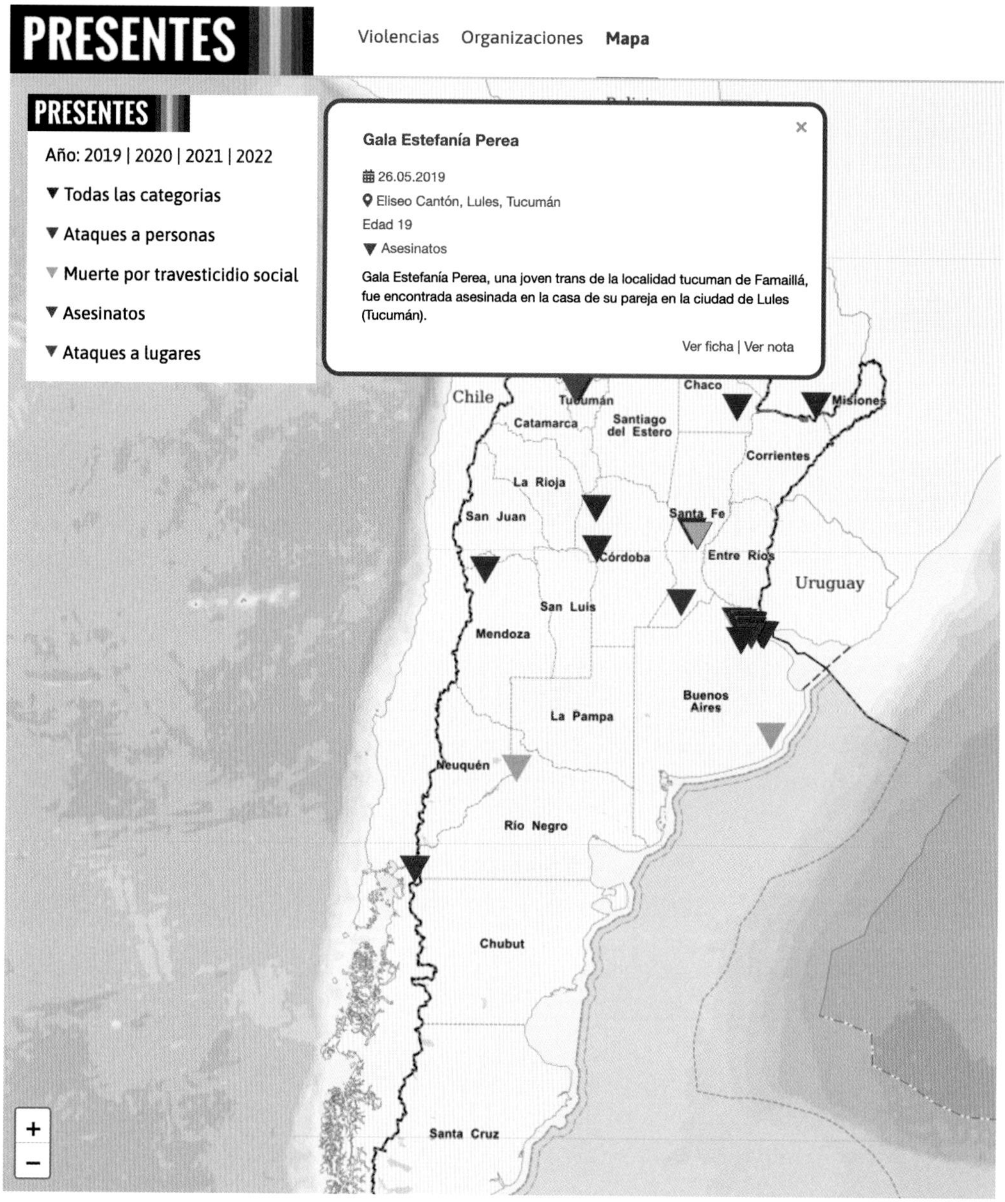

FIGURE 6.6
Mapa Periodístico Crímenes de Odio Contra LGBT+ (Journalistic Map of Hate Crimes Against LGBT+ People), 2019—present. Courtesy of Agencia Presentes.

FIGURE 6.7
An explainer by Agencia Presentes about the importance of the ruling of travesticidio in the case of Diana Sacayán. Courtesy of Agencia Presentes. Translation by Valentina Pedroza Munoz. Graphics production in English by Tiandra Ray.

DIANA SACAYÁN IT WAS A *TRAVESTICIDIO*

DIANA SACAYÁN WAS A *TRAVESTI* ACTIVIST AND ARGENTINE HUMAN RIGHTS DEFENDER, PROMOTED THE GENDER IDENTITY LAW, AND CREATED THE *TRAVESTI-TRANS* WORK QUOTA, AMONG OTHER STRUGGLES.

ON 10/11/2015 SHE WAS KILLED IN HER HOME, IN THE CITY OF BUENOS AIRES. DIANA WAS BEATEN, GAGGED, HANDCUFFED, AND STABBED.

UNLIKE MOST *TRAVESTI* MURDERS, DIANA'S MURDER WAS INVESTIGATED WITH A GENDER APPROACH. WITH THE JOINT WORK BETWEEN A SPECIALIZED PROSECUTOR AND THE JUSTICE COMMISSION FOR DIANA SACAYÁN, THE CASE ADVANCED TO A HEARING WITH A SUSPECT.

THE TRIAL BEGAN ON 03/12/18. THE PROSECUTOR'S OFFICE AND THE DEFENSE ALWAYS USED THE WORD *TRAVESTICIDIO*.

INSIDE AND OUTSIDE THE COURTS, THE FAMILY AND LGBTI ACTIVISM WERE ALWAYS PRESENT.

ON 06/18/18, A HISTORIC RULING WAS ACHIEVED. FOR THE FIRST TIME IN LATIN AMERICA THE ASSASINATION OF A *TRAVESTI* WAS RULED A HATE CRIME IN RELATION TO GENDER IDENTITY.

#NO MORE *TRAVESTICIDIOS*

prosecutor repeatedly used the term *travesticidio* in the hearings to frame it as a murder that was motivated by her gender identity. Travesti, trans, and LGBTI+ activists played a large role in drawing attention to the case, and Agencia Presentes devoted a great deal of coverage. "We chronicled each hearing," explained Fornaro. "So, there is also something like having put ourselves there, to show that we were there, physically present with our bodies, which built another type of trust." This is, in fact, the point of having *presentes* in the group's name—they are present, physically. They show up to document and narrate, rigorously and carefully, those events that have to do with ensuring gender rights, and to contextualize and disseminate them to a broad public. In a monumental decision in 2018, the Argentine court ruled that Sacayán's murder did indeed constitute travesticidio.

Fornaro and Luduena decided to create a visual artifact to explain and frame the historic import of the court's decision for nonexperts. "Because it is something that is difficult to communicate. Why is it important that the judicial system take up the word *travesticidio*? It is somewhat technical," recounts Fornaro about what they were up against. They worked with activist illustrators to create the comic shown in figure 6.5, which ultimately went viral. For the two journalists, reaching people who may never have heard of the case or the trial or the word *travesticidio* was one of their proudest moments, since such issues rarely get attention in the mainstream press in Latin America.

Agencia Presentes's work is emblematic of a third function of activist data circulation: to *reframe* feminicide and fatal gender-related violence. This work seeks to produce narrative change in the realm of media and culture. It often draws from data but uses storytelling to challenge stigma and stereotypes and weave a broader and more contextualized story around structural violence. The goal is typically to raise consciousness in the broader public. Thus, while Diana Sacayán's case was in Agencia Presentes's registry and map, it was their in-depth reporting, live videos, and viral comic that enabled the data points to go beyond the map and to assert a new, structural frame into the public conversation: it was travesticidio.

Reframing happens by creating and distributing alternative narratives. Data activists refuse the narrow interpretations of violence coming from the police, the state, and mainstream media; they uplift stories of grassroots and community struggle for justice; and they insert feminist framings of the violence into public circulation, often using their own data as key evidence.

Reframing doesn't only happen via the data artifacts—infographics, reports, maps, visualizations, stories, and even comics—that groups produce. It also comes from activists strategically inserting their data and statistics into mainstream media narratives and from their efforts to educate journalists and other communications professionals.

A key way that activists leverage data to reframe public narratives about feminicide is to use their own databases to position themselves as authorities on the matter for the media. Observatories, like Mumalá's, put out press releases at strategic moments and make themselves available for interviews and public events. Activists leverage the fact that state data on feminicide are missing, and, possessing their own counterdata, become a key credible source for news outlets. "They call us from the press, from television, from everywhere, because the state does not have that information systematized," says Estefanía Rivera Guzmán from Red Antimilitarista Feminista in Colombia.

For activists, this is a crucial way of interrupting the toxic and harmful narratives around individual cases of feminicide, where the press, seeing the state as the only legitimate source of information about a criminal case, tends to solely interview law enforcement and judicial officials and thus reproduce those narrow views. Some groups have even studied this media pattern. The US-based organization Jane Doe did a qualitative study of over a hundred news articles. "We found that few provided any information that would be valuable to survivors, like where to look for help, or what the warning signs were, anything like that," related Toni Troop, their director of communications. Thus, through the legitimacy that producing data and quantitative information confers, activists are able to intervene in this dynamic and insert both their data and their framing of feminicide into the public conversation. Silvana Mariano, from the Brazilian group Néias—Observatório de Feminicídios Londrina, stated, "Our contribution has been in the sense of producing another narrative about these crimes that is not the police narrative, of offering to the media an interpretation of the phenomenon that is not a police interpretation. I think this is a very important collective gain for us."

A final way that activists use data in the service of reframing feminicide has to do with offering trainings and workshops to other journalists and communications professionals. In particular, groups out of journalism, like Agencia Presentes and Cuántas Más, leverage the visibility and "clickability" that maps and data artifacts provide to their organizations to push for larger structural change in the media ecosystem through running workshops and trainings. Advocacy groups with strong links to networks of feminist journalists, such as the Casa del Encuentro, also do this. These educational sessions are not about data analysis, but rather about feminist and queer analysis. They provide conceptual tools to analyze the problem of feminicide and fatal gender-related violence, as well as to help disseminate creative and ethical approaches to communicating about it. As Irma Lugo Nazario from the Observatorio de Equidad de Género Puerto Rico describes it, such capacity building sessions are "a space to guide and educate other people who are approaching these themes."

Reframing actions carry ethical risks and are not without conflict. For example, in attempts to reframe the phenomenon of feminicide as structural and widespread, a great many groups produce maps.[32] But Cuántas Más decided to take their feminicide map offline when they realized that rather than bringing justice, the spatial patterns it showed might further stigmatize low-income and Indigenous regions in Bolivia. This illustrates activists' awareness of data's power to portray deficit narratives—that is, narratives that communicate what Palawa scholar Maggie Walter calls the five Ds: the "difference, disparity, disadvantage, dysfunction and deprivation" of minoritized groups.[33] In the case of Cuántas Más, they shifted to aggregated statistics and stories as a refusal to participate in such deficit narratives, but it remains a tension for those groups that use maps for communication.

The takeaway here is not "don't make a feminicide map," but rather that feminicide data producers must navigate some of the representational tensions that are inherent to data communication about any issue in which there are social and political disparities due to structural oppression. Such ethical questions around representation could apply equally to other domains, such as educational outcomes, health disparities, or food security: How do you visually demonstrate group-based differences while also providing a nonstigmatizing and structural interpretation of those differences? Likewise, because activist data have become a source of credibility and authority, with their data and their voices being widely circulated in the media, this can produce competition and conflict among groups, as evidenced in the story of Mumalá's data loss from chapter 4. Thus, though I have emphasized how activists work collectively and in concert with movements, this work is not always happy, multicultural sunshine. In fact, much of the pluralistic coordination is forged through conflict, deliberation, and difference.

REFORM

"We were born in times of war," stated Giovanna Lemus matter-of-factly. The Guatemalan Civil War lasted from 1960 through 1996. It was instigated and prolonged in no small part due to US government and corporate involvement in the country. The decades-long war used strategies of forced disappearances, razing the villages of peasants, and weaponizing gender and sexual violence, in particular against the Maya peoples. In this context of extreme violence, the Grupo Guatemalteco de Mujeres (Guatemalan Group of Women) was founded by Lemus and colleagues in 1988 to support women survivors of violence and to work—primarily through laws and policies—to ensure a life free of violence for women and girls in Guatemala. The group engages in a wide range of activities, including coordinating with feminist organizations around the country, running shelters, engaging with international scholars and organizations, and producing comprehensive data about femicide in the country.

The first legal win the group secured was in 1996—a law about domestic violence, but painfully limited in its scope. When Grupo Guatemalteco de Mujeres and other feminist groups pushed for more comprehensive protections and for a broader definition of violence against women, political institutions told them that they couldn't do more because of the lack of data and information to support their demands. Thus, that same year, Grupo Guatemalteco de Mujeres began producing comprehensive data and analysis about femicide in Guatemala. When they started, there were no nationally systematized statistics or death records, so they had to count themselves. They had piecemeal access to police reports and regularly went to the morgues to document the women who had died and their causes of death. They released public statements with their findings to try to push the media to report more widely about the problem. In 2006, after intense lobbying from Grupo Guatemalteco de Mujeres and others, the government established the Instituto Nacional de Ciencias Forenses de Guatemala (INACIF), Guatemala's national institute of forensic medicine, and it started publishing death records. The creation of this national institution made the group's work vastly easier—they were able to triangulate official data with press reports to scrutinize individual cases and produce their femicide data, a system that they continue to this day (see chapter 4 for more on triangulation). Grupo Guatemalteco de Mujeres communicates the results of their research with reports, analysis, and conventional charts—forms of data communication that are readily understood by the government. For example, figure 6.8 is a bar chart from their 2018 report and shows six months of monitoring work in which they compare counts of violent deaths of women against those deaths they consider to be femicide.[34]

The work of Grupo Guatemalteco de Mujeres has played a significant role in institutional reform around femicide. Some of their achievements include helping to establish, in 2000, a national commission for the prevention of violence against women (La Coordinadora Nacional para la Prevención de la Violencia Intrafamiliar y Contra la Mujer, CONAPREVI), which the group has a seat on; the establishment of the INACIF in 2006; and the passage of a 2008 law defining femicide in the Guatemalan penal code. This was one of the first laws in the region codifying femicide. These were not solo achievements. Grupo Guatemalteco de Mujeres emphasizes that their work is always in coordination with other women's groups, with the larger feminist movement in Guatemala, and with international networks. "It was hard at first getting institutions to give us information," recounts Lemus, "but as we systematized that information we saw the possibility that all of us, including members of the Latin American network

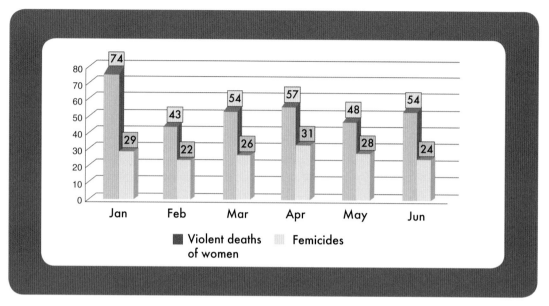

FIGURE 6.8

Chart of femicides as compared with violent deaths of women from a report by Grupo Guatemalteco de Mujeres in 2018. Courtesy of Grupo Guatemalteco de Mujeres. English graphic produced by Wonyoung So.

we worked with, could show the seriousness of violence against women by using these strategies."

Like all groups we interviewed, for Grupo Guatemalteco de Mujeres it is never solely about the data. Emphasized Lemus, "It's not about having data for simply having it. It's about putting it into action in such a way that it can generate a response from the state." Putting data into action, for their group, meant undertaking "constant analysis"; it meant circulating their data in reports naming and framing the problem of femicide; it meant articulating specific political demands that first *created* institutions and then pushed those institutions toward recognition—using feminist analysis—of the problem of femicide.

Reform encompasses the ways in which activists use counterdata to push for changes in laws, policies, institutions, and/or official data collection practices around feminicide. Since the target of reform efforts is typically the state, activists design their data communications in aesthetic forms that are understood as "evidence-based" by the state, including reports with charts (like figure 6.8) or policy recommendations backed by statistics. Reform actions include activists leveraging their counterdata (a) to get a seat at the table with the state, (b) as evidence to demand new laws or reform existing laws, or (c) to shift and strengthen state data collection. Activist data can also take on a life of their own and circulate independently of activists themselves.

In the case of reform, activists often seek the state's recognition of feminicide as a legitimate public problem, and their counterdata can be a way to gain a seat at the table. The Alianza in Ecuador has monthly meetings with state officials to compare cases, and the state speaks publicly and proudly about their collaboration.[35] Numerous data production groups that we interviewed had been invited into committees in the municipal, state, or federal government, especially when progressive parties are in power who seek to build alliances with civil society groups around human rights issues. Others, like María Salguero, have given testimony to national congresses and commissions.

These exchanges, as well as news articles that cite activist data, have led to state actors relying directly on counterdata as evidence for legislative reform. This was the case in Puerto Rico, for example, where the report and data described in chapter 2 were directly quoted in the island's 2021 law that updated the legal framework for feminicide and transfeminicide. The Argentine observatory Ahora que sí nos ven has seen their statistics cited in proposed legislation. Activist data have also been used to debate laws and policies that reach beyond codifying feminicide. For example, in 2018, Law 27452 passed in Argentina concerning reparations from the state for children orphaned by femicide, and activist data around the number of children of femicide victims played a role in the public debate around implementing the law.[36] Coverage and maps produced by Agencia Presentes played a role in Argentina's 2020 passage of a law outlining a travesti and trans quota for public sector employees. And evidence and advocacy from the Grupo Guatemalteco de Mujeres played a significant role in the government's establishment of the Isabel-Claudina Alert, which mandates better government response times in cases of missing women.

Counterdata also travel into officialdom, sometimes without the knowledge of their producers. In Ecuador, in 2018, the Alianza was surprised to see their feminicide map used as evidence by an assemblywoman in debates around updates to the national law to prevent violence against women. Data produced by the Organization of Salvadoran Women for Peace (ORMUSA) are regularly quoted in US State Department's human rights reports for El Salvador.[37] These statistics, in turn, are then used as evidence to make determinations in individual cases of women seeking asylum in the United States. "The immigration courts lean heavily on these State Department reports," described Mneesha Gellman, a political scientist who studies El Salvador, and provides expert review of asylum cases.[38] This is not a use previously conceived of by ORMUSA, nor do the activists have any role in writing or framing reports from the US State Department, but it illustrates how counterdata are viewed by institutional actors as the best available data in an imperfect information ecosystem, and thus it may be picked up and employed in high-stakes contexts. This may, in fact, be the most enduring way that activist data are used for reform—not necessarily through directly protesting and lobbying in order to get specific bills passed, but through gradual, informatic infiltration into the official understanding of a problem.

Activists also use their data to influence official measurement and classification practices by the state. This can be an informatic strategy for achieving official state recognition of the problem. "There was no data. There was no acceptance that there was this problem of violence," says Fabiola Ortiz of Grupo Guatemalteco de Mujeres about the early days of their data production. Many activists related to us that their work is primarily about this *visibilization*, particularly to the state. Once counterdata exists, "it's a tool to argue with the authorities about official figures," related Geraldina Guerra Garcés of the Alianza. Their group is in regular monthly dialogue with state agencies and uses their data to push the state to expand their criteria for what counts as a femicide. The Alianza counts induced suicides, transfemicides, and girls' murders as femicides, for example, while the state does not. "We want the criminal code to be reformed so that these related femicides are included," states Guerra Garcés. The Fórum Cearense de Mulheres in Brazil has a similar strategy: the group monitors whether the Secretary of Security of the state of Ceará has designated a case as homicide or feminicide, and uses its counterdata to challenge the state's classification. These are examples of what Helena Suárez Val describes as *discordant data*, in which activists use counterdata to push for different definitions, measurements, and classifications of a particular issue. The discrepancies then provide the basis for discussion, exchange, and comparison.

Numerous ethical issues and contradictions arise in using activist data for reform. For example, at the same time that Grupo Guatemalteco de Mujeres has worked for decades producing counterdata about femicide, they also refuse the idea that this is their responsibility. Lemus declared, "This was our fight: to not do this work and to make the state do its job." There is an inherent contradiction, then, in producing counterdata for reform. On the one hand, the state will take no action without proof of the problem, but it will not invest in such proof when the problem is invisibilized and normalized. Public health scholars have called this the "no data, no problem" problem.[39] At least half of the activists we interviewed wrestle with the contradiction that they are engaging in work that they do not want to do, that they refuse to accept as their responsibility, that they are not resourced to do, and that can, in fact, result in severe undercounting of feminicide because their sources of information are limited and they don't have the same reach as government institutions.

This leads to another ethical tension that relates to the visual communication of counterdata for reform. To be legible to the state, activists seeking reform employ some of the most familiar conventions of data communication: neutral titles, dry reports, clean graphic layouts, and geometric shapes and lines (like figure 6.8). As digital studies scholar Helen Kennedy and coauthors have asserted, such visualization conventions work to "imbue visualizations with a sense of objectivity, transparency and facticity."[40] Such apparent objectivity helps counterdata achieve legitimacy in the eyes of the state, and certainly helps them travel and spread, going on to be cited in legislation, media reports, and international policy documents. And yet activists are deeply aware of the limitations of their data. They know that they are underreported and subject to media bias, and they also want to carefully communicate those limitations. As Lara Andres from Ahora que sí nos ven stated, "If we present a report where we say that we have three transfeminicides, we are missing reality. We know that it is not so. So, it is about being able to look for ways to be able to present the data, and always mentioning that our analysis is of media reports . . . Understanding that what we present is always an underrepresentation." This, again, represents a kind of ethical contradiction in which activists both use and refuse data—they engage the visual politics of data neutrality, leveraging those conventions to achieve recognition with the state, their primary audience, and yet simultaneously reject such neutrality and try to emphasize the underreporting, partiality, and incompleteness of the data.

Amid these various efforts at reform, it is striking how fuzzy the lines start to get between counterdata and official data. Indeed, perhaps the greatest achievement of counterdata is when it is used by state institutions themselves as the evidentiary basis for taking official action and strengthening official data collection. In some cases, such as that of Grupo Guatemalteco de Mujeres, the state even comes to rely on activists to do the research and produce the statistics for which the state does not have the capacity (or the political will). There are many cases of dialogue between activists and state agencies, even leading in some cases to friendship and shared vision. This can happen because the line between activists and public sector employees is also not so hard and fast. Any seeming binary of activist versus state is constantly disrupted by the flows of people into and out of public-sector employment. For example, the woman who founded Ahora que sí nos ven spent two years working on a national violence observatory for the Ministry of Women in Argentina, and then departed to once again direct the activist observatory. In another instance—the Office of the Prosecutor for the Investigation of the Crime of Feminicide in Mexico City—a municipal agency has initiated their own database of feminicide that combines official data with news media reports.[41] Here we see an example of inside-government counterdata production. Or is it official data production using counterdata methods?

Whatever we call such state-adopted data, it is clear that the state is not monolithic. Indeed, as Pine and Liboiron point out in their paper on the politics of measurement, data activism can often come from insiders and experts who seek to use quantification to reframe matters of health and harm.[42] Necroresistance itself may not always come from outside of institutions but from concerned professionals within institutions who draw counterdata into official data settings, who develop mechanisms for collaboration with activists, journalists, and nonprofits, and who leverage the authority and legitimacy of the state itself to transform counterdata into institutional action. Helena and I saw this in the active and committed participation of many public sector employees who showed up to take our online course, Data Against Feminicide: Theory and Practice, in Spring 2022. And activists themselves are not always operating at cross-purposes from the state. At the end of our interview with her, Ortiz stated that what kept Grupo Guatemalteco de Mujeres going was their firm conviction that "we are building something better for this country."

REVOLT

A group of ten women showed up early on International Women's Day 2020 in Mexico City's emblematic public square: the Zócalo. They brought buckets of paint, long rolling brushes, and four-foot-tall stencils. From the early hours of the morning through the surge of crowds in the afternoon, they stenciled the names of women killed in feminicide onto the large gray concrete expanse (figure 6.9). Jacqueline. Mercedes. Sirenia. Olimpia. Veronica. Catalina L. Araceli R. Christy C. Gertrudis S. Laura M. Martha M. Josefina. Agustina. Amada. The group of ten was joined by over two hundred more women painters that morning. The names were arranged in quadrants just like the concrete squares of the Zócalo, spiraling out from the center (figure 6.10). At the edges of the arrangement of names, the painters stenciled the movement's slogans in bright pinks and purples:

NI UNA MENOS

NI UNA MÁS

NOT ONE LESS (WOMAN)

NOT ONE MORE (WOMAN)

FIGURE 6.9
Members of Colectiva SJF stenciling their large-scale installation of names of victims of feminicide in the Zócalo, Mexico City, for International Women's Day (8M) 2020. Courtesy of Colectiva SJF. Photo by Santiago Arau.

The Colectiva SJF is a group of artists, filmmakers, and graphic designers based in Mexico who do large-scale creative interventions into public space around human rights issues. Explains Marcela Zendejas Lasso de la Vega, "The main objective is how to translate the message and the fight and, well, *la lucha*. This very important work that human rights activists do. How can we translate it into a language that more people know about, or that has the power to influence in a stronger way?" For example, in 2021, the group coordinated an action that spelled out "Dónde están?" (Where are they?) in front of Mexico's national palace. The letters were formed by clothing belonging to some of the thousands of people—of all genders—who are missing and whose families continue to search for them.

Colectiva SJF decided to take action on feminicide one afternoon at a social gathering at one of their member's houses. They were discussing the inadequacy of data, said Mónica Meltis Vejar, and "we started talking about how it's crazy that the numbers don't validate the importance of the person that we are missing in a feminicide case.

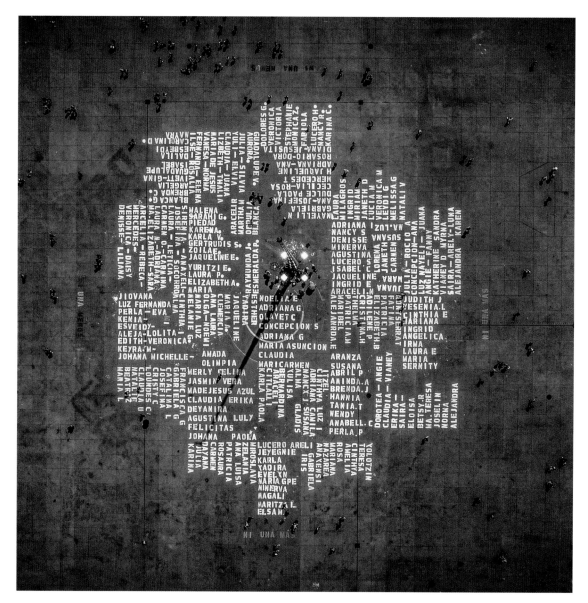

FIGURE 6.10
Aerial view of Colectiva SJF's installation in the Zócalo, with over two hundred names. Courtesy of Colectiva SJF. Photo by Santiago Arau.

So, we started to say, 'Yeah, we have to do something. We'd need to find a way for people to feel closer to the problem.'" But to do a large-scale intervention, they needed the names of feminicide victims. Meltis Vejar also happens to be the director of a Mexican NGO called Data Cívica, which provides technical assistance to groups who monitor feminicide and gender-related violence against LGBTTTIQ+ people.[43] Data Cívica wrote a web-scraping application to review the last four years of news articles about feminicide and extracted the names of thousands of victims.

Ultimately, on March 8, 2020, the activists were able to paint more than two hundred of the names provided by Data Cívica before masses of protesters swarmed the Zócalo with signs and slogans and chants that reverberated the words from the ground: "Ni Una Menos," calling the named women back into presence for the collective political body. It was a massive aesthetic takeover of public space.

The federal government tried to take it back the next day. In line with the Mexican president's preoccupation with protecting public property over listening to citizens, government employees showed up promptly the next morning for cleaning.[44] Although they tried to efficiently erase the names and suppress the record of the action, it was fitting that they were unable to fully do so, leaving large smudges and smears of white across the plaza (figure 6.11). The traces remained, not only materially through the white paint lingering in the Zócalo, but also through the spectacular documentation, produced by the activists in collaboration with aerial photographer Santiago Arau. The images are intentionally and strategically designed for digital circulation, part of a wave of antimonument monumentalism in Mexico that seeks to document, preserve, and circulate feminist actions in public space.[45] And indeed the action received a great deal of national and international media coverage, with articles in mainstream Mexican media and international art magazines, and thousands of retweets on Twitter.

Revolt is focused on the ways that feminicide activists mobilize data for public protests and spectacular collective actions aligned with specific political demands. Revolt actions usually take place in public spaces and in concert with larger mobilizations like marches and protests. In using data for revolt, activists build on a long history of Latin American protest art and artivism to make particularly symbolic uses of both space and time in order to register their dissent and make political demands. For example, Colectiva SJF chose to take action in the Zócalo, an emblematic public space at the center of Mexico City that served as a sacred space in pre-Hispanic times and, later, as a backdrop of official ceremonies and of political protest during the colonial period and then thru independence. Activists also align their data actions to symbolic dates and times—such as March 8, November 25, or the anniversary of Ni Una Menos—so as to maximize public attention and the circulation of documentation on social media.

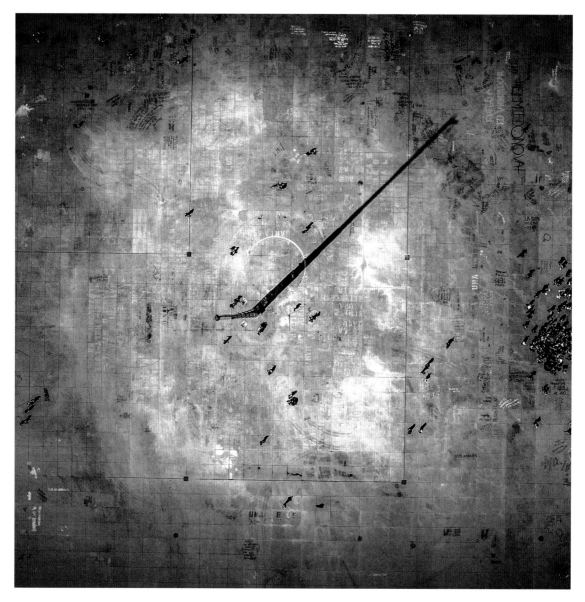

FIGURE 6.11
Image of the Zócalo plaza on the day after Colectiva SJF's intervention when government employees tried to scrub it clean. Courtesy of Colectiva SJF. Photo by Santiago Arau.

FIGURE 6.12
On the International Day for the Elimination of Violence against Women, in 2021, the Mujeres de Negro Rosario group placed 250 empty chairs outside the provincial court building in Rosario to symbolize the 250 women lost to feminicide that year. Courtesy of Mujeres de Negro Rosario.

In some cases, the same activists who produce feminicide data are also the ones who are staging revolts in public spaces. As Silvana Mariano, from the Brazilian collective Néias—Observatório de Feminicídios Londrina, stated, "We have seen that when we are physically in these spaces—and this has to be with a protest, with a demonstration—this enhances the visibility of our work, our message." Another example includes the regular public installations by Mujeres de Negro Rosario, based in the Argentine city of Rosario. For example, on November 25, 2021, they placed 250 chairs outside the provincial court building in Rosario (figure 6.12). Each chair was labeled with the first name of a victim of femicide from that year and some information about her case. Cofounder Marta Pérez wanted to represent the lost lives not only through names but through bodily space (or rather, the absence of it): "Seeing people's faces . . . I mean, one says 250, and it's a number, but when you see that represented with the space that

that woman occupied, this has another impact on people." The court building was chosen as the site of their public action because of the group's work demanding justice and reparations for children and families left behind.

In other cases, data activists operate as data intermediaries and provide information to other activist groups so that they may stage physical or digital interventions. Mariana Mora, in Costa Rica, described how she gets data requests each year from activists in the weeks leading up to November 25. The Ahora que sí nos ven observatory has provided data to family-led groups so that they could produce large-scale banners for marches they were organizing. And revolt actions do not only happen in physical spaces. Utopix, in Venezuela, has provided data to activist groups coordinating large digital *tuitazos* or tweet-a-thons in which activists seek to temporarily take over certain hashtags on social media with messages against feminicide and gender-related violence.

There are also many cases in which counterdata travel and circulate of their own accord. For example, Paola Maldonado Tobar, from the Alianza in Ecuador, described how she encountered a group at a march on November 25 who had printed an enormous version of the Alianza's map and carried it as a protest banner. Memorably, she called it a "giantography" and noted her realization that "for me, this demonstrated that this map had a life of its own." Nerea Novo, from Feminicidio.net, recounted how she had, more than once, experienced "going to a demonstration and seeing the names of the victims that you have documented, seeing them with the classification that you have given." She was filled with pride to see her group's data on display and catalyzing debate at mass mobilizations.

The ethical questions for activists using data for *revolt* have to do with whether and how much to disclose regarding the identity and stories of victims while trying to protest structural violence. For activists, using names can be a key strategy to humanize victims and remind the public that they are more than numbers. These were two frequently stated goals, which represent—again—a refusal to let people's lives be subsumed into numbers. Yet using names and stories can revictimize families and communities, especially when used without consent. Colectiva SJF, like most groups using data for revolt, chose to use only the first names of victims, sometimes followed by the initial of their last name (figure 6.10). The installation of empty chairs undertaken by the Mujeres de Negro Rosario (figure 6.12) has a similar ethic, where the group only uses first names and does not include a photo of the person unless the family has sought them out and given their explicit permission. In the case of the three women in the foreground of the photo, such consent was granted, but the vast majority of chairs have no photo or surname because the family was not in contact with the group.

A second ethical tension lies in the aesthetic strategies used to depict individual victims within aggregate groups. Activists refuse that individuals would be seen as numbers, and yet they also refuse to keep individual cases isolated and separated. When
aggregated and connected, using chairs or names or other visual strategies, these collections of unjust deaths illustrate the larger structural phenomenon of feminicide. Here,
activists insist on representing data in a multiscalar way, similar to our discussion of
scale in chapter 5, which addressed how they use data as countermemory and data as
megaphone. Activists simultaneously focus on parts and wholes of the phenomenon of
feminicide; asserting that the scale of the individual cannot be forgotten and the scale
of the structural cannot be ignored. Both scales are essential and both scales are not
enough on their own.

Data actions for revolt refuse the state-sanctioned silence around gender-related violence by claiming space and time to insert new necropolitical narratives. These actions
claim physical public spaces associated with the state in large-scale demonstrations of
dissent using data. They also claim discursive and digital spaces by circulating spectacular documentation, which persists even after the protest is over and despite the state's
attempted erasure, symbolically and literally, of the phenomenon of feminicide. They
claim time by capturing strategic moments of public attention. Using data for revolt
often functions as what scholar-activist Joy Buolamwini has termed an *evocative audit*:
"an evaluation of a system through a combination of human experience and documented evidence to viscerally relay harmful behavior like discrimination that has real-
world impacts."[46] For example, in the Zócalo action we see individual names (human
experience, human scale) aggregated into a massive artwork (documented evidence,
structural scale). This combination of scales functions as a powerful and evocative audit
of both structural harms and their interpersonal effects.

Nevertheless, in using data for revolt, activists continue to refuse the data themselves. They refuse to allow names and people and lives to be subsumed into aggrega-
tions and counts—even as they produce counts for families to carry as banners or on
maps handed out as flyers at protests. Andres from Ahora que sí nos ven stated that,
as they plan public interventions, they are always discussing how to put a name and
a face to the numbers that they have collected. Guerra Garcés, from the Alianza in
Ecuador, mentioned many times the concept of "nourishing the data point" (*nutrir el
dato*), connecting it back into its full context and into the full lifeworld from which it
emerged. Data actions for revolt thus embody this apparent contradiction of using data
while (doubly) refusing it: using data for massive actions to refuse feminicide while
simultaneously, persistently, refusing the reduction that accompanies aggregation and

abstraction. These tensions are productive—they invite us to think and feel and act, simultaneously, across scales of the matrix of domination.

LESSONS FOR A RESTORATIVE/TRANSFORMATIVE DATA SCIENCE

Feminicide data activists circulate their data in many forms and to many audiences. If there is one major lesson for a restorative/transformative data science, it is this embrace of a multiplicity of uses of data, audiences to reach, and modalities of data communication. This constitutes an activist epistemology of data that arises because activist data circulation is not about the data themselves but about what kinds of political impacts and effects that data may support. This is to say that data play a supporting role, but they are not center stage in any of these public data actions and circulations. This lifts the constraint, inherited from more positivist notions of data science, that data should be neutral, rational, raw, and primary material for decision-making by status quo institutions. Pluralism in communication formats opens up possibilities for elevating emotion and embodiment, for speaking to more diverse audiences, and for inviting evocative and human experiences into the equation.

Groups and individuals that produce counterdata about feminicide told us repeatedly in interviews: "No somos números"—"We [women] are not numbers." This resonates with a widely circulated statement of feminist refusal called the Feminist Data Manifest-No: "We refuse to understand data as disembodied and thereby dehumanized and departicularized."[47] Yet data activists told us this while also speaking about their rigorous methods of quantifying, enumerating, and aggregating feminicide. Data activists feel this tension in attempting to shift power using exactly those strategies that are associated with the power of the patriarchal, hegemonic state to allocate the differential disposability of women. They are using necropolitical tools to enact necroresistance. This relates to the famous argument advanced by Audre Lorde, who argued that "the master's tools will never dismantle the master's house."[48]

Extending this argument to data, we might assert that data and quantification cannot eliminate feminicide and the gender-related oppression that undergirds it. As Lorde states in the same speech, "What does it mean when the tools of a racist patriarchy are used to examine the fruits of that same patriarchy? It means that only the most narrow parameters of change are possible and allowable."[49] In their article "Numbers Will Not Save Us," information scholars Roderic Crooks and Morgan Currie build on Lorde's words to document the ways in which data-driven evidence of oppression may not serve liberatory ends. This is particularly true, they argue, for grassroots community organizers who face high emotional burdens in documenting harms and have fewer resources to enable them to extract value from data.[50] Where the coauthors see some liberatory potential is in the concept of agonistic data practices, the idea of using community data in affective and narrative ways (such as the performances, vigils, installations, and social media posts discussed in this chapter).

Yet no activist that we interviewed thought that more data *would* save us, or that it *could* dismantle the master's house; that is to say, for activists, data on their own would never be enough to eliminate feminicide. "We do not count the dead," emphasized Lorena Astudillo from the Red Chilena contra la Violencia Hacia las Mujeres. "Rather, the registry is something political." The political purpose of the registry, for the Red Chilena, is to make violence against women more visible and to deepen public understanding of such violence. As Maldonado Tobar from the Alianza explained it, data activists see their work as part of a larger chorus of other actors and actions, embracing pluralism to achieve transformative social change. This was, in fact, actually the point of Lorde's "master's tools" speech: it was first given in 1979 to an academic conference organized by white feminists about the lives of American women. Lorde noted how the conference did not consider race, nor sexuality, nor class, nor age. *Using the master's tools* in this context meant repeating the racism, heteronormativity, patriarchy, and ageism inherent to academic structures. It meant failing to recognize difference and multiplicity. Lorde was calling the attendees into collective action, into pluralism, forged from exactly such difference. In the same speech, she stated: "Without community there is no liberation, only the most vulnerable and temporary armistice between an individual and her oppression. But community must not mean a shedding of our differences, nor the pathetic pretense that these differences do not exist."[51]

Thus, the point of the master's tools argument shouldn't be to posit a binary—whether data science will or won't liberate us. This isn't a useful question in the abstract. It's a rather tall order and a silly idea for a tool, a tactic, or an aesthetic form to either foreclose or achieve liberation on its own, disconnected from the context it's inserted into. It's similar to saying "placards at protests will overthrow the patriarchy" or "documentary films will dismantle the prison industrial complex" or "maps will restore sovereignty to Indigenous people." Of course, they will not. But they will also not always support the patriarchy, nor always uphold the prison industrial complex, nor always support extraction and profiteering on Indigenous lands. The more important question is rather around whether and how reformative/transformative data science makes pluralism and collective action possible and under which constraints. Thus, it may be more useful to look at where Bronwyn Carlson lands in her discussion of using social media to document Indigenous femicide: "Social media, however, despite its conspicuous corporatist—and colonialist—objectives, provides us with platforms we are able to seize and utilize for our own ends. These sites are where we will continue to document our deathscape archive. *This is not a solution, but it is part of a solution* for so many of us who have been silenced and are now refusing that (im)position."[52] The italics in that quote are mine, added to emphasize the point that no data activist would see social media platforms or data or infographics or a data-driven artwork as a "solution." The technosolutionist realm of magical thinking is exclusively the territory of Silicon Valley; activists themselves refuse to accept such nonsense.

Activists use data because they have been able to develop creative and rigorous methods to produce it and because they strategically exploit the legitimacy that quantification and data confer on their creators. But they also refuse the idea that data would dehumanize women killed in feminicide (by reducing human lives into counts) and they acknowledge the tension that data always carry that risk and sometimes do have that effect. Once a body has been abstracted into rows and columns, it is hard to revivify. Activists are antiabstraction and antireduction while abstracting and reducing. They do not fully resolve this tension because it isn't resolvable. Instead, they both use and actively refuse data.

I advance that this tension constitutes a key feature of restorative/transformative data science. Similar questions around using and refusing data have been described by scholars writing about data activism in other domains. For example, Lourdes Vera and colleagues described how, in the process of trying to archive environmental data that the US federal government was taking offline, their group's aspirations toward environmental data justice "are embroiled in the very extractive logics that we aim to critique and replace."[53] In a different locale, technology scholar Dmitry Muravyov details a case study about the DTP Map in Russia. This activist map combines official data on traffic accidents with other data as a way of mobilizing public will to reduce accidents. While activists engage deeply with the data, they simultaneously express doubts about its veracity and comprehensiveness because traffic accident mortality is one of the metrics by which Russian governors are assessed, which leads to systemic incentives to manipulate numbers. Muravyov calls this the *epistemological ambiguity of data*—meaning that activists occupy a middle ground between complete trust in data (and, consequently, alignment with the positivist approach of hegemonic data science) versus the total dismissal of data as a source of knowledge.[54]

For feminicide data activists, the epistemological ambiguity derives from the fact that they are intimately aware of the uncomprehensiveness of their data (because they themselves produce the data) as well as the risks of reducing human lives to lists and statistics. They consistently navigate these ambiguities by inserting and reasserting emotion and embodiment back into their data communications. "No number is adequate, because it's not about numbers, it's about lives," stated Geraldina Guerra Garcés from the Alianza in Ecuador. Many of the data communications produced by activists echo this sentiment. For example, a social media post announcing Mumalá's report leads with the all-caps statement:

THEY ARE NOT NUMBERS, THEY ARE LIVES THAT WERE TAKEN FROM US.[55]

That said, at the same time that data activists refuse the fiction that data are neutral representations of reality, they still strategically leverage such popular, positivist conceptions of data science in order to garner authority for their organizations and legitimacy for their struggle. Particularly in the case of reframing and reforming actions, activists leverage the high epistemological status of data—as a form of knowledge perceived by the media and the state to be authoritative—in order to amplify their message through the press or get a seat at the table in dialogue with the state. This is not without contradictions. Both Roderic Crooks and Morgan Currie, as well as Helena Suárez Val, note that activists' embrace of data and quantification can simultaneously lead to devaluing those forms of knowledge that they most prize and want to elevate: lived experience, oral history, and testimony.[56] Yet from surveying the diverse ways that feminicide activists circulate their data in this chapter, it is apparent that they are tremendously resourceful in reincorporating those voices and testimonies in their "data-driven" communications. Activists' data practices complicate the idea that using data means that one is endorsing quantification over lived experience and community voice. This is a final lesson for a restorative/transformative data science: viewing data not as an authoritative truth claim but rather as a passageway to building power and amplifying the voices of grassroots groups and impacted families.

CONCLUSION

Refusing and using data is the last stage of a reformative/transformative data science project in which activists communicate their data in a variety of forms to a variety of audiences. Feminicide data activists circulate their data in order to *repair* and heal communities, to *remember* people lost to gender-related violence, to *reframe* media and cultural narratives around feminicide, to *reform* state policies and practices, and to *revolt* in large public demonstrations. I have argued in this chapter that activists use their data to refuse: to refuse feminicide; to refuse gender subordination; and to refuse state neglect, media bias, and public ignorance. Activists also refuse data themselves—repeatedly refusing to reduce people's lives to numbers and living with the epistemological ambiguity of data about feminicide. They refuse data while using data to enact their refusals.

They refuse quantification while providing numbers to families. They refuse to forget a person's life while representing it as an empty chair, an absence. It is not that activists are inconsistent or that they are ignorant of these apparent contradictions. It is rather that they have developed complex ways to reclaim the "master's tools" and bend them toward necroresistance. And part of this bending involves staying with the tensions of refusing while using data and using while refusing data. A restorative/transformative data science, then, advances a fundamentally nondeterministic view of data and technology. This is the idea that it is possible to reclaim hegemonic methods of knowledge production (such as databases) and hegemonic forms of information presentation (such as data visualization) for ends that are aligned with the twin goals of restoration—of rights, dignity, life, living, and vitality—and transformation—elimination of feminicide and gender-related violence.

While refusing and using data is the ostensibly "final" stage of a reformative/transformative data science project, as illustrated in the diagram in figure 2.4, the reality is that communicating and circulating data is always taking place in parallel with researching and recording cases (chapters 4 and 5). What activists learn from their labor in these three stages doubles back to inform their analysis of power and the theories of change that I described in the resolving stage (chapter 3). These stages are helpful to differentiate characteristics of data activist labor, but they are not linear and they are often all unfolding simultaneously in a given restorative/transformative data science project. The other reason this stage is truly not "final" is because the work continues. As Ana Alvarez from the Casa del Encuentro stated, "The work never ends. Not in the numbers and not in the law." This is especially true for the many activist observatory projects that aspire to monitor continuously—and ostensibly without end. For these projects, the end is when the violence has been eliminated. As Audrey Mugeni from Femicide Count Kenya stated: "One of the things I realized is that, for us, counting was also a form of saying no to violence, that this all needs to stop, because it's what is leading to all of these deaths. It was a form of us saying, 'No, this can't continue. We cannot continue talking to one another like this. We cannot continue with corporal punishment. We cannot continue with the violence in our homes. This all needs to stop.'"

III ACTION-REFLECTION

7 CO-DESIGNING FOR RESTORATIVE/ TRANSFORMATIVE DATA SCIENCE

What does it look like to design technology in the service of healing and liberation? Let us start with a counterfactual: what it does *not* look like. Here is an imagined brief news article for an alternate version of this project that I could have created:

> MIT PROFESSOR USES AI TO SOLVE GENDER VIOLENCE
> MIT Professor Catherine D'Ignazio has developed an advanced automated system using artificial intelligence and machine learning to sift through massive amounts of news and social media data to detect articles about *feminicide*—the gender-related killings of women and girls. Based at MIT, her project creates a centralized and comprehensive global archive of feminicide at a scale never before seen. According to D'Ignazio, the machine learning classifier she trained can detect news articles about feminicide with an accuracy of 92 percent.
> —Fictional High-Tech News Blurb

What is the takeaway from this (fictional) blurb? The hypothetical system is, from a technical perspective, incredibly sophisticated. The database is large—"massive," in fact. It is global, and thus ostensibly captures "all the data" in a comprehensive fashion. It is framed as an authoritative central repository. The project is credited to me, individually, and my affiliation with an elite institution is mentioned several times. This narrative works well for my own social capital; as an academic, I'm incentivized toward individual impact, and collaboration is viewed suspiciously by tenure committees and funding agencies. This narrative works well for my institution: *MIT saves the world again! Thank you, MIT.* This narrative works well for funders who are eager to "solve" complicated social problems like gender-related violence with efficient and scalable technologies rather than investing in people and relations. This narrative works well for white supremacy (as well as settler colonialism, patriarchy, and imperialism) as it reinforces white saviorism from both a benevolent white individual and a predominantly white institution in the Global North.[1] And this narrative indirectly bolsters the widely held misconception that feminicide is a problem that only happens in the Global South.

So what does this narrative erase and exclude? The labor of the organizations—often women-led and Indigenous-led and Black-led and queer-led—already doing the actual work of producing these data, usually in a more situated and culturally appropriate way. It erases and excludes the asymmetry of resource allocation in the space. The agency of families, communities, and social movements. The technological and social difficulty of monitoring feminicide across culture, language, and race/ethnicity/class. It obscures the scale of the missing data that we produce by monitoring from afar, without grounded context. In other words, it erases all of the rich complexity of grassroots data activism practices that I've spent the past six chapters telling you about. It's important to remember that the academy, like the data economy, is set up for and incentivizes extraction.[2] So how do we design against (or in the midst of) these structural barriers?

This chapter is a case study in participatory design for restorative/transformative data science. Throughout this book, I have talked about the Data Against Feminicide team and our work interviewing and learning from grassroots data activists. In this chapter, I want to explore one very specific aspect of that work: our attempt to co-design and deploy tools that support activists' informatic practices. How can we design interactive technologies to support and sustain the activist labor of restorative/transformative data science about feminicide? This chapter details our team's attempt to answer that question, in community, working collaboratively with the activists themselves. We co-designed and continue to maintain two digital tools that support the production of feminicide counterdata: the Data Against Feminicide Highlighter and the Data Against Feminicide Email Alerts. Guided by data feminism, we made many intentional decisions along the way to center its principles, such as *elevating emotion and embodiment, embracing pluralism, considering context,* and *challenging power*.

I led this chapter with the heroic counterfactual because such hero stories are often what we academics and technologists are pressured to tell—in academia, at conferences, and to funders. In these stories, the community is always lacking in one way or another. The academic or the technologist, in turn, always fills that void, usually with some kind of tech. The reality of what our team did is much more humble, more relational, and also more fraught. It was a sound process, and a process I am proud of, and a community with whom I continue to be in relation. But it was riddled with learnings and fumblings, and it isn't finished. It has left me with many questions about the role of academics and technology in activist labor. Thus, this chapter also constitutes a reflection on both the possibilities and the barriers that exist to putting research and technology development in the service of movements for liberation.

HUMAN–COMPUTER INTERACTION AND PARTICIPATORY DESIGN FOR DATA ACTIVISM

As a case study, this chapter is in dialogue with the fields of participatory design (PD) and human–computer interaction (HCI). Participatory design, also called co-design, originated in collaborations between university researchers and Scandinavian organized labor in the 1970s regarding the design and use of computer applications in the workplace, and was explicitly geared toward building worker knowledge and power.[3] HCI is a subfield of computer science that studies how people use digital systems as well as the design of novel interactive technologies to meet the needs of individuals and groups. While PD has explicitly political roots, HCI has more recently turned to addressing questions of equity, social justice, and the role of technology in social change. Scholars and designers have been trying to integrate critical and urgent ideas—such as feminism, critical race theory, intersectionality, feminist solidarity, Afrofuturism, or decolonial methods—into the process and products of design.[4] For example, feminist HCI, a framework originally elaborated by Shaowen Bardzell in 2010, draws from feminist theory to suggest key qualities that can challenge patriarchal approaches to computing and contribute to more liberatory design practices.[5] HCI designer-researchers don't only study what elements should be on a screen or how a digital workflow should go. They also look at theories and values, process concerns, and the impacts of designs on communities.

There is a small but growing body of work in HCI that addresses gender-related violence. A number of projects have focused on designing safe and supportive digital spaces for women and LGBTQ+ people, on violence prevention, and on designing with impacted populations.[6] Jill Dimond examined the relationship between data activism and co-designing feminist technology in relation to street harassment in her 2012 thesis.[7] One study in this vein of healing, prevention and direct support to impacted individuals, has addressed the topic of feminicide. In response to high rates of feminicide in Brazil, Giulia Bordignon Silveira and her colleagues designed a web application to aid Brazilian women in abusive relationships in getting support.[8] In a 2021 paper, sociologist Renee Shelby analyzed the rise of safety technologies that seek to prevent gender-related violence by mobilizing responses from peers rather than from law enforcement. One example is a wearable panic button that notifies friends and community when pressed by a person in danger of sexual assault. While Shelby lauds the "abolitionist sensibility" of such tech to bypass the state, she critiques these tools for "perpetuating gendered rape myths, commercializing assault, and disproportionately placing the burden of prevention on women."[9] Thus, there are tensions in using tech to address gender-related violence, including whether (1) they may introduce *carceral creep*—that is, fortify punitive and state-based responses to violence that perpetuate mass incarceration—and (2) focus on individual behavioral changes that, in order to cope with unjust systems, reduce rights and access.[10] While these technologies might provide some immediate harm reduction to individuals excluded or abandoned by the state, they do not address the more structural aspects of gender-related violence.

Other HCI studies have focused less on individual-scale designs and more on digital methods that support activism and collective action around gender-related violence. For example, Morgan Vigil-Hayes and colleagues showed that missing and murdered Indigenous women (MMIW) consciousness-raising represented a significant amount of Native American activist use of Twitter in 2016.[11] Several years later, Angelika Strohmayer and colleagues collaborated with sex workers in North East England to commemorate lives lost to violence by co-organizing a march and reflecting on the use of digital technology during it and thereafter.[12] The case study I describe in this chapter is situated in this latter vein of work, which does not provide direct services to survivors, but rather aims to support activists who are already taking collective action with digital technologies to combat structural violence. In our case, these are the data activists who are working in concert with social movements to visibilize feminicide as a public issue through the production of counterdata. For us, it was an intentional design decision to support the people already doing the work.

Data activists have specific informatic needs. Recent case studies, both inside and outside of HCI, have highlighted the growth in data activism as well as the informatic needs and practices of citizens and residents undertaking it. The growing literature on data activism, discussed in chapters 1 and 2, shows that counterdata production may be undertaken by activists, journalists, nonprofit organizations, librarians, citizens, and other groups. It does not happen only outside of mainstream institutions, but also with insiders and experts looking to reframe political problems. Research on violence as the object of data practices has explored how citizen organizers mobilize the affective and narrative potential of data through agonistic data practices, and scholar-activists recording fatal violence have also reflected on their own practices, including the data challenges and vicarious trauma of recording homicide.[13] In a 2017 paper, Adriana Alvarado García and colleagues discuss data about feminicide and sex crimes against children as areas where human rights organizations attempt to combat data gaps through community-based data practices in Mexico.[14] They describe existing data practices and speculate about how HCI may work to support these practices using design, including addressing infrastructure concerns, designing for safety, and supporting community data production and circulation.

Work in HCI that explores social justice, participatory methods, and feminist and decolonial approaches to design has also had to navigate the ethics—and ethical pitfalls—of design that aims for transformative change. One salient contribution to this literature is a 2022 paper titled "On Activism and Academia," in which Débora de Castro Leal, Angelika Strohmayer, and Max Krüger untangle some of the tensions and possibilities of working at the intersection of academia and activism. The authors warn against what they call *community fetishism*—the tendency for academics to reap career benefits from working with marginalized groups. They also enumerate the various ways in which academics may fail to do the "good" they aim for—and, in fact, may perpetuate harm against activist communities through extractive practices, invalidation of community know-how, and a focus on narrow research products over process. These and other potential harms arise, no surprise, because of the ways in which oppressive forces—white supremacy, patriarchy, and colonialism, to name a few—permeate all aspects of Western academia. As a result, the authors discuss how researchers working with activists can recognize those forces as well as challenge such forces within academia itself. Moreover, there are the pragmatic matters of funding, maintenance, and sustainability that need to be addressed for tools that emerge from academia, such as the basic and urgent question posed by Max Krüger and coauthors in a separate paper: "What happens when the funding ends?"[15]

One goal of the present case study is to think through how to use co-design in the service of restorative/transformative data science. This requires considering not only which specific digital tools may support data activism, but also which ethical and epistemological frameworks may help us navigate some of the complexities of working across sectors (such as activism and academia) and contexts (geographic, cultural, racial). To that end, I will show how we integrated data feminism's principles into designing for restorative/transformative data science, but also how principles—just like good intentions—can only get you so far. Practice and design and relationships are messy and specific and don't always conform to general principles. That said, the friction is something to embrace because the friction is where the actual work happens.[16]

RESOURCES AND RELATIONSHIPS ARE DESIGN QUESTIONS

This case study emerges from the ongoing work of Data Against Feminicide, the South-North collaboration led by myself, Silvana Fumega, and Helena Suárez Val. In 2023, Isadora Cruxên, who had worked with us since the beginning as a researcher, joined our leadership team. As I described in the introduction, Data Against Feminicide has three goals: to foster an international community of practice; to develop digital tools to support and sustain the work of activists; and, where appropriate, to help standardize the production of feminicide data.[17] This case study is about the second pillar of our collaboration—the development of digital tools and technologies to support the widespread counterdata production work of activists, journalists, and civil society organizations.

Participatory design often gets discussed in terms of process: what activities were conducted and what people created during them. Recently there has been a shift to look at infrastructure and institutions and their roles in enhancing or inhibiting the emancipatory goals of participatory design.[18] This work makes the point that resources, governance, and social relations are also designed and can be designed for. Engineering healthy setup conditions for a co-design project includes intentional decisions made around these infrastructural and relational aspects. These are like the soil from which co-design for liberation may grow (or be stunted). Here I draw from Mariam Asad's formulation of "prefigurative design" in which she asserts that redistributing resources and transforming social relationships are key opportunities for academics to engage in justice-oriented research.[19] In that spirit, I want to tell you about three foundational decisions regarding resources and relationships that we made, guided by the principles of data feminism.

First, we worked collaboratively and in community with grassroots activists. This aligns with the *embrace pluralism* data feminism principle, which asserts that the most complete knowledge comes from synthesizing multiple perspectives, with priority given to local, Indigenous, and experiential ways of knowing. This was a very intentional decision that runs counter to where the money is concentrated for technologies related to gender-related violence—namely, in carceral technologies that tend to focus on punitive "solutions" like incarceration and that bolster police and law enforcement budgets.[20] These technologies tend to be deployed without community consultation, effectively centralizing the singular, monolithic perspective of state surveillance. For example, the Markup recently reported on police around the globe who are looking to predict domestic violence with algorithms.[21] And, in a stunning example of carceral fortification around gender-related violence, US-based corporation Honeywell partnered with the Indian city of Bengaluru on a $67 million project to implement a command-and-control center, facial recognition, drones, and apps in the service of detecting and prosecuting sexual harassment and gender-related violence.[22] There are so many alarming aspects of this project, but one question I will posit here is this: Instead of handing over tens of millions to police—who have consistently proved themselves not only ineffectual and ignorant in responding to gender-related violence but who are also often perpetrators of it—what if such funds had been used to resource the grassroots organizations that are already providing expert community defense, preventative services, and access to support and healing for survivors?[23]

The Data Against Feminicide project sought to place our own modest time and resources in coalition with such community defense, not least because Helena is herself a data activist and brings this political orientation and lived experience into the collaboration. Thus, in contrast to typical research, the "we" that undertakes this project is not disinterested. That is, we "have allowed [ourselves] to become interested" and to use "solidarity as a method" to engage with activist data practices around feminicide and feminicide data.[24] This is in line with recent calls from senior scholars in participatory design for researchers to "team up with partners to fight for shared political goals."[25] Our shared commitment is to collaborative knowledge production and technological development, but also to actively caring for the already existing community of data activists by supporting practitioners' work and connecting activists to each other. At a more personal level, this involves *elevating emotion and embodiment* as tools to foster caring relations. Angelika Strohmayer wrote that her academic training had not prepared her for how "the boundaries between research partner, colleague, and friend started to blur. I started to care for and love those with whom I collaborate."[26] In the Data Against Feminicide project, we have embraced the love and friendship that we hold for each other, and we have supported each other in numerous ways inside and outside the collaboration—traveling together, cowriting together, venting together, snarking together, crying together, reflecting together.[27]

Second, more about money. Decisions about funding structure all downstream relationships and products. A key way to *challenge power* is to think upstream about funding—that is, to reengineer flows of social and financial capital through an institution or a research project, before it begins. This is part of what Asad describes as *redistributing resources*, which are disproportionately concentrated in the Global North, at elite institutions, at predominantly white institutions, and in fields with more industry sponsorship (i.e., profit-making potential) such as computer science or business.[28] To get Data Against Feminicide off the ground, we did not seek highly structured grants from industry, foundations, or federal agencies. We used operational funds provided by ILDA to underwrite our community events and courses, and funded the majority of the research and tool development with my start-up funds from joining MIT in 2020. This was a way to maximize flexibility and autonomy as the three coleaders developed our relationships with each other and as we developed relationships with data activists.[29] Trust takes time and requires pivoting and reflection, especially when navigating differences of language, culture, geography, ethnicity, and race. Funders don't often see the value in such open-ended exploration. The risk of seeking large grants, especially at the incipient phase of a project before you are working in community, is that the seeker sets up the goals, timeline, and outcomes, which may in fact not be desirable to the community. These lead to urgency, a key feature of white supremacy culture and a sure way to damage relationships with a community before they are well-formed.[30] Moreover, we recognized that data activists' labor is gendered, racialized, always under-resourced, and almost never paid, and that participation is work. Thus, we compensated all interviewees and co-design partners for their time.[31]

Finally, at no point did Data Against Feminicide request to see, seek to use, or aim to centralize activist data. Why do I state this so explicitly? First, grassroots efforts like MundoSur are already underway to harmonize activist data (see chapter 5) in sensitive and participatory ways in Latin America. And second, Silvana, Helena, and I were aware of the extractivist tendency in academia: the all-too-common pattern whereby researchers take knowledge and stories and experiences from communities; they analyze them and write them up in paywalled academic papers; and they do not return to the community to share results and impacts. This pattern has led to deep and very warranted distrust of academia by marginalized communities, evidenced by Linda Tuhiwai Smith's (Ngāti Awa and Ngāti Porou, Māori) affirmation that the word *research* "is probably one of the dirtiest words in the Indigenous world's vocabulary."[32] We did not enter the community with the (arrogant!) idea that we could take activists' data and synthesize their work better than they could. We entered with admiration for their labor and commitment (and in the case of Helena, lived experience of doing the work herself), and an offer to explore together how to support and sustain it. This represents a way of *making labor visible*—seeing and appreciating the tremendous collective efforts that result in feminicide counterdata and offering a vehicle to create social and technical infrastructure around them. Our aim was to manifest a design-as-service orientation rather than a design-as-hero orientation.

Together these three setup decisions—collaborating with activists, keeping funding flexible while still paying participants, and deliberately not requesting activist data—paved the way for the co-design process that followed.

THE CO-DESIGN PROCESS AND PRODUCTS

Designing digital tools with data activists began in parallel with our interviews with them in 2020, described in more detail in chapter 2. From these forty-one interviews, we developed the process model that I have used throughout this book, which describes the four workflow stages of a restorative/transformative science project: resolving, researching, recording, and refusing and using data (see figure 2.4 or table 7.1).

These interviews also surfaced numerous informatic challenges that activists face in producing data about feminicide. Two of these challenges emerged again and again, and so we tried to directly address them in the co-design process. First, all groups have to reckon with missing data, which is inevitable due to the negligence, inaction, and bias of the state and media, and which makes the researching stage of counterdata work very challenging (as described in chapter 4). In the face of such missing information, many activists sift through news media articles to source cases that they may then triangulate with other sources. The informatic hurdles at this stage are more acute for activists monitoring violence against Black women, Indigenous women, women in rural areas, and/or LGBTQ+ people. It seemed clear to us that digital tools might at least help to surface some of the information that activists are seeking. We soon started talking about some sort of case-detection system that could scan news or social media, especially hyperlocal media, to detect cases of feminicide; and/or notification systems that could alert activists to new cases.

Second, all groups face resource constraints in terms of time, money, mental health burden, and emotional labor. The majority are volunteer-led efforts with no funding source, though some were supported by small grants or crowdfunding. Across all projects, activists noted how counterdata work was time-intensive and also emotionally challenging because of the continuous exposure to violence. Tools that help groups anticipate and plan for some of these challenges might therefore be useful in the resolving stage of work. From these challenges, we can extrapolate some design implications, such as that any new or adapted digital tool should be free, easily maintainable, collaborative, and easy to learn by newcomers to accommodate activists' volunteer and sometimes ad hoc labor. Another approach for tools in the researching stage could be the creation of case-detection systems that seek to reduce the number of nonrelevant violent results that counterdata activists are exposed to and that they need to filter manually. This could reduce overall time spent seeking cases as well as some of the emotional burden of the work. At the same time, activists do not want to fully eliminate this labor—they see the emotional labor of caring for murdered people's lives, stories, and families as an essential part of public witnessing and memory justice.

The more we talked with activists, the more ideas surfaced. It became clear that design could be useful in many ways—that there were likely tools that would aid with all four stages of a restorative/transformative data science project (see table 7.1). It seemed only fitting to workshop these design possibilities directly with activists. Because the project was on a limited budget, we decided to draw from the activist expertise already on the leadership team—Helena Suárez Val of Feminicidio Uruguay, whose experience comes out of Latin American conversations about feminicide—and then to invite the first North American activist that we interviewed, Dawn Wilcox of Women Count USA. We ran six co-design sessions with Dawn, Helena, myself, and a

Table 7.1

Stages of a restorative/transformative data science project and designs that could be useful at each stage

	Resolving	Researching	Recording	Refusing and using data
Activist activities	Starting a monitoring effort	Seeking and finding cases and related info	Information extraction and classification	Where data go, who uses them
Design example	*Tools to map out feasibility of counterdata effort*	*Tools to detect relevant cases (e.g., Data Against Feminicide Email Alerts system)*	*Tools to record and categorize cases (e.g., Data Against Feminicide Highlighter)*	*Tools for visualizing and humanizing feminicide data*

number of student researchers between June 2020 and February 2021—roughly one session a month. Because this was during the height of the COVID-19 pandemic, and we were all in different locations, these sessions took place on Zoom, and there were usually around five to seven of us on the call. Our sessions were scheduled for an hour but often went longer, as we got to brainstorming, reacting to wireframes, and proposing hand-sketched changes (figures 7.1a, b). Design ideas fell into five categories: detecting and recording cases, enabling collaboration, storing sources, archiving databases, and data analysis and visualization (figure 7.1c).

From our six brainstorming sessions, we emerged with a list of more than fifty potential ideas, which ranged from the relatively simple (a bar chart with photos) to the more technically challenging (a data repository where activists share the data they produce and get recognized for their efforts) to the nontechnical (activist events to share knowledge around managing volunteers). From these ideas, the activists and our team decided which to carry forward into development based on two questions: According to our interviews and our partners' knowledge of other data activists' work, how widespread was the need that this tool met? And how technically feasible was this tool to develop? We chose to move forward with two tools that help activists detect and record feminicides—that is, tools that address the researching and recording phases of a project. These are the two stages in which activists spend the most time and exert the most emotional labor.

Our team developed and built the first version of the tools in early 2021 and then piloted them later in the year.[33] For the pilot, we wanted to test both a Spanish and English language version of the tools, so we recruited seven groups from Argentina,

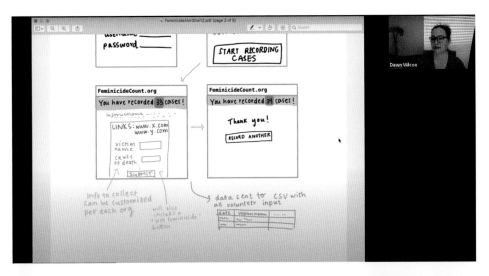

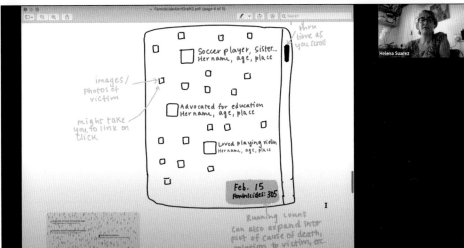

FIGURE 7.1

(a, b) Screenshots from our co-design sessions, which took place on Zoom and often involved reviewing hand-drawn sketches and generating new ideas. (c) The final Miro board cataloging and categorizing ideas generated during six months of co-design work with data activists. Courtesy of the author.

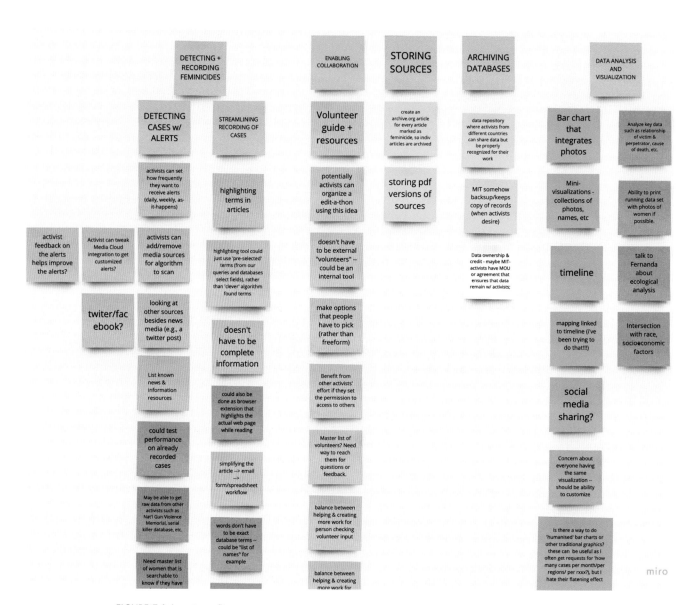

FIGURE 7.1 (continued)

FIGURE 7.2

Screenshot of the Data Against Feminicide Highlighter, a browser extension for the Chrome browser. (a) When a reader loads a news article, the tool highlights people, places, ages, and custom words to facilitate activist scanning. (b) The interface of the Highlighter provides a selection of toggleable options and an area for custom term entry. Courtesy of the author.

Uruguay, and the United States (see appendix 1) from those that we had previously interviewed. Groups participating in the pilot filled out a weekly survey and participated in several two-hour focus groups over a period of two months.

THE DATA AGAINST FEMINICIDE HIGHLIGHTER

The Data Against Feminicide Highlighter is an extension for the Chrome browser that aids in the *recording* stage of a restorative/transformative data science project. This is the stage when activists have found relevant information and need to extract it into structured spreadsheets or databases. When an activist opens a news article about feminicide or fatal gender-related violence, the extension autohighlights names, places, dates, and numbers with different highlighting colors (figure 7.2a). This is an example of what is called *named entity recognition*. Data activists and journalists can also put in custom words to highlight, such as "gun", "husband", or "boyfriend" (figure 7.2b). The Highlighter has an Open Database link that activists can customize to open their own database or spreadsheet for easy copying and pasting between the browser and the spreadsheet.

The idea for the Highlighter emerged from an early participatory design session with our co-design partners where we were discussing how, for a given case, activists need to scan a huge amount of information in order to fill all of the fields for a single case in their database. Our interviewees estimated that they needed to read anywhere between three and fifty news articles per case to get the information they needed. Activists move back and forth, copying and pasting between the browser and their databases. The process is incredibly time-consuming, and also emotionally intensive. The goal of the Highlighter is to reduce activists' overall time spent scanning violent news articles by visually highlighting the key pieces of information they are seeking for their databases.

Graduate student Harini Suresh led the development of the Highlighter, from early sketches through multiple rounds of feedback—first with the two activist partners, and then with seven groups in our pilot in spring 2021. She was aided by undergraduate Amelia Dogan and by Rahul Bhargava, a professor at Northeastern University, who is a collaborator and partner on the Data Against Feminicide project. During the pilot, activists found the Highlighter very easy to use and also suggested many additions to the Highlighter to support their existing workflows. These included the ability to email an article to a colleague (to support collaboration), multiple custom word fields (to highlight key terms they search for, such as specific mode of death—like "gun" in a separate color from words that describe the relationship between the perpetrator and victim like "ex-husband"), and a feature for choosing which colors to use for which highlights. Our team implemented these in a second round of development following the pilot.

THE DATA AGAINST FEMINICIDE EMAIL ALERTS SYSTEM

The Data Against Feminicide Email Alerts system supports activists in detecting new cases of feminicide and fatal gender-related violence, and in following the development of existing cases. This is part of the researching stage of a feminicide monitoring project, wherein activists are seeking information about relevant cases or observations for their database (see table 7.1). Our system is designed to be similar to Google Alerts, but with a few significant tweaks. An activist sets up a project in a particular geography. Then they input keywords for finding news media articles related to feminicide or gender-based killing and pick the frequency with which they wish to receive email alerts. Many groups we interviewed had attempted to use Google Alerts to find cases, but most had stopped because the search results were too broad, the system returned cases from outside their geography of interest, and/or the system repeated a single case or article many times over, making it hard to distinguish between new and old information. In addition, Google didn't index the most useful media sources in their country (and also wasn't transparent about the sources it did include). These drawbacks related by activists provided us with some high-level design goals: provide more geographically relevant results, increase the proportion of relevant articles, group news articles by feminicide case, and provide transparency and agency to activists regarding media sources.

In the co-design sessions, we also discussed the topic of automation—everything from full automation (the system would monitor the media, extract relevant information from articles, and put all results into a database) to partial automation (the system surfaces alerts and the activist chooses which are relevant for their database). Helena pushed back on the idea of full automation and described the central importance of her emotional labor of witnessing and caring for the people whose cases she logged in her database. Because of this, as well as the fact that definitions of feminicide vary across cultures and contexts, we stayed with partial automation, where it is ultimately up to the data activists to decide whether a case is relevant for their database or not. The overall goal of the system is to reduce activists' time spent searching for new cases as well as to reduce the emotional burden of reading violent news articles that are not relevant for their databases.

The idea of the system is simple—it surfaces news alerts for relevant cases of gender-related violence. Yet what's happening behind the scenes is a bit more complicated. The system draws news content from Media Cloud, our partner project and an open-source platform for media analysis (and also an academic research project I have participated in[34]). An organization using the Email Alerts System can customize a search query and set of place-based media sources to best suit their project needs. Media Cloud then retrieves matching articles from its continually updated database of global news stories, which are run through a machine learning model we developed that predicts the probability that the article will be relevant to the organization (i.e., the article describes an instance of feminicide). Articles above a particular probability threshold (which defaults to 0.75) are sorted by the probability of feminicide and delivered in a daily email digest (figure 7.3) and can also be viewed in an online dashboard. Articles are grouped together if they share many of the same entities (people, places, organizations). For example, in figure 7.3, under the first headline, "Judge to Decide Wednesday

**Women Count USA: Femicide Accountability Project
Recent News Articles**

05/27/2021

Title	Date	Count
Judge to decide Wednesday whether James Prokopovitz can ever be released from prison in wife's homicide	05/26/2021	4
└ Kewaunee County Star News	05/26/2021	
└ Door County Advocate	05/26/2021	
└ greenbaypressgazette.com	05/26/2021	
└ Oconto County Reporter	05/26/2021	
Chad Daybell and Lori Vallow indicted on murder charges for the death of Vallow's children	05/25/2021	1
└ nbc-2.com	05/25/2021	
Oxnard man charged in stabbing death of wife	05/26/2021	1
└ keyt.com	05/26/2021	
Person of interest won't be charged in Arkansas woman's overdose death: Sheriff	05/25/2021	1
└ truecrimedaily.com	05/25/2021	
Husband charged with murder of wife he dumped in Edwardsville	05/26/2021	2
└ fox2now.com	05/26/2021	
└ kplr11.com	05/26/2021	

FIGURE 7.3
Sample email alert delivered from the Data Against Feminicide Email Alerts system to Women Count USA.

whether James Prokopovitz Can Ever Be Released from Prison in Wife's Homicide," you can see that there are four articles from different news outlets about an upcoming sentencing in one specific case.

The first machine learning models involved participatory annotation of training datasets—meaning that our team and activist partners manually labeled several hundred news articles about feminicide in both English and Spanish.[35] This is because we needed a reliable set of news articles describing feminicide in order to teach the machine how to detect it via natural language. These data were used to train two language-specific logistic regression models to predict the probability of feminicide from the text of an article. The English and Spanish models achieved 84.8 percent and 81.6 percent accuracy, respectively. This means, for example, that when we ran our English-language model on test data, it was able to correctly predict whether the article described a feminicide in 84.8 percent of cases. Most of the errors were false positives, meaning the model guessed that a news article described feminicide when it did not. This felt like the appropriate direction of error, because activists told us they would prefer to get alerts about nonfeminicide cases than to miss an actual case of feminicide. Further details about data collection, annotation, and model performance for this initial iteration can be found in our paper for the MD4SG community from 2020.[36] Since then, we have developed models for different types of fatal gender-related violence (which I discuss further in the next section) as well as begun work on participatory data annotation and a Portuguese language model with Brazilian activists.

THE TENSIONS OF CO-DESIGN

For the most part, these two tools received positive reviews during our two-month pilot. More than half of the groups in the pilot reported that the tools saved them time, helped them detect new cases, and made their work easier. Rosalind Page, who runs Black Femicide US, expressed that the best part about the Highlighter was that she didn't have to read the whole news article to see if it was relevant for her database. Members of Mumalá said that while not all articles delivered by the Email Alerts tool were relevant, the system was delivering several cases a week that they would not have otherwise found out about. They were also receiving alerts about feminicide attempts, which they also monitor, so it made those easier to detect and track. Five of the seven groups in the pilot continued using the two tools after the pilot ended, so this is a good indicator of overall performance and utility. After doing a round of improvements following the pilot, we have continued to maintain the tools and, in November 2021, launched them for use by the Data Against Feminicide community.[37] At the time of this writing, there are forty organizations who have set up one or more projects in the system.

While these were relatively positive results, there were three tensions that surfaced in the design process, pilot, and maintenance periods of the project. These warrant some reflection because they point to larger structural concerns for other reformative/transformative data science projects and raise questions that HCI, PD, and allied fields will need to navigate as we aspire to support data activism with informatic tools and infrastructure.

TENSION #1: MODELING POLITICALLY CONTESTED CONCEPTS

We have seen, again and again, the significant variation in how the concepts of femicide and feminicide are elaborated in laws, in the media and by civil society. This poses challenges for the Data Against Feminicide Email Alerts system. What constitutes a feminicide? How does the news media describe such an event in a given context? How can a machine learning model learn to detect that event using only the text of the article? In the Americas, there is significant language variation encompassing, at the very least, media in Spanish, Portuguese, French, English, Quechua, and Guaraní. Beyond language, there are variations in legal definitions of feminicide at the country level (see table 1.1). And beyond legal variations, there are significant differences in media ecologies at the country level. This is to say that feminicide is reported with different language in Peru than in Argentina, though both countries' dominant language is Spanish. In some places, such as the United States and Canada, the terms are very rarely used in the media and terms such as *MMIW* and *domestic violence* have more media prevalence. As discussed at length in chapter 5, even within a given country, different groups—government agencies, journalists, nonprofits, activists—may define and count feminicide differently.

This amount of variation poses a significant challenge for a technical system seeking to detect news articles describing cases of feminicide. Computer scientist Rediet Abebe and her colleagues argue that one role for computing in social change is to act as a *formalizer*, a way to codify the definition and understanding of a social problem, as well as its measurement.[38] Yet because of the contextual variation, at multiple scales, across multiple stakeholders, such formalization is premature: it would constitute an attempt to bypass the social and political contestation taking place around feminicide and preemptively bake it into a one particular model. Often, computer science, being a somewhat conservative field, is inclined to side with larger, credentialed, status quo institutions for definitions and formalizations. But doing this is at odds with the Data Against Feminicide project's design goals of challenging power, centering the expertise of grassroots feminist activists, and using computing for liberation. So how do we operationalize the data feminism principle of *considering context*, especially given that there is so much context to consider?

The way we navigated this in the Email Alerts tool is twofold. First, we made the design decision to *not* try to fully automate case detection and information extraction. This was for two reasons. First, full automation was not technically feasible without having a much more formalized and rigid definition of feminicide. Second, as mentioned earlier, Helena and other activists pushed back on the idea of full automation due to their commitment to witnessing and caring for the individual people in their database. Instead, our system involves the activists and grassroots groups—human expertise and judgment—as the ultimate arbiter of whether a case is relevant for their database. Thus the system surfaces probable feminicide cases, and the humans monitoring in different contexts make the ultimate judgment about what counts. Such a system may still need different models for different languages or different types of violence, but it can then support wide variation in definitions and circumstances.

The second way that we considered context in the design of the Email Alerts system has to do with how we assembled and annotated training datasets. For our initial prototype, we collected and annotated two datasets of around four hundred articles each: the first in English, in collaboration with Women Count USA, and the second in Spanish, in collaboration with Feminicidio Uruguay.[39] Each article was annotated by three separate people attempting to determine whether it described a feminicide and, if so, what type of feminicide. Where there were discrepancies, we held deliberation meetings to debate each case and determine the ultimate judgment. This participatory data annotation process is a way of *embracing pluralism*, supporting the integration of multiple knowers and positionalities into a machine learning model. This also mirrors activist processes of deliberation on cases as I described in chapter 5, wherein activists often use WhatsApp channels to debate and make a collective determination as to whether a case should be included in their database. As I write, we are in the process of systematizing this participatory data annotation approach and developing a Portuguese feminicide detection model in collaboration with activists who represent five different feminicide monitoring groups across Brazil.

Thus, while computing and machine learning often play the role of formalizing and reducing and centralizing, they do not *always* have to be forces in that direction. We did not have to choose one country's or one group's definition of feminicide to build into our system. Despite the fact that feminicide is a politically contested concept, we were able to use computing and machine learning to support multiplicity and contextual difference across stakeholders, media ecologies, countries, and languages.

TENSION #2: COLIBERATION REQUIRES INTERSECTIONAL ML MODELS

"It's not working for us," reported Michaela Madrid, a staff member at Sovereign Bodies Institute at the time of the pilot. "We are getting so many cases every day but we check every single one and none of them are Indigenous women." It was spring 2021 and our team was conducting a pilot study for the Email Alerts system. We were in a Zoom focus group where ten activists spanning four feminicide monitoring organizations based in the United States were sharing their perspectives.

Among the four groups, we received dramatically different feedback about the Email Alerts tool's performance. Women Count USA, which monitors all US femicides, reported that the results were overall very relevant and useful. Another organization, Black Femicide US, monitors femicides of Black women and reported mixed but still useful results. However, the system did not source relevant results for two organizations in particular, both of which monitor specific, racialized forms of feminicide: (1) Sovereign Bodies Institute, a group that tracks missing and murdered Indigenous women, girls, and two-spirit people (MMIWG2/MMIP); and (2) the African American Policy Forum, which monitors police violence against Black women as part of the Say Her Name campaign. Feedback from these two groups consistently showed a lack of relevant articles being returned by the system, despite modifying their search queries to add relevant terms.

Why is this? As described in prior chapters and table 4.1, racialized groups face more missing data from the state and more underreporting from systemic bias in the news media. In addition, it is well-documented that the state systematically misclassifies Indigenous victims' race.[40] News articles often do not report the race or tribal identity of a killed person, and they regularly misgender trans people. Not surprisingly then, it is very difficult for a machine learning model to try to distinguish news articles based on language that is either (1) absent or (2) incorrect. This is, in fact, one of the primary ways in which trying to count feminicide from news articles is deeply fraught, because if you systematically collected information from news articles and analyzed the patterns, victims would disproportionately be white, cisgender, middle or upper class, and killed in intimate partner violence. This demographic profile does not reflect who is most vulnerable to gender-related violence. Rather, it reflects the underreporting, misclassifying, and misgendering practices of the media and the state.

Navigating these biases and gaps in the information ecosystem surrounding racialized feminicide is part of a commitment to *considering context*—that is, operating with the knowledge that "data are not neutral or objective. They are the products of unequal social relations, and this context is essential for conducting accurate, ethical analysis."[41] With agreement from the two groups, we went back to the drawing board with the machine learning model and tried to develop new classifiers to meet their needs.

This is work that was led by computer science PhD student Harini Suresh, with significant contributions by undergraduate Rajiv Movva. The two collaborators came up with the data annotation and training pipeline depicted in figure 7.4. Round 1 is what we used in the pilot and what did not work for the two groups monitoring racialized feminicide. Round 2 supplants the original training data with feminicide cases contributed from the organizations themselves, who typically store links to news articles about cases in their databases. When we tested the model developed from round 2, it returned somewhat more relevant articles. However, there were also more

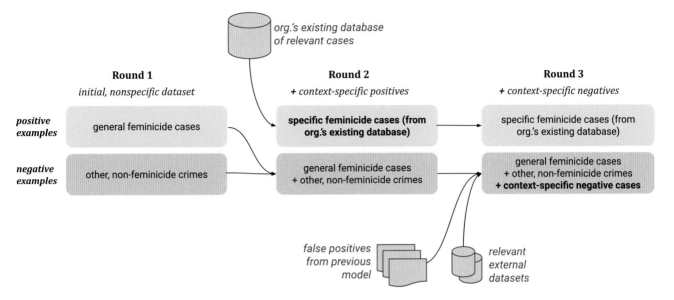

FIGURE 7.4

Our data-collection process involves iteratively collecting context-specific positives (e.g., by sourcing ground truth articles from organizations' existing databases) and context-specific negatives (e.g., by identifying and collecting types of negative examples close to the decision boundary, such as articles describing cases of Black men killed in police violence). Courtesy of Harini Suresh and coauthors.

false positives. For African American Policy Forum, we found that the list of returned articles was often dominated by police violence against Black men, which is more commonly reported in the media than Black women killed by the police. Thus, round 3 augments that further by supplementing training data with "negative" cases—training the model on which cases are *not* relevant. Harini and Rajiv explored several ways of combining and chaining training data into the machine learning model and determined that a method they called *contextual hybrid* worked the best. In the case of the African American Policy Forum, this meant first running a classifier to determine whether a news article described any kind of police violence and then, as a second pass, training a feminicide classifier to recognize articles about police violence specifically against women.[42]

Here, I cannot offer you a hero story that "we fixed it" and successfully navigated the deeply biased information ecosystems for racialized feminicide. As a method for embracing pluralism, our team considers the final test for these models' effectiveness to be the judgment of the groups themselves. But before burdening groups with evaluating new models (because, again, participation is work), we are performing two stages of internal evaluation.[43] The first involves internal cross-validation: our team will do quantitative evaluation of the model against a test dataset drawn from our hand-annotated data. The second involves an internal monitoring phase in which our team will deploy these models in the Data Against Feminicide Email Alerts system. So far, our internal tests show far more relevant results, but the final proof will be in the groups' own assessment of relevance for their work in a next stage pilot.

I want to highlight one final design decision we made in the course of navigating this tension. We did not try to infer the racial identity of the victim described in a news article. While we annotated the African American Policy Forum's data with *police violence* and *feminicide*, we did not include a race annotation even though their focus is specifically on Black women. Although technologies exist to infer race (e.g., from names or photos or language), they are often empirically wrong and ethically fraught. Race and gender are not essential properties of an individual body but rather are their own kind of classificatory technology operating at the structural level—hence why a person may be white when they are in El Salvador and Latinx/Hispanic when they are in the United States.[44] A person doesn't "have" a race; rather, a person is racialized. Moreover, the way in which different populations are racialized differs. Being Indigenous is an identity that is racialized by settlers—for example, the US census racial category "American Indian and Alaska Native"—but it is in fact a political and cultural identity.[45] Thus, because race is inherently social and political and historical, racial classification technologies that model it as an individual property of an individual body are operating with a fundamental misunderstanding of what race is and how it works.[46] It is a misunderstanding that holds the potential for reproducing racial violence. The data feminism principle of *rethinking binaries and hierarchies* requires us to challenge the gender binary, along with other systems of counting and classification that perpetuate oppression. In this case, challenging those systems meant refusing to use reductive and erroneous racial inference tech.

TENSION #3: SUPPORTING AND MAINTAINING TOOLS IN THE REAL WORLD

It was January 2022, and the Email Alerts system was in a stable state. There were around thirty monitoring projects in the system created by activists, nonprofits, and artist collectives. And then the system went down.

Well, the whole system didn't go down, but a key piece of its infrastructure did: Media Cloud, our main source of news articles by geographic region. Without a database of news articles to query against, our Email Alerts tool just couldn't function.

Activists, especially those who had participated in the pilot and who had come to rely on our tools, were emailing us asking when the service would come back online.

Unfortunately, the technical glitch at Media Cloud coincided with staffing issues and an urgent server migration and the service didn't come back online for almost three months. During that time period, Rahul Bhargava, our partner at Media Cloud, and Wonyoung So, lead developer, searched for other large-scale news sources that we could query against. Google's news APIs had query length limitations (and activists have long and complicated queries) and severe geographic limitations (no news sources for a number of Latin American countries, so that was a showstopper). After a couple months of testing and searching, we found NewsCatcher, a start-up company with a friendly founder who gave us a low-cost monthly subscription. We integrated NewsCatcher right as Media Cloud was coming back online and we now draw from both sources so that we have some redundancy in news databases in case of future outages.

Together with Helena, Silvana, Rahul, and Wonyoung, I have reflected a lot on this outage and the ways in which it highlights some fundamental tensions in designing and deploying "real-world" tools from academia. First, novelty in design and research is incentivized in HCI as it is in academia more broadly. This leads to prototype proliferation in which there is funding for early-stage tools, and academics reap social and career benefits for doing that work with communities. Yet following such prototypes, there are neither structural incentives, nor funding, nor accountability to ensure that relationships grow and deepen nor that prototypes blossom into services. This is what de Castro Leal and colleagues call *community fetishism* and Suzanne Bødker and Morten Kyng, from PD, call the *least-effort strategy*—that is, the best way to quickly publish research papers.[47] This constitutes a kind of nominal adoption of the *embrace pluralism* data feminism principle, but only when it serves an individual's career or the larger enterprise of academia. It's the same old academic extraction, but with a veneer of "participation-washing."[48] Moreover, the end result is a proliferation of presentist thinking, which "leads to too many processes and products with no utility and no impact."[49]

Second, when our team tried to move from academic prototype to real-world tool, we made a fundamental mistake of optimism, which, in retrospect, seems absurdly basic. The first lesson you learn as a software developer is that technology always breaks. Probably 80 percent of software development is building infrastructure to anticipate and address the ways in which it could stop functioning. In practice, our services went down and we didn't have a large professional team ready to address outages and respond to users. As Lilly Irani and M. Six Silberman wrote over a decade ago in their work on critical infrastructure, "Repair and maintenance fall out of view in most design discourse."[50] Most HCI work remains in the prototype and evaluate-the-prototype stages of design. How do we ensure that the design of an artifact also includes the design of its deployment, infrastructure, and maintenance?

One answer to this question might be to anticipate these needs and hire staff. In the review process for this book, J. Nathan Matias, founder of the Citizens and Technology Lab at Cornell, offered that the lab's policy is that if they are going to develop a tool with communities that requires ongoing maintenance, then they also commit to fundraising for professional staff to maintain it (rather than relying on consultant or student labor). That may be a high price tag for a community tool, but it actually gets closer to a true-cost accounting for the labor involved. Another answer to this question is proposed by HCI scholar Oliver Haimson and colleagues. In a survey of trans technologies—technologies created to help address challenges that transgender people face—they find that the most successful and inclusive design processes took place in settings like academic research and classrooms where the actual tools were not deployed. They recommend establishing more links between trans tech designs created by and for trans people and the infrastructure to implement, deploy, and maintain those designs so that they actually reach the people that need them.[51]

Finally, work on HCI to support grassroots social movements has highlighted fundamental tensions between building alternative technologies and deploying them at a scale that can work for movements. As Sucheta Ghoshal and colleagues write, "The most popular and 'effective' solutions that 'work' (with significant implications in technological power) in movement settings are centralized technologies made and marketed by corporations like Google or Facebook."[52] While there have been brilliant critiques of scalability as an unquestioned value in computing, it would seem that no alternative approach could get off the ground without buying into such Silicon Valley logic—since who among us has the resources for ensuring uptime and infrastructural soundness other than the Facebooks and the Googles of the world?

Interestingly, our partners at Media Cloud are facing the same set of issues as they try to develop and maintain a key piece of digital public infrastructure. This is how Ethan Zuckerman, cofounder of Media Cloud, frames projects that seek to provide informatic services to communities and broad publics.[53] Currently, I am sitting with these tensions, talking with our collaborators and partners and activists, and pursuing pathways for more sustainable, reliable delivery of feminist technology services to human rights activists. One path we are exploring is a strategic partnership with a justice-oriented software development agency. Another path is outlined by Julian

Oliver, who developed self-hosted infrastructure for the Extinction Rebellion climate movement. He advocates for open-source infrastructure in the service of global civil disobedience movements.[54] A final path is the call by seven participatory design scholars to "defund Big Tech and refund community."[55] They outline concrete actions for different actors, at different scales, to take in order to work toward the profound reallocation of Big Tech profits earned through surveillance capitalism. Here, I appreciate the focus on funding, because, as with most things, these are fundamentally questions of expertise, labor, and money.

For me, these questions of infrastructure are a fundamental challenge to designing and deploying alternative technologies with restorative/transformative values and visions, particularly when those visions are not profitable (and may even be anticapitalist in nature). And yet, I take inspiration from Bødker and Kyng's insistence that "high technological ambitions are necessary in order to influence our technological future."[56] The response to infrastructural challenges cannot be to always think small and focus on beautiful and ethical participatory workshops that don't yield products. It also cannot be to require perfection in uptime and services (which means we hand over tech to Big Tech and venture capital). I refuse those outcomes, and call on HCI and PD to *challenge power* by thinking together with grassroots social movements, and open-source software movements, about novel ways to infrastructure data activism and restorative/transformative data science.

CONCLUSION

In this case study I have described how the Data Against Feminicide team co-designed two specific tools to support and sustain the work of grassroots data activists. From this experience, I see many more opportunities to draw from HCI and PD to design for data activism and reformative/transformative data science. The descriptive model outlined in table 7.1 may help us structure design efforts across various stages of project activities. The two tools our team built fell into the researching and recording stages, but our co-design process generated ideas across other workflow stages. For example, in the resolving stage, there could be useful tools to collaboratively map out the data and information ecosystem and help evaluate the feasibility and labor required of any restorative/transformative data science effort. Activists we worked with generated many ideas for the refusing and using data stage, which related to novel tools for visualization, communication, and memorialization.

This chapter has also detailed some of the ethical decisions that our team made, guided by the principles of data feminism, which led me to make a case that resources and relationships should be considered some of the most important elements of design. Even with strong guidance, numerous tensions arose around our design interventions: how to model politically contested concepts, how to design for intersectionality, and how to infrastructure academic projects that aspire to go beyond parachuting prototypes into communities. I end on these tensions—and some emerging responses to them coming from HCI and PD—because I want us to be able to tell design stories that are not hero stories, that leave us with open questions, and that open a door to both critically and practically challenging power, across our technologies and our institutions and our larger societies.

8 A TOOLKIT FOR RESTORATIVE/ TRANSFORMATIVE DATA SCIENCE

This final chapter is a toolkit for anyone who may be contemplating starting or joining a restorative/transformative data science project. Perhaps you are a data activist or a journalist, or maybe you are an urban planner or a librarian, a digital humanities scholar or an academic, the head of a nonprofit or a member of a community group or a designer. Or somebody else entirely. My hope is that these pages will offer both reflection and practical guidance.

As I sat down to map the contours of this toolkit, I realized again and again that all of the ideas I thought I had were actually things that activists had told me or showed me or taught me. One of the most humbling and inspiring aspects of writing this whole book has been the continual realization of the depth and the breadth of what our partners have to teach the rest of us who aspire toward transformative social change. Feminicide data activists center care, memory, and justice without sacrificing rigor. They don't imagine data as a "solution" but rather see data as one tactic in a larger, networked movement of social and political actors. They stay close to the communities and families impacted by feminicide, in some cases providing direct healing and support. They are highly creative in acquiring information from a deeply biased and unjust information ecosystem. They develop ways to circulate feminicide data for diverse impacts, ranging from policy reform to narrative change to mass mobilizations.

This toolkit is a first step toward drawing out some of these lessons and speculating on how these may be useful to others using data in the service of restoration and transformation, healing and liberation. While much research produced in the Global North about the Global South ends up inaccessible behind paywalls, this toolkit, which is included in a book published in a free, open-access format, is an attempt to counteract that dynamic and contribute knowledge back to grassroots communities, activists, journalists, academics, and others.

The questions and activities in the toolkit are drawn from themes that surfaced in our work on feminicide, but they are purposely written to be broadly applicable. My hope is that these examples, questions, and activities will be helpful to anyone using data-driven methods for monitoring, auditing, and inquiry, and especially projects related to structural inequality. I am confident that at least some of the many lessons learned from counting feminicide will be helpful for other efforts working to restore rights and to transform systems. These might include projects that document forms of violence other than feminicide—for example, the already existing efforts to document police killings of Black Americans, LGBTQIA+ hate crimes, or the murders of Indigenous land defenders. Yet there are also counterdata efforts occurring that are related not only to physical violence but to other forms of economic violence or structural inequality. Recently there has been a remarkable growth, for example, in nonprofit and activist groups that count and monitor evictions in the United States. Here the rows of data represent cases of eviction rather than cases of feminicide, but the intent is similar: first, to use quantification to provide direct services and legal aid to impacted people. And second, to transform policies and systems that treat housing as a commodity rather than as a human right. Here there is great resonance with the feminicide data activists' examination of power and use of data to reframe and remake personal problems into structural patterns. To inspire connections to work in other domains, I have included a list of restorative/transformative data science examples in the toolkit.

That said, because the toolkit arose in the context of feminicide data activism, it is highly likely that not all activities in the toolkit apply to *all* restorative/transformative data science projects in *all* domains. I encourage people to use what is useful and, if you are moved to do so, contribute comments, questions, and critiques to the evolving open toolkit located at https://mitpressonpubpub.mitpress.mit.edu/pub/restorative-data-toolkit.

GLOSSARY

Coliberation: This is the idea that all of us are harmed by systems of unequal *power* and that we, working together, can free ourselves of its multiple burdens—material, psychic, spiritual, and intergenerational. There is a well-known quote from Aboriginal activists in Queensland, Australia, that best represents this idea: "If you have come here to help me you are wasting your time, but if you have come because your liberation is bound up with mine, then let us work together."[1]

Counterdata: Data that are produced by civil society groups or individuals in order to challenge unequal *power*. Producing counterdata is not (only) about countering *missing data* or inadequate *official data* from institutions but is also used to challenge state bias and inaction, to galvanize media and public attention, to reframe political debates, to work toward policy change, and to help heal wounded communities. Counterdata production, like *data activism* more broadly, is a citizenship practice. It is an informatic form of enacting democratic dissent, prompting protest and insisting on political engagement. But not all activist data is counterdata: Indigenous scholars and data activists emphasize that their work is about sovereignty and not about countering the settler state.

Data activism: The use of data and software to pursue collective action and exercise political agency. Producing *counterdata* and engaging in *restorative/transformative data science* are specific ways—but far from the only ways—of engaging in data activism.

Data epistemologies: Theories and approaches to knowing things about the world with data. Mainstream data epistemologies are heavily positivist—seeking to use data to find universal truths over a consideration of context. This results in *hegemonic data science*. Many scholars and activists have highlighted how mainstream data epistemologies replicate violent and extractive and colonial modes of knowledge generation. Emerging alternative data epistemologies include data feminism, feminist data refusal, emancipatory data science, environmental data justice, decolonial AI, Indigenous data sovereignty, and queer data (see the next section for a short guide).

Discordant data: An idea from Helena Suárez Val that describes the fact that *official data* and *counterdata* often deliberately do not coincide; they are *discordant* because they intentionally use different definitions, measurements, and classification strategies.[2]

Emotional labor: Producing and working with data related to social inequality almost always involves being witness to trauma and violence, whether the data are about feminicide or evictions or environmental harms. The psychic and emotional burdens of this witnessing work should not be overlooked, particularly for survivors who may have firsthand experience of such violence. Projects engaging in *restorative/transformative data science* can develop strategies for self- and team care and/or they may choose to "flip the script" and take an assets-based approach to producing data, where the focus is on mapping communities' strengths and joys, not their deficits and traumas.

Hegemonic data science: Mainstream data science that works to concentrate wealth and *power*; to accelerate *racial capitalism*, perpetuate *patriarchy*, and sustain *settler colonialism*; and to exacerbate environmental excesses and social inequality.

Information ecosystem: A dynamic constellation of actors that includes infrastructure, tools, technology, producers, consumers, curators, and sharers of information about a particular topic. The metaphor of the ecosystem is designed to capture the dynamic nature of information—it moves and flows across scales and sites and actors as it is produced, curated, transformed, and used.

Intersectionality: The idea from Black feminism that systems of *power* compound and combine and cannot be understood in isolation. For example, a single-axis analysis that looks only at *patriarchy* will miss the ways in which *patriarchy* intersects with *white supremacy*, with *settler colonialism*, with *ableism*, and so on. Intersectionality comes from theorizing the experiences of Black women in the United States, and substantial contributions to it have been made by the Combahee River Collective, Kimberlé Crenshaw, and Patricia Hill Collins, among many others.

Memory work: Practitioners of *restorative/transformative data science* understand that producing and circulating *counterdata* is a way to assemble and care for stories from the past in order to develop new visions for the present and future.

Missing data: Data that are neglected by institutions, despite political demands that such data *should be* collected and made available. Missing data may include data that are entirely absent but also data that are sparse, neglected, poorly collected and maintained, purposefully removed, difficult to access, infrequently updated, contested, and/or underreported.

Official data: Data that are produced by the state, international governing bodies, and/or other mainstream institutions such as large corporations or professional associations.

Power: The current configuration of structural privilege and structural oppression, in which some groups experience unearned advantages—because various systems have been designed by people like them and work for people like them—and other groups experience systematic and violent disadvantages—because those same systems were not designed by them or with people like them in mind.[3] Specific systems of privilege and oppression include but are not limited to the following:

- **Ableism:** The systemic privileging of ability that results in the oppression of disabled people based upon real or perceived impairments. It "others" disabilities, chronic illnesses, and neurological or mental illness.[4]
- **Cisheteropatriarchy:** (Synonymous with *patriarchy* in this book.) The social and political system that elevates cisgender, heterosexual men and oppresses those with

minoritized gender identities (women, trans, travesti, nonbinary, two-spirit people, and more) and minoritized sexual orientations (lesbian, gay, bisexual, and more).

- **Colonialism:** Refers to some combination of territorial, cultural, linguistic, political, and/or economic invasion and subsequent domination of one group of people by another group of people.[5]

- **Economic violence:** Economic policies and practices that systematically deprive groups of people of their human rights to life, food, clothing, housing, and medical care. This might be through exclusion from labor markets, underwaged or unwaged work, exclusion from education, privatization of public goods, and more. *Neoliberal* policies result in economic violence, and so do *colonialism*, *patriarchy*, and the other systems of *power* referenced here.

- **Extractivism:** Positions land, air, water—and increasingly also data and digital information—as free resources to be mined, wasted, privatized, and profited from.[6] Often results in *economic violence*.

- **Neoliberalism:** Refers to economic policy that favors free-market capitalism and tends to oppose government regulation that intervenes in markets. Approaches include privatization of public goods, deregulation, and restriction of public spending. Neoliberalism strengthens corporate power, extracts private profit from public resources, externalizes private costs (such as environmental harms) to the public sector, removes social protections for the most vulnerable, and undermines representative democracy by creating a vicious cycle of super rich oligarchs who manipulate the political machinery, doubling down on neoliberal policy to get super-richer. Neoliberalism produces *economic violence*.

- **Patriarchy:** See *cisheteropatriarchy*.

- **Racial capitalism:** The idea from Cedric Robinson that capitalism is founded upon racial stratification. In other words, capitalism and racism are inextricable: free markets are founded upon unfree labor that arises from slavery, colonization, *economic violence*, and *extractivism*. This system is historic and ongoing.

- **Settler colonialism:** A form of historic and ongoing colonization in which outsiders come to land/air/water/subterranean earth inhabited by Indigenous peoples, dispossess them of that land, and then insist—through institutions, laws, culture, and violence—that settlers are sovereign over the stolen land.[7]

- **White supremacy:** A historically based, institutionally perpetrated system of exploitation of continents, nations, and peoples of color by white peoples and nations of the European continent, for the purpose of maintaining and defending a system of wealth, power and privilege.[8]

Restorative/transformative data science: An approach to working with systematic information that seeks, first, to heal communities from the violence and trauma produced by structural inequality and, second, to envision and work toward a world in which such violence has been eliminated. *Restoration* involves the use of data for restoring life, living, and vitality to the individuals, families, communities, and larger publics harmed by unequal systems of *power*. It also seeks the restoration of rights—the right to live a life free from violence, for example, or the right to adequate housing, or the right to ancestral homelands. *Transformation* involves the use of data to dismantle and shift the structural conditions that produced the violence in the first place. It is both visionary and preventative.

Triangulation: Using multiple sources of information about the same event or phenomenon to cross-reference and verify details. This is often necessary when *official data* are suspect or sparse or when there is not a single authoritative source of information in the *information ecosystem*.

WHAT IS YOUR DATA EPISTEMOLOGY?

Practitioners of restorative/transformative data science mobilize alternate epistemologies of data that challenge the extractive and violent regimes of hegemonic data science. In the last decade, the number of alternate data epistemologies has multiplied. Each one offers at the very least some foundational texts and principles, or else some flagship projects that can serve as models to follow. Some also offer supportive communities, and others, like the global movement for Indigenous data sovereignty, have robust theoretical foundations, thriving communities, and policy and governance frameworks.

Many of these data epistemologies come from theorizing directly from the experiences and histories of specific groups and the data harms that they have experienced. For example, Data for Black Lives aims to use data "to create concrete and measurable change in the lives of Black people." Data epistemologies are not interchangeable ethical checklists: moving forward with a particular epistemology carries responsibilities for who your work serves, who you are committed to being in dialogue with, and who you are accountable to. If you seek to mobilize an epistemology that comes from theorizing experiences that you have never had (e.g., you are in a cisgender heterosexual team who wishes to draw from a queer data perspective) then you will need to think carefully about how you can do that without causing harm and without engaging in appropriation. In some cases, the best answer may be to step aside and make space for a group who does have that lived experience to lead the project and your team can get them coffee. That's a joke—but only a half-joke! The real issue is thinking carefully about how you can support and center leadership by people who have the knowledge and cultural grounding that comes from lived experience.

For example, one troubling thing I have witnessed recently is grants being given to settler people who propose to draw from Indigenous data sovereignty and work with Indigenous communities in a participatory way. These projects receive funding without having established partnerships with tribes or nations or Indigenous-led groups, so at no point in the project has an Indigenous person or community weighed in on the validity and utility of the idea. This sets up the structural conditions for harm to occur because once funding has been secured, the settler person is most accountable to the settler funder and has to retroactively find a community partner amenable to an idea created in a settler vacuum.

To seek an appropriate data epistemology for your project, you may want to reflect on your own positionality as well as the community that you want to serve and be accountable to with your data project. See the data epistemology activity in the Getting Started section ahead for a starting point for this kind of reflection. And ahead is a list of data epistemologies that showcases those approaches with which I am most familiar and have been participating in, reading about, or following online.

DATA EPISTEMOLOGIES

DATA FOR BLACK LIVES

A movement of activists, organizers, and mathematicians based mainly in the United States that aims to use data "to create concrete and measurable change in the lives of Black people." It is organized into regional hubs. See in particular https://d4bl.org/ and their 2021 report on *Data Capitalism and Algorithmic Racism*.

⇒ **Especially relevant for:** Projects by/with Black Americans, projects in solidarity with Black Lives Matter, projects seeking a regional community.

DECOLONIAL APPROACHES

❖ **Tierra Común**—A global community of scholars and activists working to decolonize data. See https://www.tierracomun.net/, the scholarship of Paola Ricaurte, and the book *The Costs of Connection: How Data Is Colonizing Human Life and Appropriating It for Capitalism.*

❖ **Ubuntu ethics for AI**—Scholars and technologists such as Sabelo Mhlambi and Serena Dokuaa Oduro have advanced the idea of incorporating Ubuntu philosophy into AI systems and AI policy. See Mhlambi's paper *From Rationality to Relationality: Ubuntu as an Ethical and Human Rights Framework for Artificial Intelligence Governance.*

❖ **AI Decolonial Manyfesto**—A collaborative statement by two dozen scholars across computer science, social sciences, humanities, and human rights, to challenge the "Western-centric biases" being baked into AI. See https://manyfesto.ai/.

⇒ **Especially relevant for:** Projects focused on colonialism, imperialism, and dispossession, projects from the Global South, projects seeking South-South community, and projects mobilizing non-Western ethics.

THE DESIGN JUSTICE NETWORK

A network of practitioners with an important set of principles grounded in intersectional feminism, which is focused on using design to support social justice. While the Design Justice Network is focused more on design, the principles have great relevance and applicability to undertaking data science projects in community. See https://designjustice.org/ and the book *Design Justice* by Sasha Costanza-Chock.

⇒ **Especially relevant for:** Projects coming out of design disciplines, projects drawing from intersectionality, projects using participatory methods, and projects seeking to connect with a robust community of practice.

EMANCIPATORY DATA SCIENCE

Developed by professor Thema Monroe-White, emancipatory data science draws from emancipation theory, critical race theory, and critical quantitative theory (see #QuantCrit ahead), to theorize how to use data science in the service of "uplift and empowerment." Monroe-White outlines three functions emancipatory data science can undertake: diagnosis and critique (of data harms), viable futures (related to more equitable data practices and policies), and transformation (of the data science community itself). See the paper "Emancipatory Data Science: A Liberatory Framework for Mitigating Data Harms and Fostering Social Transformation."

⇒ **Especially relevant for:** Abolitionist projects, projects drawing from critical race theory, and projects seeking to describe data harms for minoritized people.

ENVIRONMENTAL DATA JUSTICE (EDJ)

An emerging approach that joins environmental justice (elaborated in the 1990s by scholar Robert Bullard and principles adopted at the First People of Color Environmental Leadership Summit) with data justice and critical data studies. Environmental data justice develops ways of working with environmental data, using participatory methods, in the service of community self-determination. To learn more, see the EDJ working group of the Environmental Data Governance Initiative (https://envirodatagov.org/environmental-data-justice/), projects by the Technoscience Research

Unit (https://technoscienceunit.org), and the work of Lourdes Vera, Sara Wylie, and colleagues.

⇒ **Especially relevant for:** Projects using environmental data, projects drawing from environmental racism and environmental justice (EJ) principles, and people seeking to be in community with activist-scholars from a variety of disciplinary backgrounds.

FEMINIST APPROACHES

Publications, programs, and emerging networks that mobilize feminist theory and activism to work with data and AI have been proliferating in recent years.

❖ **Data feminism**—Discussed in the next section.

❖ **The Design Justice Network**—Discussed earlier in this list. Grounded in intersectional feminist principles.

❖ **Feminist Data Manifest-NO**—A collaborative statement from scholars across numerous disciplines that "refuses harmful data regimes and commits to new data futures" (https://www.manifestno.com).

❖ **Data Feminism Network**—A learning community focused on equitable and feminist approaches to data. They run reading groups and produce events (https://www.datafeminismnetwork.org).

❖ Spanish language networks include Feminismo de Datos (https://www.facebook.com/FeminismoDeDatos), La Red Mexicana de Feminismo de Datos (https://www.redmexicanadefeminismodedatos.org) and La Red Latinoamericana de Feminismo de Datos (in formation, see https://www.datagenero.org/).

❖ **<A+> Alliance**—A global, multidisciplinary, feminist coalition of academics, activists, and technologists working to use artificial intelligence and IT to accelerate gender equality (https://aplusalliance.org).

❖ **Data Feminism Program**—By Data-Pop Alliance. Draws from data feminism to undertake international development projects (https://datapopalliance.org/program_data_feminism).

⇒ **Especially relevant for:** Projects focused on cisheteropatriarchy, projects focused on gender and intersectionality, projects by/with women, queer, and/or trans people, projects using participatory methods, and projects that use qualitative and creative methods with data (art, storytelling, etc).

INDIGENOUS DATA SOVEREIGNTY

If you are looking to put data in service of Indigenous nations and people, the global Indigenous data sovereignty movement has produced excellent scholarship, practical guidelines, and policy:

- Foundational texts include *Indigenous Statistics* and the edited volume *Indigenous Data Sovereignty: Toward an Agenda.*

- Many Indigenous data sovereignty regional and global networks exist to link practitioners and scholars. For a list, see https://indigenousdatalab.org/networks/.

- The First Nations Information Governance Center developed the ownership, control, access, and possession (OCAP) principles. See https://fnigc.ca/ocap-training/.

- Indigenous data sovereignty scholars have added the collective benefit, authority to control, responsibility, and ethics (CARE) principles to the FAIR principles for open science.

- Indigenous data sovereignty research labs include the Data Warriors Lab and the Collaboratory for Indigenous Data Governance.

⇒ **Especially relevant for:** Projects by/with Indigenous people, projects focused on sovereignty, projects about dispossession and land rights, projects drawing from Indigenous worldviews, and projects that use oral history and storytelling methods.

#QUANTCRIT

Also known as quantitative criticalism and quantitative critical race theory, this is an approach developed in education, ethnic studies, and the social sciences that aspires to unite critical race theory with quantitative methods. Principles include the centrality of racism and oppression, the acknowledgment that numbers are not neutral, the idea that community voice is central, and the affirmation that statistical analysis can play a role in struggles for social justice. For a starting point, see the handy resource guide assembled by chemistry professor and educator Paulette Vincent-Ruz, available at https://sites.lsa.umich.edu/pvincentruz/quantcrit-resources/.

⇒ **Especially relevant for:** Projects from education, social work, and the social sciences, projects using quantitative and statistical methods, projects drawing from critical race theory, and projects seeking a robust academic conversation.

QUEER DATA

Recent publications aim to show how data may be used (or refused) to center the lives and well-being of queer, trans, nonbinary, and LGB+ people. See the essay "Counting the Countless" by Os Keyes on radical data science, the book *Queer Data* by Kevin Guyan, and the volume titled *Queer Data Studies*, edited by Patrick Keilty. And make sure to read Dean Spade and Rori Rohlfs's powerful critique of "gay numbers" and the ways in which counting the LGBTQ+ population can perpetuate white supremacy.

⇒ **Especially relevant for:** Projects by/with LGBTQ+ people, projects focused on gender and sexuality, projects engaging theories and practices of refusal, and projects engaging queer theory.

PRINCIPLES OF DATA FEMINISM

Data feminism is one epistemology of data, and the one that comprises the conceptual backbone of this book. In *Data Feminism*, Lauren Klein and I outlined what a feminist approach to data science might look like—an alternate data epistemology to challenge the standard operating procedures of hegemonic data science. We draw from intersectional feminist theory, activism, and writing to outline seven principles for working with data in a feminist way. I offer these principles here in the hopes that they will be a useful part of your toolkit, just as they have guided my own research and writing throughout this book. But I offer them with the acknowledgment that these are far from the only principles that you could use in your own data-driven work. The frameworks emerging from other data epistemologies might be more relevant and useful, depending on who you are, where you are located (geographically, socially, and spiritually), and what your project is about.

1. **Examine power.** Data feminism begins by analyzing how power operates in the world.
2. **Challenge power.** Data feminism commits to challenging unequal power structures and working toward justice.
3. **Elevate emotion and embodiment.** Data feminism teaches us to value multiple forms of knowledge, including the knowledge that comes from people as living, feeling bodies in the world.
4. **Rethink binaries and hierarchies.** Data feminism requires us to challenge the gender binary, along with other systems of counting and classification that perpetuate oppression.
5. **Embrace pluralism.** Data feminism insists that the most complete knowledge comes from synthesizing multiple perspectives, with priority given to local, Indigenous, and experiential ways of knowing.
6. **Consider context.** Data feminism asserts that data are not neutral or objective. They are the products of unequal social relations, and this context is essential for conducting accurate, ethical analysis.
7. **Make labor visible.** The work of data science, like all work in the world, is the work of many hands. Data feminism makes this labor visible so that it can be recognized and valued.

RESTORATIVE/TRANSFORMATIVE DATA SCIENCE EXAMPLES

What does a restorative/transformative data science project look like? Across my research into feminicide data activism practices and my participation in various communities of practice, I have made some basic and preliminary observations about characteristics that many restorative/transformative data science projects share.

A reformative/transformative data science project

- has a theory of power and a theory of change;
- intentionally uses a data epistemology that focuses on liberatory goals (e.g., healing, liberation, emancipation, sovereignty, refusal, self-determination);
- can be done with minimal computing resources and basic data literacy skills;
- can be done with small, medium, or big data;
- can be done *without* an advanced degree from a fancy institution;
- often (but not always) produces or uses counterdata;
- engages in ethical, long-term relations of care with the communities most impacted;
- engages in pluralistic and culturally appropriate conceptions of rigor and truth; and
- holds the responsibility of caring for the data and the people and the stories and the relations assembled in the database.

All the grassroots feminicide monitoring projects I have discussed in this book are examples of restorative/transformative data science (indeed, they are the motivation for theorizing the concept in the first place). Examples from other domains abound. Ahead are projects across domains ranging from housing to health to the environment to civic engagement. Each one does not necessarily share all of the characteristics listed previously, but each has something to offer in terms of epistemology, thematic focus, community engagement, research methods, or outputs. I offer these as models to serve as inspiration for your own projects.

HOUSING
Anti-Eviction Mapping Project

https://antievictionmap.com/

This activist-academic project defines evictions more broadly than the legal definitions and uses their discordant data to advocate for a more structural framing of the root causes of eviction in specific areas. They acquire official eviction data from court records and other eviction data from surveys and collaborations with housing clinics. They have produced maps of evictions, landlord monitoring tools, murals, oral histories, and a book called *Counterpoints: A San Francisco Bay Area Atlas of Displacement & Resistance.*

"The Secret Bias Hidden in Mortgage-Approval Algorithms"

https://themarkup.org/denied/2021/08/25/the-secret-bias-hidden-in-mortgage-approval-algorithms

The result of a year-long investigation by the Markup, this data journalism story demonstrates systemic racial bias in home mortgage loan approvals in the United States. The journalists used publicly available official data triangulated with academic research studies, but there are still key missing data hindering a comprehensive analysis—notably credit scores—which the mortgage industry has successfully lobbied to keep secret.

LAND USE

The Detroit Geographic Expedition and Institute

https://medium.com/nightingale/gwendolyn-warren-and-the-detroit-geographic-expedition-and-institute-df9ee10e6ad2

The Detroit Geographic Expedition and Institute (DGEI) was a collaboration between Black young adults in Detroit led by Gwendolyn Warren and white academic geographers that lasted from 1968 to 1971. The group worked together to produce data about aspects of the urban environment related to children and education, and produced numerous maps from that work, including the widely circulated map *Where Commuters Run Over Black Children on the Pointes-Downtown Track.* The DGEI collected many other types of counterdata and published them in reports with analysis and recommendations.

Waorani territorial mapping / Mapeo Territorial Waorani

https://waoresist.amazonfrontlines.org/explore/

The Waorani people's lands are in the upper part of the Amazon River in Ecuador. Threatened by oil companies, the Waorani began a process of mapping their knowledge of and relation to the land in order to "defend our way of life and protect our future from threats like oil exploitation, mining impacts and invasions."[9] They use a participatory process that involves training different groups to use GPS devices, produces large paper maps that can be discussed and annotated in community, and finally leads to publication of the resulting maps on the open-source platform Mapeo. Some maps are public, others are not, and "everything included on the map is the cultural property of the Waorani."

"Land Grab Universities"

https://www.landgrabu.org/

High Country News produced an original report and unique database documenting how land grant universities across the United States were funded with expropriated Indigenous land via the 1862 Morrill Act. Data sources included land patent records, congressional documents, historical bulletins, historical maps, and more. Many data needed manual entry, and the database took months to assemble. It is publicly available for download and research.

HEALTH

Mapping Police Violence

https://mappingpoliceviolence.org/

Run by advocacy organization Campaign Zero, this project has tracked and mapped fatal police violence—and its systemic racial injustice—in the United States since 2013. Similar to feminicide data activists, the project relies on media reports as a primary source and triangulates those with official data and other counterdata sources. The database is open and publicly available.

"Lost Mothers"

https://www.propublica.org/article/lost-mothers-maternal-health-died-childbirth-pregnancy

Data on maternal mortality in the United States have been characterized as "an unreliable mess" by *Scientific American*.[10] In 2016, ProPublica set out to identify every single mother or parent who died from pregnancy-related causes in the United States (estimated to be between seven hundred and nine hundred people). They used social media, crowdfunding sites where funds had been set up for families left behind, public records, and obituaries. The journalists discuss how their crowdsourcing approach ended up overrepresenting the stories of dominant groups—the white, educated women who were more likely to respond to their call for stories—and underrepresented Black women in particular. Adriana Gallardo wrote about these methods in "How We Collected Nearly 5,000 Stories of Maternal Harm." This is an important reminder of the limitations of counterdata tactics and the ways in which they may reproduce the matrix of domination.

COVID Black

https://covidblack.org/

An organization founded during the COVID-19 pandemic out of a national campaign for people to call and demand that US state and federal agencies collect and publish racial data. Founded by historian and digital humanities scholar Kim Gallon, COVID Black not only gathers and publishes Black health data, but also does trainings and produces data stories, visualizations, and other public interpretations to combat racial health inequities (and also to push back against the relentless, racialized deficit narratives depicted by the mainstream media). They state on their website, "Data is more than facts and statistics. Black health data represents life."

The Qanuippitaa? National Inuit Health Survey

https://nationalinuithealthsurvey.ca

Qanuippitaa? National Inuit Health Survey (QNIHS) is an ongoing longitudinal survey of the health and well-being of the Inuit—the Indigenous peoples of the Arctic. It is Inuit-owned and Inuit-determined, and works in partnership with four major Inuit land claims organizations to ensure that the survey, data collection, data analysis, and research outputs are owned by Inuit people and informed by Inuit knowledge, values, and worldview.

ENVIRONMENT

The Global Atlas of Environmental Justice

https://ejatlas.org/

Initiated in 2012 by a team of researchers at the Universitat Autónoma de Barcelona, in Spain, the Environmental Justice Atlas undertakes systematic collection of global ecological conflicts in partnership with activists, civil society organizations, and social movements. They source conflicts from media reports, crowdsourcing, and local partnerships with impacted communities. The project has developed its own typology of ecological conflict and publishes its data openly.

Environmental Data & Governance Initiative

https://envirodatagov.org/

Environmental Data & Governance Initiative (EDGI) is a research collaborative that was formed in 2016 in the United States as the Trump administration threatened to take open federal environmental datasets offline (a notable example of intentionally producing missing data). The group started organizing "data rescues" where they downloaded and archived federal datasets on university and civil society servers. Now EDGI monitors US government action (and inaction) related to environmental information, organizes campaigns, and educates communities who want to use data to lobby the government. They have also led the development of environmental data justice principles (see the data epistemology section).

Land and Environmental Defenders Campaign

https://www.globalwitness.org/en/campaigns/environmental-activists/

An advocacy campaign, open dataset, and series of reports undertaken by Global Witness to record the unjust deaths of people—largely Indigenous people—killed while defending their land and environments. Similar to feminicide data activists, Global Witness sources cases from social media, news reports, and trusted local partners and networks. They have a strict information verification methodology to determine if a case should be included in their database. They outline that their numbers are "only a partial picture" because of the difficulty of obtaining comprehensive information about these cases.

TRANSPORTATION

Assaults on ride-hail drivers

https://themarkup.org/newsletter/hello-world/tracking-tens-of-thousands-of-assaults
  -on-ride-hail-drivers

Since 2021, the Markup has been monitoring assaults and carjackings on Uber and Lyft drivers, some of which end in fatalities. Reporter Dara Kerr sourced hundreds of these assaults from phone calls, police reports, public information requests, and local news articles. The data are published in an open, searchable database. Affected individuals and labor organizations are using these numbers to call for more gig worker safety protocols and protections.

The DTP Map / Карта ДТП

https://dtp-stat.ru

An activist map and ongoing monitoring effort in Russia that combines official government data on traffic crashes with weather, streets, participant types, and more, as a way of instigating civic engagement and social change to reduce accidents. Here the activists are not producing their own counterdata but rather reframing official data as a way of building a public sense of urgency around traffic fatalities. Yet the activists have mixed feelings about the veracity of the official data, as described in a case study by Dmitry Muravyov, "Doubt To Be Certain: Epistemological Ambiguity of Data in the Case of Grassroots Mapping of Traffic Accidents in Russia."

ARTS AND CULTURE

National Monument Audit

https://monumentlab.com/audit

A study by the arts-based nonprofit Monument Lab found that the monument landscape in the United States is overwhelmingly white and male and elevates themes of war and conquest. There is no authority in the United States that keeps records on monuments, so to undertake the national audit, they assembled records from dozens of federal, state, local, tribal, and institutional sources, many of which have different criteria and definitions of "monument." They held participatory analysis sessions to develop their analytical categories and themes. The data can be explored in their interface on the project's website.

Whose Heritage? Public Symbols of the Confederacy

https://www.splcenter.org/whose-heritage

Since 2015, the Southern Poverty Law Center (SPLC) has maintained a database and map of Confederate-related monuments and place names. They periodically reaudit the list to monitor removals, relocations and renamings. For example, following George Floyd's murder, the SPLC found that almost one hundred Confederate symbols were removed, relocated, or renamed. They publish reports as well as a community action guide that allows users of the map to take direct action by providing instructions on how to build a campaign to remove monuments and/or rename streets.

Book Censorship Database

https://www.everylibraryinstitute.org/book_censorship_database_magnusson

A project by Dr. Tasslyn Magnusson in partnership with EveryLibrary Institute and EveryLibrary to monitor book bans and book challenges across the United States since 2021. The open spreadsheet is organized by school districts, books challenged/banned in school districts, public libraries, and books banned/challenged in school libraries. Data are sourced from news reports, online forums, and social media.

Abortion Onscreen Database

https://www.ansirh.org/research/abortion/pop-culture

The Advancing New Standards in Reproductive Health (ANSIRH) organization compiles and publishes an open database of all film and television depictions available to viewers in the United States that discuss abortion, from 2016 to the present. ANSIRH publishes annual reports on media portrayals of abortion, and the database has been used in a range of media studies.

CIVIC ENGAGEMENT

Feminindex

https://winguweb.org/en/caso-impacto/feminindex/

A civic media project by Ecofeminita and Wingu that documented and visualized where political candidates in Argentina stood on gender and LGBTQ+ issues, including reproductive rights, femicide, care work, and trans rights. The first version was released in 2017, with subsequent versions in 2019 and 2021. Data on politicians' views were collected through media reports, candidates' public statements, and policy documents and through surveys administered by the organization.

First Nations Information Governance Centre

https://fnigc.ca

First Nations Information Governance Centre (FNIGC) is a nonprofit organization leading the establishment of Indigenous data sovereignty for all members of the Assembly of First Nations in Canada. They undertake a variety of ongoing First Nations population surveys, run trainings and capacity-building sessions for tribal partners, and develop data governance strategy aligned with the goals of sovereignty and self-determination.

HOW TO USE THIS TOOLKIT

RESOLVING	**RESEARCHING**	**RECORDING**	**REFUSING + USING**
Developing a theory of change	Finding + verifying information	Information extraction + classification	Where data go, who uses them

Although there are many ways the toolkit could be used, here are three possibilities:

1. **Strategic project planning and visioning**

    Use the toolkit at the beginning of a project to undertake planning sessions, establish a shared vision for the project, who it is serving, and how to sustain it through the different stages of work. In this model, the project team would take the time to go through the questions and activities in this restorative/transformative data science toolkit, as individuals and as a group, align around their answers at each stage, and incorporate them into their plan and their vision (and their budget!).

2. **Equity pauses and recalibrations**

I learned about the idea of *equity pauses* from Jenn Roberts, who runs VersedEd and the Colored Girls Liberation Lab. This is the idea of regularly stepping back from the intense day-to-day work, say, of researching and recording counterdata, and pausing to evaluate your process and whether you are meeting your equity goals. In this model, teams would take a short period of time to engage with the questions at one stage of work, discuss their answers, and surface shifts and recalibrations to make in data practices to better meet their goals. These equity pauses could happen at regular, scheduled intervals—for example, at one meeting a month.

3. **Ethics crisis moments**

There may be moments in a restorative/transformative data science project that provoke an equity pause that the team did not foresee. A community may come forward and express that they have been harmed. An individual's information may have been made public in a traumatizing way. You or your team may have included a story or a case or some information without permission. These are moments where a more profound recalibration—of data practices and of relationships—becomes necessary. This toolkit could aid in that recalibration by providing a structured set of ethical questions and activities for the team to use to draw out their analysis of what happened, how to redress it in the short term, and how to prevent such harm in the longer term.

GETTING STARTED

To get started, I suggest teams first do *Start-up Activity 1: What is your data epistemology?* to reflect on who they are, who they serve, and who they are accountable to with their data work. This will help you understand which emerging alternative data epistemologies may match with your team's backgrounds, relations, goals and values. It will also help you begin to reflect on and refine your theory of change for your work. *Start-up Activity 2: Map the information ecosystem* guides your group to create a map of the information ecosystem for your topic of interest. Understanding the information ecosystem will be invaluable as you begin to think about how your group can mitigate biases, address missing data, build coalitions, and use data and information for healing and liberation. The information ecosystem map that you make during this activity is referenced in a number of later activities, so it is handy to keep around as a guide for your project.

START-UP ACTIVITY 1: WHAT IS YOUR DATA EPISTEMOLOGY?

Time required: 60–75 minutes

Materials: Pens and paper/sticky notes

Preactivity homework: Review the list of data epistemologies provided in this toolkit (or other data epistemologies you may be considering).

Activity: Choosing an appropriate data epistemology involves locating yourself and your team and your organization/institution in relation to the topic. These questions can serve as a starting point and are designed to be answered as a group. For each question, take five minutes to quietly freewrite answers, and then ten to fifteen minutes to share responses with the group.

1. Who are you (individually and as a team and as an organization) in relation to the topic? Do you bring lived experience of the topic? How and why were you brought to the topic?

2. Who are you producing data for? Which communities or publics do you serve, or aspire to serve, by doing this work?

3. Who are you accountable to? Which communities or publics should have a direct say in influencing the course of the project?

Final discussion (15 minutes): Review your responses in relation to the list of data epistemologies in this toolkit. Which one or ones are best aligned with your team's responses?

START-UP ACTIVITY 2: MAP THE INFORMATION ECOSYSTEM

Time required: 2½–3 hours

Participants: 5+. Can be done with fewer than five people but it will take more time. If you are able, try to recruit participants with lived experience, legal experience, data experience, and movement experience.

Materials: Computers with Internet access, multicolored sticky notes, six large posterboards or papers

Preactivity homework: Organizers and team members should determine the geographic scale of interest and the time period of interest for the project and write it into a concise mission statement in this form: "We are mapping the information ecosystem for TOPIC in PLACE from START TIME to END TIME."

Activity setup:

1. Write out your mission statement (see preactivity homework) on a blackboard, whiteboard, or big paper posted for everyone to see.

2. Choose one color of sticky notes that will represent "missing data" and one color that will represent "bias" and communicate those to the participants.

Activity:

1. 15 min.: Introductions—participants go around the room and say their name, pronouns, and one personal or professional reason they are in the room today.

2. 10 minutes: Organizers read the mission statement and outline the purpose for gathering today. Divide participants into groups and give each group a large surface to work on (posterboard or wall) and sticky notes.

3. 60 minutes: Group work.

   • **Group 1—Legal inventory:** Place a sticky note on your surface for each law relevant to your topic. Note the year it was passed and whether it is a municipal/state/federal/tribal law on the sticky note. Include laws that relate to (1) definitions of the phenomenon, (2) government or official monitoring of the phenomenon, and (3) public disclosure of information about it. As you examine each law, place a "bias" sticky note if you see a source of bias.

   • **Group 2—Official data producers inventory:** Place a sticky note on your surface for each agency or group that produces official information about the topic of interest. Around it, place another sticky note for each relevant dataset that the agency produces. Note on the sticky note whether that dataset is open or closed, aggregated or disaggregated. If the dataset is open, download it and examine it. As you examine each agency and dataset, place a "bias" sticky note if you see a source of bias and a "missing data" sticky note if you see missing datasets, rows, features, or variables.

   • **Group 3—Counterdata producers inventory:** Place a sticky note on your surface for each agency or group that produces counterdata or activist data about the topic of interest. Note on the sticky note what sector they are from—e.g., activism, journalism, nonprofit, academia, government, etc. Around it, place another sticky note for each dataset that that group produces. Note on the sticky note whether that dataset is open or closed, aggregated or disaggregated. If the dataset is open, download it and examine it. As you examine each group and dataset, place a "bias" sticky note if you see a source of bias and a "missing data" sticky note if you see missing datasets, rows, features, or variables.

   • **Group 4—Data users inventory:** Place a sticky note on your surface for each agency or group that *uses* data (official data or counterdata) about the topic of interest. Note on the sticky note what sector they are from—e.g., government, activism, journalism, nonprofit, academia, etc. As you examine each data user, place a "bias" sticky note if you see a source of bias.

   • **Group 5—Larger landscape inventory:** Place a sticky note on your surface for other actors in this information ecosystem that are influencing policy, advocacy, and public conversation. They may not produce or use data, but they are individuals, organizations, government agencies, and/or social movements that are doing agenda-setting on the topic. As you examine each individual or group, place a "bias" sticky note if you see a source of bias.

Stretch break! (5 minutes)

Discussion & reflection (60 minutes):

- 30 minutes: Each group shares back their results (5 min. per group)
- 20 minutes: Large group power analysis. Facilitators move participants through the following questions:
  - What data remain missing from this mapping? What data should exist (according to your group or to other advocacy groups) but do not exist in either official or counterdata efforts? Why?
  - What are the biases in this information ecosystem? What are their root causes? Can the biases and inequalities be mitigated informatically?
- 10 minutes: Close out and designate participants who can help document the work.

Post-activity documentation: Consider creating a large visual map synthesizing groups' work that can be posted in your team's space (digital or physical space).

Once you have done the two start-up activities, your team is ready to pick and choose from the rest of this toolkit to see which of the activities and discussions might be relevant for your project. Throughout this book, I have described the different workflow stages of a restorative/transformative data science project: resolving, researching, recording, and refusing and using data. These stages are derived from our team's interviews with grassroots data activists working to challenge feminicide, predominantly from Latin America. These stages form a four-stage process model, and I offer that model here in the hopes that it may be useful to other practitioners working on restorative/transformative data science projects.

Different ethical concerns and questions arise at each stage of work in a restorative/transformative data science project. For example, during the resolving stage of a project, data practitioners are developing their analysis of the problem, their theory of change for how and why counting and data analysis might be useful, and their data epistemology. Here it is important to think about your own positionality in relation to the phenomenon, who the beneficiaries of a project may be, and how to work collectively, in networks. In contrast, during the research stage of a project, it is important to reflect on the systemic biases in the information ecosystem, creative ways to source information, and how your project will handle missing data about minorities and subgroups. The rest of the toolkit is structured around activities to do and questions to ask during these four stages of a restorative/transformative data science project.

As you will see ahead, each discussion or activity is mapped to the data feminism principle that it aligns with. This is my attempt to demonstrate how one's data epistemology can translate into concrete matters of discussion and action for people working with data. Following the publication of *Data Feminism*, many people have asked Lauren and me for practical guidance on how to use the data feminism principles—to move them from general guidelines into something applicable in specific contexts. This mapping is an attempt to do that, as well as a way of inviting scholars and activists to do this kind of mapping work for other data epistemologies.

RESOLVING	RESEARCHING	RECORDING	REFUSING + USING
Developing a theory of change	Finding + verifying information	Information extraction + classification	Where data go, who uses them

Resolving is the stage of a restorative/transformative data science project in which an individual or group seeks to address a problem of structural inequality and determines how and why counting and registering data will be an effective method to do so.

Examine power	Look at the information ecosystem map you generated in Start-up Activity 2. How can you *join* rather than *initiate* your own effort? [ACTIVITY]
	Whose job is it to count? Whose job are you doing? How will you remind them that it is their job? [DISCUSSION]
	Are you from the community you are counting? Are you working within institutional structures that incentivize knowledge extraction (academia, journalism, nonprofit) or profit (industry)? Are you the right person or group to be doing this work? Try to have the courage to *not* do the project. [DISCUSSION]
	Which ideas—from theorists, activists, and/or communities—will you draw from for your structural analysis of power? How does your counting work help to name and frame the phenomenon as a structural problem? [DISCUSSION]
Challenge power	Who are you counting for? [DISCUSSION]
	What is your theory of change? How and why do you think measuring and monitoring will challenge power? [DISCUSSION]
	What is your data epistemology? See Start-up Activity 1 in this toolkit if you don't know yet. Once you do know, create a collaborative document with resources and guidance about your chosen data epistemology to help get new team members up to speed with its foundational ideas and methods. [ACTIVITY]

What are ways that counting may harm the people and communities you want to serve (e.g., by making them visible to institutions that want to target them)? [DISCUSSION]

Is this a one-time study or an ongoing observatory? How does that match your available resources (people and money)? How does that match who you want to serve or influence? [DISCUSSION]

Consider context	How can your project center the lived experience of those who have been impacted by the issue? Without exploitation, extraction, or tokenism? [DISCUSSION]
	What is specific to the geography or community that you are counting? What differences do you need to highlight? Do you need to develop new names, frames, concepts, and/or categories for that context? [DISCUSSION]
Embrace Pluralism	How can you avoid hoarding, whether data or credit? How can you count in community—leveraging collectives and networks of solidarity? How can you build partnerships and work in networks instead of trying to do everything on your own? [DISCUSSION]
Elevate emotion and embodiment	What kind of emotional labor is involved in this work? How will you care for yourself and your team as you measure injustice? [DISCUSSION]

RESOLVING	**RESEARCHING**	RECORDING	REFUSING + USING
Developing a theory of change	Finding + verifying information	Information extraction + classification	Where data go, who uses them

Researching is the stage of a restorative/transformative data science project in which an individual or group seeks and finds data observations and related information to add to their database. This can include sourcing existing datasets, discovery and detection of new observations, triangulation of information across sources, and ongoing research to add information to existing observations.

Examine power	Discuss the following questions with your team before beginning research: • What harm does collecting counterdata potentially incur for the groups that you want to influence or serve? • Whose trauma does it make visible and what is your (individual/ team's/institution's) relationship to that trauma? • Could bad actors use your data to target minoritized groups? • Do the data perpetuate a deficit narrative about minoritized groups—painting them as in need of saving by dominant groups? [DISCUSSION]

	Return to the biases surfaced on your information ecosystem map (Start-up Activity 2). What are their root causes? How can your work name and challenge those biases? [DISCUSSION]
Consider context	Return to missing data surfaced on your information ecosystem map. Discuss with your team members the following question: "why don't these data exist?" [DISCUSSION]
	Augment your information ecosystem map by placing a sticky note to denote creative ways that you can navigate, mitigate, and triangulate missing data. Don't forget to consider mass media, hyperlocal media, social media, private chat groups, relationships, partnerships, friendships, and crowdsourcing. [ACTIVITY]
	What groups, especially those at the intersection of multiple forms of domination, will still be missing, underreported, erased, or neglected by your methods of counterdata research? How can you address those limitations or, at the very least, acknowledge them? [DISCUSSION]
Embrace pluralism	How can you cultivate human networks of counterdata research predicated on ethical, authentic, nonextractive, and enduring relations? These might be relations with individuals, social movements, coalitions, journalists, nonprofits, or service organizations from your information ecosystem map. [DISCUSSION]
Elevate emotion and embodiment	How emotionally challenging is the research? Consider what it might be like for survivors or people with first-hand experience. How will you handle self-care and team care for secondary trauma? What does a trauma-informed approach to the production of this data look like? [DISCUSSION]
Make labor visible	How can you make the labor of researching counterdata visible and for whom? Are there strategic reasons for hiding the labor of your counterdata research? Are there ways to acknowledge labor and care internally, even when it may be strategic to conceal them externally? [DISCUSSION]

RESOLVING	RESEARCHING	**RECORDING**	REFUSING + USING
Developing a theory of change	Finding + verifying information	**Information extraction + classification**	Where data go, who uses them

Recording is the stage of a restorative/transformative data science project that involves extracting unstructured data from various sources into structured datasets (text documents, spreadsheets, and/or databases); classifying cases according to diverse typologies; and managing data—including ethics, access, and governance of the database.

Examine power	What does consent look like for the type of data you are collecting? How will you obtain consent and also support individuals and groups who decide to withdraw their consent? [DISCUSSION]
	Take a look at the legal landscape inventory from your information ecosystem map in Start-up Activity 2. How are legal protections falling short? Do laws adequately capture and classify the scope of the phenomena? [DISCUSSION]
Challenge power	How can you count and classify in order to exceed and/or challenge those standards? How might you demand that the phenomenon be conceptualized and measured differently? [DISCUSSION]
	Look at your information ecosystem map. What important variables and categories are missing from existing data? How can you incorporate those into your recording work? [DISCUSSION]
Embrace pluralism	How can your counterdata project operate as a *megaphone* for amplifying the voices, power, knowledge and agency of the people closest to the harms that you are trying to challenge? [DISCUSSION]
	How will you engage multiple and diverse stakeholders in the development of your data variables and categories? How will you participate in building community—the essential, ongoing social and technical infrastructure that can sustain this work? [DISCUSSION]
	Who else is recording counterdata about the issue? How can you scale your impact by *harmonizing* with them—that is, sharing definitions, categories, dialogue, and recording tips, and even potentially pooling data for greater impact? Look at your information ecosystem map for groups to start with. [DISCUSSION]
Rethink binaries and hierarchies	Whose experiences are sidelined, erased, or marginalized by your schema and categories? Whose experiences will be sidelined because there will be quantitatively fewer of them in the dataset and/or because of known biases in the data sources? How do you bring these experiences back in? [DISCUSSION]
	If necessary to the project, how can you collect identity categories such as race, gender, and ethnicity without naturalizing and essentializing them? Which categories might you avoid collecting because to collect them would be to do harm (e.g., trying to record someone's race or gender from a photo of them)? [DISCUSSION]
Elevate emotion and embodiment	How will you care for and respect your data? How will you develop your team's intimate knowledge of and relationship with the data? [DISCUSSION]
	How do your columns and categories communicate certain narratives about the issue and the people involved? (For example, is a woman always named as a "victim," defining her life and her agency by a single event?) How can you push back on that essentializing tendency? [DISCUSSION]

	How is your database a memorial to structural trauma—a *cultural countermemory*? Whose lives and whose pain is represented therein? How are you accountable to them and how are you in relationship with them? [DISCUSSION]
	If your database is a memorial, how does this shift your thinking about ethics and access to the database? [DISCUSSION]
Make labor visible	What is the minimal computing infrastructure you need for your team to do the work easily, safely, and reliably? How can you balance minimal computing and easy-to-use tools with data security and redundancy? [DISCUSSION]

RESOLVING Developing a theory of change	RESEARCHING Finding + verifying information	RECORDING Information extraction + classification	**REFUSING + USING** **Where data go, who uses them**

Refusing and using data is the stage of a restorative/transformative data science project in which individuals and groups circulate data in order to push specific actors toward thinking, feeling, or acting differently. The goals of these data actions and circulations may include to repair, to remember, to reframe, to reform, and/or to revolt.

Challenge power	Freewrite or free-draw about *refusal* and the issue area that you are working on. What are you refusing? Who is refusing? What is the affirmative, generative vision forged from your refusal?
	(For example, for feminicide data activists, the affirmative vision is a world that has erradicated gender-related violence and its causal forces of oppression: cisheteropatriarchy, settler colonialism, white supremacy, racial capitalism, and more.) [ACTIVITY]
	Data may refuse the status quo in a variety of ways. Here are some questions to start thinking about the ways your project may use data for refusal:
	• **Repair:** How can data support the provision of direct services, support, and healing to the people most impacted by the issue?
	• **Remember:** How can data support the holding of collective space around a structural problem? How can data reshape collective memory and communal interpretation of an issue?
	• **Reframe:** How can data's authority be leveraged to amplify grassroots voices and analysis in media and culture? To use narrative change to do mass public consciousness-raising?

	• **Reform:** How can data be leveraged to design new or reform existing institutions, laws, policies, and official practices? How can data's authority be leveraged to get a seat at the table for impacted communities? • **Revolt:** How can data be used in support of massive mobilizations in public spaces or online? How can data become spectacular and evocative, physically or digitally, to claim space, to claim time, to demand public attention, and to enact dissent? [DISCUSSION]
Embrace pluralism	Review your information ecosystem map and make a list of the different groups or audiences that you want to move to action and brainstorm multiple forms of data communication tailored for each audience. (For example, if you want to move policy makers to action, one form might be a report with data visualizations. Another form might be oral testimony from impacted communities. Another form might be a visual slideshow with photos, quotes, and statistics. Another form might be a protest outside their offices.) Each form is an opportunity to involve different groups in the communication and the circulation of data. [ACTIVITY]
Elevate emotion and embodiment	When is it politically advantageous to communicate data neutrally and minimally, as if from an omniscient observer? When is it more appropriate to center emotion and embodiment in data communication? [DISCUSSION] What emotional impact will your data artifacts have when circulated publicly? Is the impact different for different groups (say, survivors or impacted communities)? How will you care for the impact on those most affected and give them avenues for healing and action? [DISCUSSION]
Consider context	How can you recontextualize your data points and recuperate them from their abstraction into rows and columns? How can acts of data communication and circulation connect each data point back into the fullness of the lifeworlds from which it emerged? [DISCUSSION]

A LIVING TOOLKIT

This is a first step toward a toolkit for restorative/transformative data science. I welcome feedback and dialogue. Given that many of the examples I provide are from the United States, I would especially love to learn about restorative and transformative data science projects outside of the US context. Please post all comments, questions, and critiques online to the evolving open toolkit located at https://mitpressonpubpub .mitpress.mit.edu/pub/restorative-data-toolkit.

CONCLUSION: PUTTING DATA SCIENCE IN ITS PLACE

As I have described in this book, feminicide data activism, advocacy, and journalism is widespread and growing in the Americas and globally. Its practitioners mobilize alternative data epistemologies and approaches to data science that disrupt many of the hegemonic notions of how data science is done and who it is for. Grassroots data activists center care, memory, healing, and justice. They work collectively, in concert with social movements, networks, and coalitions. Their data circulate in diverse ways and produce a variety of social and political impacts, from supporting families to raising consciousness to reforming policy to mobilizing publics. Generalizing from common practices across activist workflows, and in dialogue with activists themselves about how to name and frame this work, I have described this approach as the practice of a *restorative/transformative data science*. This is a data science that aspires to restore life, living, vitality, rights, and dignity to those families and communities and publics from whom those things were taken. And it is a data science that aims for structural change—shifting the balance of power in the world so that such violence and inequality is no longer possible.

As I have been writing and revising this book, I found myself coming back to specific conversations with various members of the Academic-Community Advisory Board for the book. There are two points in particular that have stayed with me from our dialogues.

First, we have discussed the idea of necropolitics multiple times in this book: this is Achille Mbembe's idea of the unequal distribution of death, the conversations around whose deaths matter, and the contested interpretation of what the widespread violent deaths of women mean. And with a title like *Counting Feminicide*, it would be easy to think that you have just finished reading a book about death. And yet, during the review process, as I was in dialogue with Geraldina Guerra Garcés from the

Alianza Feminista para el Mapeo de los Femi(ni)cidios en Ecuador, she encouraged me to emphasize that this work is not about counting the dead. Rather, it is about defending life itself. It is about restoring and reclaiming and preserving life, living, and vitality in the face of colonial and patriarchal violence. This resonates with what activists told us over and over again. As Laura Hernández Pérez from Coordinadora Nacional de Mujeres Indígenas (CONAMI) remarked, "Our aspiration is to eradicate the violence. It's not about prevention nor about handling it. Our dream is to eradicate all of the violences." This work is about defending the right to life in order to realize a generative vision of a new world order: the remarkably nonradical idea that all women can and should live a life free from violence.

How do we then situate the role of counting and data science in this defense of life? How do we critically (but generatively) reflect on the role that information, artificial intelligence (AI), and machine learning have to play in restoring vitality and transforming the structural conditions of inequality?

In the last few decades, mainstream Western society has developed a shocking degree of faith in the power of technology to solve problems. In the face of neoliberal austerity, technology appears to be a cost-saving fix. Governments are adopting automated systems to allocate social services, to determine who gets a loan, or to judge who should be imprisoned. Corporations are racing to automate whole industries, offer free services so they can sell off consumer data, and develop technologies that exacerbate gendered and racialized violence (e.g., facial recognition). These efforts are having devastating effects on minoritized populations—expanding corporate and government surveillance, concentrating wealth and power, exacerbating inequality, pillaging the planet, and fortifying mass incarceration.

For those of us who practice data science and data communication, it is important to resist the technosolutionist snake oil from Silicon Valley: data and AI are not going to "solve" matters of social inequality. More data doesn't readily translate to more action or more justice. On the contrary, the demand for more proof—more data collection, more analysis, more research—is often a terribly effective delay tactic employed by mainstream institutions to avoid taking meaningful action on social justice.

Against this backdrop, it is crucial for us to recognize that data science—even feminist data science or restorative/transformative data science—is not a "solution" and it is not saving anyone or anything anytime soon. Rather, it is something much more humble. Restorative/transformative data science is a specific tactic of knowledge generation and consciousness building deployed amid deeply unjust information ecosystems. Just as neither a documentary film nor a new policy nor an advocacy campaign can "solve" structural inequality, it is silly for us to expect that a database or data system—even a counterdatabase or counterdata system—might do the same. What we might rather hope for (and tangibly organize for) is that a restorative/transformative data science project might participate in a collective concert of social action toward structural change.

Thus, the second point that I find myself coming back to over and over again is one that Paola Maldonado Tobar, also from the Alianza in Ecuador, articulated. This, too, was echoed by many other activists across countries and contexts. She told us: "We play a minimal role in this process, which is basically a defense, a fight for rights and for a dignified life for women and for the eradication of violence, but this is collective work, networked work, work that it has to continue expanding upwards, downwards, towards all the edges that we can give it."

Maldonado Tobar is wise. She helps us see the multiplicity of meanings in our efforts. The Alianza's work producing feminicide data in Ecuador is both minimal and centrally important. It is a community defense—a way to refuse violence and work toward its eradication. But she positions their work relationally—within the many collectives and networks that work on the same issue but who may use different methods and serve different communities. All of that work is necessary. And all of the work that each of us does is necessary; individually insufficient, and yet increasingly powerful in aggregate. This is certainly true of the Latin American feminist movements, whose popular strength across the continent is a model for what can be accomplished by collective political action.

Thus, as I come to the end of my writing process, I realize that this has not been a book about counting. And this has also not been a book about death. It has, in fact, been a love letter to the data activists, data journalists, nonprofits, political collectives, and academics that work to defend life in small, networked, relational, transnational, grassroots, informatic ways across the Americas. Their work models an approach to data science that foregrounds life, living, vitality, rights, and dignity, along with an insistence on structural change. *Counting Feminicide* has narrated the behind-the-scenes labor of this data activism: the workflow process of resolving, researching, recording, and refusing and using data. We have seen how the Data Against Feminicide team undertook participatory technology development to support such a restorative/transformative approach to data science. And the book concluded with a toolkit designed to draw out and build on lessons from this tremendous labor so that others may undertake restorative/transformative data science in other domains. My hope, through these contributions, is to enable and inspire more of us to enact a vision and a practice of a *humble* data science that plays a small—yet powerful—part in the larger constellation of efforts to restore and transform the world.

ACKNOWLEDGMENTS

I am grateful to so many people for getting *Counting Feminicide* out into the world. At the MIT Press, Gita Manaktala supported this project in its earliest stages and provided crucial editorial guidance, and Suraiya Jetha provided editorial assistance. At PubPub, Catherine Ahearn and Allison Vanouse supported the digital production of the book through the PubPub platform, without which the open community review would not have been possible. I want to especially acknowledge David Weinberger, editor of the Strong Ideas series at the MIT Press. He is the person who originally pointed me to a Global Voices article about María Salguero's work, which led me to learning about and then writing about the global landscape of feminicide data activism.

Many colleagues have given me feedback and input as this project has taken shape over the past several years. I am deeply indebted to and so grateful for the friendship of my collaborators on Data Against Feminicide: Silvana Fumega and Helena Suárez Val. Our Zooms were a source of laughter and community during the COVID-19 lockdowns and I'm so glad to get to see them more in person now. This book simply wouldn't exist without the conversations, event organizing, writing, and teaching that we have done together since 2019. This project also would not exist without the collaboration and friendship of Lauren Klein, coauthor of *Data Feminism*. I build on and draw extensively from *Data Feminism* in this work. I am grateful that she has confidence that I can take our collaborative work forward in this new direction.

It was important to me that this book honor the collaborative nature of this prior work, as well as be accountable to the feminicide data activists that we have been working with since 2019. Toward this end, I formed an Academic-Community Peer Review Board comprised of Lauren, Silvana, and Helena, along with four others: Geraldina Guerra Garcés and Paola Maldonado Tobar (ALDEA & Alianza Feminista para el Mapeo de los Femi(ni)cidios en Ecuador), Debora Upegui-Hernández (Observatorio de Equidad de Género Puerto Rico), and Annita Lucchesi (Sovereign Bodies Institute). I am so grateful for this board's feedback because it significantly shifted the framing of the book and the development of particular concepts—such as restorative/transformative data science. This board and its members are described further in appendix 2.

In addition to feedback from the Academic-Community Peer Review Board, I received generous comments and questions from three anonymous peer reviewers. They pushed me to deepen theoretical discussions, broaden my ideas for who the audience could be for the book, and to make some of the pragmatic contributions, like the toolkit, even more accessible and useful. And, just as Lauren and I did with *Data Feminism*, I worked with the folks at PubPub to put the first draft of this book online for a wider community review. I am grateful for the many (hundreds) of comments I received on the book, ranging from copy edits to reference suggestions to larger structural and conceptual comments. I want to particularly acknowledge the comments from Alexis Henshaw, Bronwyn Carlson, Carl DiSalvo, Cynthia Bejarano, Débora de Castro Leal, Derya Akababa, Eric Gordon, James Scott-Brown, Jimena Acosta, Julia Monárrez Fragoso, Matt Bui, Milagros Miceli, Mimi Ọnụọha, Paola Ricaurte, Patricia Garcia, Renee Shelby, Saide Mobayed, and Susana Galan. I am also grateful to the many data activists who sent me comments and fact corrections as part of the community review process.

I wish to thank the senior scholars whose work I did *not* cite from fields such as urban planning, STS, HCI, PD, computer science, the sociology of classification, critical data studies, and feminist theory. I build on your work, but given the limited space I have for my own words and for references to the work of others, I have chosen to prioritize uplifting works by activists and advocacy organizations, artists, junior and student scholars, women of color scholars, Black scholars, Indigenous scholars, and Latin American and border studies scholars. For anyone who wishes to see the extended bibliography of the book—the list of everything that I reviewed and wanted to cite in the making of this book—it is available here: https://mitpressonpubpub.mitpress.mit .edu/pub/extended-bibliography.

Apart from more formal review processes, my colleagues at the MIT Department of Urban Studies and Planning gave me extremely helpful feedback on my book proposal at our junior faculty workshop: thank you to Jason Jackson, Andres Sevtsuk, Justin Steil, and Delia Duong Ba Wendel. And a special shoutout to devin michelle bunten for her feedback at the workshop and then the many other times I called on her! My dear faculty mentors and friends Ethan Zuckerman and Eric Gordon also provided excellent suggestions and feedback on early stages of the book proposal. I enlisted my mother,

Janet Sue, and my father, Fred, for feedback, life coaching, and moral support, and they showed up more promptly than I did at all of our Zooms and boosted me when I needed it, and even when I didn't need it. Janet Sue, Helena Suárez Val, and my partner Dave also answered my urgent call for last-minute editing when I discovered my book manuscript was far over length and needed emergency trimming. My Thursday (really early) morning writing group with Helena and Isadora Cruxên was an essential source of support where we could cowork in silence but also chat and digress. Nikki Stevens, postdoc at the Data + Feminism Lab, provided generous feedback and dialogue for how to navigate TERFS in the bibliography. I also meet regularly with a wonderful, sage, grace-giving executive coach, Chris Miller. He gave me feedback on drafts of the first chapters of this book and has also been an ongoing source of support for navigating racial justice issues both in text and in real life. If you are looking for honest and accountable supervision in the areas of racial and gender justice, I highly recommend emailing Chris at chrisaverymiller@verizon.net.

Partners and student researchers have been at the heart of this book project and the larger Data Against Feminicide project from which it emerged. I want to thank project partners Jimena Acosta, Rahul Bhargava, Isadora Cruxên, Angeles Martinez Cuba, and Alessandra Jungs de Almeida for their significant contributions to the work. First, I would like to recognize the efforts of Isadora Cruxên for leading the initial design of our qualitative research and analysis process in summer 2020 and for everything that has come thereafter. What we both thought was a summer job has turned into a long-term collaboration, and Isa has played a leading role in numerous publications and spearheaded our team's participatory design work with data activists in Brazil. In 2023, Isa joined the leadership team of Data Against Feminicide. Rahul is a dear friend, long-time collaborator, and co-conspirator, and without his contributions and the use of the Media Cloud infrastructure, the tools described in chapter 7 would not have been possible. Angeles started as a master's student and was such an important force on the project, undertaking work across all aspects ranging from data analysis to information systematization to literature review to project coordination to supervision of undergraduate researchers. Once Angeles graduated, Alessandra Jungs de Almeida stepped up to provide project coordination, coleadership on our collaboration with Brazilian activists, and all-around enthusiasm for the work. A special thanks goes to designer Melissa Q. Teng for her beautiful collage work created for this book, along with data visualizations and other graphics. Thank you as well as to Wonyoung So and Tiandra Ray for their graphic design and data visualizations. Natasha Ansari brought her knowledge of disability justice to produce alt text for the images on the

PubPub website. Natasha, Alessandra, and Valentina Pedroza Muñoz contributed to literature review and demonstrated huge patience with inputting citations into Zotero using Google Docs (someone please fix this integration!). Many students contributed to the tools described in chapter 7. A special shoutout to Wonyoung So, who led the development of the Email Alerts tool and who, as technical lead for the Data + Feminism Lab, also keeps our servers running, fixes and upgrades our software, and generally makes the lab possible. Harini Suresh was lead designer and developer for the Highlighter tool and for the development of our machine learning models, and did a brilliant job of conjoining feminist principles with computer science in both process and product. Other essential contributions to the Data Against Feminicide tools, interviews, analysis, and research project have been made by student researchers Amelia Dogan, Soyoun Kang, Niki Karanikola, Patricia Michelle García Iruegas, Mariel García-Montes, Ana Amelia Letelier, Rajiv Movva, Luciana Ribeiro da Silva, and Thuận Trần.

This book would not have been possible without the funding and support of multiple groups. I received crucial start-up funds, multiple internal grants, and two semesters of junior leave time from the Department of Urban Studies and Planning at MIT. This support, along with funds secured by Silvana Fumega and ILDA, underwrote the project in its early stages as it was taking shape. Some of the travel on the project has been supported by the MIT International Science and Technology Initiatives (MISTI). A grant from the Leventhal Center for Advanced Urbanism (LCAU) has supported student researchers and data analysis. I am grateful for a 2022 award from the National Science Foundation, which has helped the project expand its focus on making contributions to human-centered computing and citizen data science (award #2213826). And the open-access edition of this work was made possible by generous funding from the MIT Libraries.

While financial support is essential, I am at the center of a caring and wonderful network of people who provided emotional, material, and caregiving support to make this project possible. Innumerable thanks goes to my best friend and partner Dave, mainly for doing the dishes when I hate them so much, but also for being a source of laughter, light, and possibility even when balancing work and life feels impossible. Thanks also to my three kids, who gave me encouragement along the way, celebrated small milestones by eating fancy chocolate with me, bought me books about #SayHerName, and who are themselves starting to stand up for reproductive justice, racial justice, and gender justice. While those kids are getting bigger, someone still has to take care of them while I'm writing. Many thanks again to Dave, who will always volunteer to pick up the sick kid from school or drive to some faraway sports game so that I'm able to meet a deadline or finish a thought. A huge debt of gratitude goes to María Lopez

Rodas, who has cared for our kids since they were toddlers and continues to do so now that the responsibilities have shifted into dressing kids for hockey and supporting their slime habits. And, last but not least, to the grandparents who are always at the ready for driving and caregiving when needed.

In short, anyone who tells you they wrote a book alone is deceiving themselves! I may have typed most of the words in this book, but so much infrastructure needs to exist for those words to be possible. I feel a deep sense of gratitude for this infrastructure and recognize that it is a privilege to be able to do this work. To all the people in my life that made this possible, I'm sending you a ginormous abrazo for being you!

APPENDIX 1: FEMINICIDE DATA ACTIVISTS INTERVIEWED FOR THIS BOOK

The Data Against Feminicide team conducted semistructured interviews with more than seventy-five participants directly involved in the forty-one projects listed ahead. Groups gave permission to be listed here and provided suggestions about which information they wanted to make publicly available about themselves. The interviews took place via Zoom between June 2020 and June 2023. They were conducted in activists' native languages and lasted between one and two hours. We aimed to understand the workflow, data production process, and conceptual categories through which activists identified and documented feminicide and fatal gender-related violence, as well as their reflections on challenges and lessons learned from their monitoring work.

The interview questionnaire was developed based on exploratory conversations with different activists and civil society groups and was tested and adjusted based on their feedback. The interviews were recorded with permission from the interviewees, and subsequently transcribed and analyzed through the qualitative data analysis software NVivo. (I do not recommend NVivo for reasons of cost and functionality, and our team has since transitioned to the feminist software platform ImpactMapper, which I do recommend for qualitative analysis.) Our qualitative coding process was iterative and collaborative, designed to allow for collective reflection and construction of shared understandings around the data. Our team is bi- and trilingual, which helped us analyze interviews conducted in Spanish, Portuguese, and English (see the acknowledgments for the many students who aided in this process). We compiled an initial codebook based on the interview questionnaire and then revised it to incorporate new categories and insights that emerged from our grounded analysis of the transcripts. Themes that emerged included the four common workflow stages—resolving, researching, recording, and refusing and using data—that are described in figure 2.4 and part II of this book. These stages of work were not explicitly included in our interview questionnaire, but they gained prominence through our close reading of the interviews.

Table A1

Feminicide data activists interviewed for this book

☆ = Co-design partner for ideation prior to and during tool development 2020–2022.
★ = Participated in co-design sessions to pilot our interactive tools in spring 2021.
✹ = Participated in co-design sessions to pilot our interactive tools in spring 2023 (Brazil/Portuguese pilot).

Organization	Project/s	Website	Country	Project focus	Sector	Year started database (self-reported)	Project status
★ Mumalá	Observatorio MuMaLá: Mujeres, Disidencias, Derechos	https://mumala.online /observatorio	Argentina	Feminicides, trans/ travesticidios, feminicide attempts, missing women, LGBTIQ+ hate crimes, suicides of feminicide perpetrators, deaths of women and LGBTIQ+ people linked to vulnerabilities and criminal economies (narcotics, crimes, etc.), deaths from unsafe abortions	Feminist and/or political collective	2015	Active
★ Mujeres de Negro Rosario		https://www.facebook .com/MujeresdeNegro Rosario/	Argentina	Femicides	Feminist and/or political collective	2012	Active
Ahora que sí nos ven	Observatorio de violencias de género	https://ahoraquesinosven .com.ar/	Argentina	Femicides	Feminist and/or political collective	2015	Active
La Casa del Encuentro	Observatorio Adriana Marisel Zambrano	http://www.lacasadelen cuentro.org/	Argentina	Femicides	Nonprofit	2008	Active

Table A1 (continued)

Organization	Project/s	Website	Country	Project focus	Sector	Year started database (self-reported)	Project status
Agencia Presentes	Mapa Periodístico Crimenes de Odio LGBT+	https://agenciapresentes.org/	Based in Argentina, compiles data across Latin America	Feminicides of Indigenous women, killings of LGBT+ and migrant people	Nonprofit	2015 (database) 2016 (organization)	Active
Mundo Sur	Mapa Latinoamericano de Feminicidios, Monitor de Crímenes de Odio LGBTTTIQ+ Marielle Franco	https://mlf.mundosur.org/	Based in Argentina, compiles data across Latin America	Feminicides and LGBTQ+ violence	Nonprofit	2020	Active
¿Cuántas Más?		https://www.facebook.com/FeminicidiosBolivia/	Bolivia	Feminicides	Journalism	2014	Paused
Néias— Observatório de Feminicídios Londrina*		https://www.observatorioneia.com/	Brazil	Feminicides	Feminist and/or political collective	2021	Active
Fórum Cearense de Mulheres*		https://www.facebook.com/forumcearensedemulheres/	Brazil	Feminicides	Feminist and/or political collective	2018	Active
data_labe	Transdados	https://datalabe.org/transdados/	Brazil	Transgender killings	Feminist and/or political collective	2016	Ended
Campanha Levante Feminista contra o Feminicídio, Transfeminicídio e Lesbocídio*	Lupa Feminista Contra o Feminicídio	https://lupafeminista.org.br/	Brazil	Feminicides	Feminist and/or political collective	2021	Active
NE10	Uma por Uma	http://produtos.ne10.uol.com.br/umaporuma/index.php	Brazil	Feminicides	Journalism	2018	Ended
Native Women's Association of Canada	Safe Passage	https://nwac.ca/ and https://safe-passage.ca/	Canada	MMIWG2S, 2SLGBTQQIA+ violence, and unsafe experiences	Nonprofit	2006–2010 and 2017–present	Active

Table A1 (continued)

Organization	Project/s	Website	Country	Project focus	Sector	Year started database (self-reported)	Project status
Canadian Femicide Observatory for Justice and Accountability		https://www.femicidein canada.ca/	Canada	Femicides	Academia	2017	Active
Red Chilena contra la Violencia hacia las Mujeres		http://www.nomasviolen ciacontramujeres.cl/	Chile	Femicides	Feminist and/or political collective	2001	Active
Red Feminista Antimilitarista	Observatorio Feminicidios Colombia	https://www.redfeminista antimilitarista.org/	Colombia	Feminicides	Nonprofit	2012	Active
Colombia Diversa	Sin Violencia LGBT	https://colombiadiversa .org/	Colombia	LGBTIQ+ violence	Nonprofit	2004	Active
Recordar-LAS		https://cartografiafemici dioscr.com/	Costa Rica	Femicides	Individual	2017	Active
Organización de Mujeres Salvadoreñas por la Paz (ORMUSA)	Observatorio de violencia contra las mujeres	https://ormusa.org/	El Salvador	Feminicides	Nonprofit	2005	Active
Alianza Feminista para el Mapeo de los Femi(ni)cidios en Ecuador		https://www.fundacio naldea.org/noticias-aldea /mapafemicidios	Ecuador	Feminicides	Nonprofit	2017	Active
Grupo Guatemalteco de Mujeres	Muertes Violentas de Mujeres y Femicidios en Guatemala	https://ggm.org.gt/	Guatemala	Femicides	Nonprofit	2000	Active
Femicide Count Kenya		https://www.facebook .com/CountingDead WomenKenya/	Kenya	Femicides	Feminist and/or political collective	2019	Active
Departamento de Estudios Culturales, Dirección General Regional Noroeste. El Colegio de la Frontera Norte	Base de datos del feminicidio en Ciudad Juárez (personal research file of Julia E. Monárrez Fragoso)	https://www.academia .edu/28683255/An _Analysis_of_Feminicide _in_Ciudad_Ju%C3%A1rez _1993_2007_pdf	Mexico	Feminicides, missing women	Academic	1998	Active

Table A1 (continued)

Organization	Project/s	Website	Country	Project focus	Sector	Year started database (self-reported)	Project status
Coordinadora Nacional de Mujeres Indígenas (CONAMI)	Emergencia Comunitaria de Género	https://www.facebook.com/mujeresindigenasconamimexico/	Mexico	Indigenous feminicides, missing women and violence against Indigenous women	Feminist and/or political collective	2013	Active
Ellas Tienen Nombre		https://www.ellastienennombre.org/	Mexico	Feminicides	Individual	2014	Active
Letra Ese	Homicidios de personas LGBT+ en México	https://letraese.org.mx/	Mexico	LGBTI+ violent deaths	Nonprofit	1996	Active
Anonymous university observatory			Mexico	Missing women, Feminicides, Women homicides	Academia	2014	Active
Yo te nombro: El Mapa de los Feminicidios en México		http://mapafeminicidios.blogspot.com/p/inicio.html	Mexico	Feminicides	Individual	2016	Active
Movimiento Manuela Ramos	El registro de los feminicidios cometidos en el Perú	https://www.manuela.org.pe/	Peru	Feminicides	Nonprofit	2018	Active
Seguimiento de Casos PR		https://www.facebook.com/seguimiento.decasos.9	Puerto Rico	Feminicides, missing women and sexual abuse	Individual	2011	Active
Observatorio de Equidad de Género Puerto Rico	Monitoreo de feminicidios y desapariciones de mujeres, niñas y personas LGBTQI+	https://observatoriopr.org/	Puerto Rico	Feminicides and missing women	Nonprofit	2019 (database) 2020 (organization)	Active
Feminicidio.net	Listado De Feminicidios y Otros Asesinatos de Mujeres en España	https://feminicidio.net/	Spain	Feminicides	Feminist and/or political collective	2010	Active

Table A1 (continued)

Organization	Project/s	Website	Country	Project focus	Sector	Year started database (self-reported)	Project status
* Sovereign Bodies Institute	MMIP Database	https://www.sovereign-bodies.org/	Based in United States, Monitors the Americas	MMIWG2, MMIP	Nonprofit	2015 (database) 2018 (org founded)	Active
Justice for Native People		http://www.justicefornativewomen.com/	United States	MMIP	Individual	2015	Active
Jane Doe	Massachusetts Domestic Violence Homicide	https://www.janedoe.org/	United States	Domestic violence, all genders	Nonprofit	1993	Active
* African American Policy Forum	#SayHerName	https://www.aapf.org/sayhername	United States	Black women and girls killed by police violence	Nonprofit	2015	Active
*⁑ Women Count USA		https://womencountusa.org/	United States	Femicides	Individual	2016	Active
* Black Femicide US		https://www.facebook.com/blackfemicideUS	United States	Femicides of Black women and girls	Individual	2018	Active
*⁑ Feminicidio Uruguay		https://www.feminicidiouruguay.net/	Uruguay	Feminicides	Individual	2015	Active
Utopix	Observatorio de Femicidios	https://utopix.cc/	Venezuela	Femicides	Feminist and/or political collective	2019	Active

APPENDIX 2: ACADEMIC-COMMUNITY PEER REVIEW BOARD

Counting Feminicide: Data Feminism in Action comes from working in collaboration, community, and solidarity with feminicide data activists since 2019. It documents the creative, intellectual, and emotional labor of data activists across the Americas and shows how their approach to data science challenges the hegemonic and extractivist logics of mainstream data science. Using these insights, the book inductively develops the idea of a restorative/transformative data science—a data science grounded in healing and liberation—and speculates about how this approach may have relevance in other domains characterized by structural inequality, such as health, housing, environment, land use, and more.

In doing this work, I draw from my prior collaboration with Lauren Klein on *Data Feminism*, as well as my current collaboration on Data Against Feminicide (https://dataagainstfeminicide.net)—a large, participatory action-research-design project I have been leading with Silvana Fumega and Helena Suárez Val since 2019. In writing this book, I feel most accountable to these two groups—my collaborators and the feminicide data activists we have been working with. For them and for me, I want to ensure that the work meets its goals around collaboration, community, and solidarity. Toward this end, I formed an Academic-Community Peer Review Board so that I could hear directly from them as part of the review process.

The board was composed of my three collaborators along with representatives from three different organizations across the Americas (discussed ahead). I called it an "academic-community" board because everyone on the board represents multiple perspectives: all have academic training; several are leading scholars of feminism, feminicide, and MMIW/P whom I admire; many lead activist and community-based organizations; many work directly with families; and some are also survivors of violence themselves. In convening an academic-community board, this was also an attempt to follow the data feminism principles of rethinking binaries and hierarchies. All too often, academia theorizes (and often romanticizes and other-izes) "the community" as something separate from "the academy." In this case, the feminicide data activism community that I have been in dialogue with over the past four years is centrally comprised of public scholars, theory-building activists, heads of advocacy organizations with PhDs, professors who drive groceries to families' houses, and more. I sought to make space for the idea that "academic" and "community" are deeply intertwined.

The process was as follows: Each board member read the book manuscript in fall 2022 and made comments, and then we discussed it at a collaborative workshop in December 2022, as well as at individual meetings in January 2023. I would like to thank the interpreters who made our multilingual book feedback workshop possible. Gracias to Pat Antuña Arbulo and Ana Barreiro. If anyone is looking for live interpretation for Spanish, English, and Portuguese, please get in touch with them at pat@mezcladas.com and ana@baameirotraducciones.com. In regard to compensation, I estimated the work at twenty hours and paid members at a rate of $100 per hour. Per their wishes, board members were compensated either individually or as a donation to their organization or else we did a labor trade. This process was made possible through grants that I received from my institution.

I am deeply indebted to the Academic-Community Peer Review Board for shaping my thinking and writing on *Counting Feminicide*. As I describe further in chapter 2, this process significantly shifted some of the central concepts of the book. For example, where I had been previously theorizing a "counterdata science," I heard from multiple board members that this framing did not align with their conception of their work. It was too binaristic—appearing to create the hard binary of official data versus activist counterdata—when in fact activist data make their way into official data and vice versa. For the Alianza Feminista para el Mapeo de los Femi(ni)cidios en Ecuador, it sounded like activists were going to the state begging to be part of the official record, when in fact they consider their work to be about restoring fundamental rights that have been denied. For Sovereign Bodies Institute, it was too focused on doing data science as a "counter" to the colonial state when they see their work as primarily about Indigenous sovereignty and kinship. From these conversations and the brainstorms that followed, the concept of a restorative/transformative data science emerged. While this was one of the largest shifts based on the board's feedback, there were many medium and small shifts that arose from the generous feedback, comments, questions, and references suggested by the board members. *Counting Feminicide* was significantly strengthened out of this process, and I am deeply grateful for the time and generosity of this board.

MEMBERS OF THE ACADEMIC-COMMUNITY PEER REVIEW BOARD

Silvana Fumega / Data Against Feminicide (https://dataagainstfeminicide.net)

Silvana is a cofounder of Data Against Feminicide together with Catherine D'Ignazio and Helena Suárez Val. She holds a PhD from the University of Tasmania (UTAS) in Australia and a master's in public policy from the Victoria University of Wellington (NZ). She served as a consultant for numerous international organizations, governments, and civil society groups as well as research director at ILDA until the end of 2022. In recent years, she has focused her work on the intersection between data and inclusion. She is currently an independent consultant while also acting as the director of the Global Data Barometer project.

Geraldina Guerra Garcés / ALDEA and Alianza Feminista para el Mapeo de los Femi(ni)cidios en Ecuador (https://www.fundacionaldea.org/noticias-aldea/mapafemicidios)

Geraldina Guerra Garcés is an activist and defender of women's rights to live a life free from violence. She represents Fundación ALDEA and the Red Nacional de Casas de Acogida para mujeres víctimas de violencia (National Network of Shelters for women victims of violence). She is part of the Alianza Feminista para el Mapeo de los Femi(ni)cidios en Ecuador and a delegate in the Red Latinoamericana contra la Violencia de Género (Latin American Network against Gender Violence) that registers feminicides in the region. She has a degree in social communication from the Universidad Central del Ecuador and a diploma in migration and development from FLACSO Ecuador. Geraldina has experience in women's human rights; a focus on working with victims of violence; and experience accompanying families whose members were victims of femicide. She is an expert in facilitation and training processes with women's groups. She has a trajectory of more than twenty years of experience working to prevent violence against women, girls, and adolescents.

Lauren F. Klein / Digital Humanities Lab, Emory University (https://lklein.com/) / **coauthor of** *Data Feminism*

Lauren Klein is the Winship Distinguished Research Professor and an associate professor in the Departments of English and Quantitative Theory and Methods at Emory University, where she also directs the Digital Humanities Lab. Her work brings together computational and critical methods in order to explore questions of gender, race, and justice, both as they emerge in the early United States and as they endure in the present. She is the coauthor of *Data Feminism* (MIT Press) and the author of *An Archive of Taste: Race and Eating in the Early United States* (University of Minnesota Press). With Matthew Gold, she edits *Debates in the Digital Humanities*, a print/digital publication stream that explores debates in the field as they emerge.

Annita Hetoevehotohke'e Lucchesi / Sovereign Bodies Institute (https://www.sovereign -bodies.org/)

Annita Hetoevehotohke'e Lucchesi is a Cheyenne survivor of trafficking, intimate partner violence, sexual assault, and police violence. She is the founder and director of research and outreach at Sovereign Bodies Institute, a nonprofit research center and service provider addressing violence against Indigenous people, especially the crisis of missing and murdered Indigenous people. Annita received her PhD from the University of Arizona's School of Geography, Development, & Environment in spring 2023. Annita resides on her ancestral homelands in southeast Montana.

Paola Maldonado Tobar / ALDEA and Alianza Feminista para el Mapeo de los Femi(ni) cidios en Ecuador (https://www.fundacionaldea.org/noticias-aldea/mapafemicidios)

Paola Maldonado Tobar is a geography specialist and environmental engineer. She has more than twenty years of experience in community mapping and social cartography. She is part of Fundación ALDEA and coordinator of the projects Mapping of Femicides in Ecuador; the Atlas of Rural Women; and Human Rights, Indigenous Peoples and Territories Monitoring System. Paola is copresident of the Group to Document the Territories of Life in the World, which is the focal point of the Indigenous Peoples' and Community Conserved Areas and Territories (ICCA) Consortium in Ecuador. Paola is a member of the Red Latinoamericana contra la Violencia de Género (Latin American Network against Gender Violence) that generates the Latin American Map of Femicide. Paola has higher education in geographic engineering and the environment, with a focus on rural territorial development. She holds an international diploma in public and socioenvironmental management of natural resources in rural areas and a diploma in local development with emphasis in natural resources management. She has experience in rural, socioenvironmental, gender and human rights issues, and community territorial planning, as well as training in GIS and spatial analysis.

Helena Suárez Val / Feminicidio Uruguay (https://www.feminicidiouruguay.net/) **and Data Against Feminicide** (https://datoscontrafeminicidio.net)

Helena Suárez Val is an activist, researcher, and producer focused on digital communications, data, feminism, and human rights. She has worked as a web developer and communications specialist, for organizations such as Amnesty International's International Secretariat, the Global Call to Action against Poverty, and Cotidiano

Mujer. Helena holds master's degrees in gender, media and culture (Goldsmiths, University of London) and social science research (University of Warwick) and completed her PhD at the Centre for Interdisciplinary Methodologies, University of Warwick in 2023. With Catherine D'Ignazio and Silvana Fumega, she cofounded Data Against Feminicide in 2019. She has been recording and caring for feminicide data for Uruguay since 2015, at https://feminicidiouruguay.net/.

Debora Upegui-Hernández / Observatorio de Equidad de Género Puerto Rico (https://observatoriopr.org/)

Debora Upegui-Hernández is a researcher and analyst at Observatorio de Equidad de Género de Puerto Rico (https://observatoriopr.org), a nonprofit organization monitoring feminicide incidence, gender violence, and gender equity on the island. She holds a PhD in social psychology (CUNY Graduate Center) and a master's in psychology (Hunter College CUNY). Her work has focused on gender violence, drug policy, drug users' human rights, and the experiences of Latinxs in the United States. She is the coauthor of *When The Earth Trembled: Violence and Women's Resistance after the Earthquakes in the South of Puerto Rico*, and *Humiliation and Abuse in Drug Treatment Centers in Puerto Rico* and the author of *Growing Up Transnational: Colombian and Dominican Children in New York City* (LFB Scholarly Publishing).

STATEMENT ON SHARING POWER

As part of my continued efforts to share power and divert social and financial capital back to where it belongs, I am directing half of the royalties from this book to Indigenous Women Rising. I have been on the board of this incredible organization since 2018.

Through IWR's abortion fund, they help Native people access abortion care, which was scarce before Dobbs and even scarcer now. Through their midwifery program, the

Indigenous Women Rising is committed to honoring Native and Indigenous people's inherent right to equitable and culturally safe health options through accessible health education, resources, and advocacy. Logo by Eloy Bida. Courtesy of Indigenous Women Rising.

Emergence Fund, they provide culturally appropriate birth care to Native people. And through their education and community power-building work, they share their vision of reproductive justice with Indigenous communities and beyond. I invite you to seek them out, engage with them on social media and in real life, and consider how you might contribute to their work.

Follow Indigenous Women Rising on Twitter at @IWRising, on Instagram at @indigenouswomenrising, and at https://www.iwrising.org.

NOTES

INTRODUCTION

1. Ortiz and Brady, "One Woman Is Behind the Most Up-to-Date Interactive Map of Femicides in Mexico."

2. This quote is from our research team's interview with María Salguero. From here forward, when the book quotes activists, the source is our research team's interviews with them unless otherwise noted.

3. Semple and Villegas, "Grisly Deaths of a Woman and a Girl Shock Mexico and Test Its President"; Salguero, "Yo te nombro: El Mapa de los Feminicidios en México" [I name you: Map of Feminicides in Mexico].

4. Official data can be found from the Instituto Nacional de Estadística y Geografía (INEGI) and the Secretaría de Salud (SSA). Feminicide is defined in article 325 of the Federal Penal Code, and there is a law titled General Law on Women's Access to a Life Free of Violence. See Mejía Berdeja and Monreal Ávila, Iniciativa que reforma el artículo 325 del código penal federal; ONU Mujeres, INMUJERES, and Conavim, *Violencia Feminicida En México*; and Ley General de Acceso de las Mujeres a una Vida Libre de Violencia.

5. D'Ignazio and Klein, *Data Feminism*, 38.

6. Driver, *More or Less Dead*, 7.

7. Temporarily renaming streets is becoming an increasingly prevalent activist tactic in Mexico. For example, the art collective Colectiva SJF produced a project called *Nombrar, No Olvidar (Name, Don't Forget)* in 2020, which renamed streets in Mexico City after the forty-three students from the Ayotzinapa Rural Teachers' College who were abducted and executed in Iguala, Guerrero, Mexico, in 2014.

8. Mobayed, "Recontar Feminicidios."

9. I have a deep aversion to the unspecified "we," so I want to make it clear that unless otherwise specified, all instances of *we* used in this book mean you, the reader, and me, the author.

10. Fregoso and Bejarano, *Terrorizing Women*, 5.

11. Feminicide is not only invisibilized in the Global North, and especially the United States and Canada, but also in English-speaking countries like Jamaica that, due to British colonialism, have similar legal frameworks to the United States and Canada. I thank Silvana Fumega for making this point clear based on her work with ILDA in the Caribbean.

12. Alvarez et al., *Translocalities/Translocalidades*.

13. UNODOC, *Global Study on Homicide—Gender-Related Killing of Women and Girls*.

14. McHugh, "Opinion | Why Aren't Women in the U.S. Also Protesting against Femicide?"

15. Semple and Villegas, "Grisly Deaths of a Woman and a Girl Shock Mexico and Test Its President"; Salmenrón Arroyo, Carrión Rivera, and Montoya Ramos, *Un Manual Urgente Para La Cobertura de Violencia Contra Las Mujeres y Feminicidios En México*.

16. Latin America includes South America (including Brazil), Central America, and Mexico. Feminicide statistics for the region come from: CEPAL, "Prevenir El Feminicidio. Una Tarea Prioritaria Para La Sociedad En Su Conjunto."

17. "Femicide Watch Initiative"; *Femicide Volume VII*.

18. Fregoso and Bejarano, *Terrorizing Women*, 25–27.

19. Taylor, "Until Black Women Are Free, None of Us Will Be Free."

20. This nomenclature is also in line with visual theorist Johanna Drucker, who reminds us that we shouldn't even be calling anything "data" in the first place but rather "capta." She writes, "The notion of data as 'given' and thus self-evident is patently false—all data are constructed, made, and should be referred to as constructa (or capta)." See Drucker, "Visualization."

21. Gramsci, *Prison Notebooks*.

22. See the glossary in chapter 8 for precise definitions of racial capitalism, colonialism, patriarchy, and other systems of power.

23. Benjamin, *Race after Technology*; Broussard, *More Than a Glitch*. For a more liberatory take on the potential of glitches to resist domination, see Russell, *Glitch Feminism*.

24. Benjamin, *Race after Technology*, 54.

25. Erin Genia was an artist in residence with the City of Boston's Department of Emergency Management in 2020–2021, and she used that time to explore techniques to approach cultural emergencies. You can learn more about her work at https://www.eringenia.studio.

26. Lucchesi, "Mapping Violence against Indigenous Women and Girls."

27. Perera and Pugliese, *Mapping Deathscapes*; Fregoso and Bejarano, *Terrorizing Women*; Wright, "Necropolitics, Narcopolitics, and Femicide"; Falquet, "Violence against Women and (De-)colonization of the 'Body-Territory'"; Segato, "Las Nuevas Formas de La Guerra y El Cuerpo de Las Mujeres."

28. Ricaurte, "Data Epistemologies, the Coloniality of Power, and Resistance," 352.

29. Milan and van der Velden, "Alternative Epistemologies of Data Activism," 63–64.

30. Cifor et al., *Feminist Data Manifest-No*; Carmi, "A Feminist Critique to Digital Consent"; Edenfield, "Queering Consent"; Lee and Toliver, *Building Consentful Tech*; Leurs, "Feminist Data Studies."

31. The Stanford Institute for Human-Centered Artificial Intelligence aims to raise $1 billion for its university-based research center. My own institution, MIT, has sought to raise over $1 billion for the creation of the Schwarzmann College of Computing, which aims to produce leaders with "the cultural, ethical, and historical consciousness to use technology for the common good." Reid Hoffman, Pierre Omidyar, and the Knight Foundation created the $27 million Ethics and Governance in AI Fund to address algorithmic discrimination. Far, far less of these funds are going to Black-led and women-led centers like the UCLA Center for Critical Internet Inquiry, coled by Safiya Noble and Sarah T. Roberts, and the Distributed Artificial Intelligence Research Institute (DAIR), led by Timnit Gebru, former colead of Google's AI Ethics group. Gebru's research on the potential harms of large language models was censored by Google, and she was then unceremoniously fired in 2020. You can listen to Noble and Gebru in conversation with J. Khadijah Abdurahman as they outline the risks of these funding disparities in terms of producing tech that is truly focused on transformative justice: https://www.youtube.com/watch?v=WqAMkmX9AuE. For the censored research paper, see Emily M. Bender, Timnit Gebru, Angelina McMillan-Major, and Shmargaret Shmitchell, "On the Dangers of Stochastic Parrots: Can Language Models Be Too Big?," in *Proceedings of the 2021 ACM Conference on Fairness, Accountability, and Transparency* (New York: Association for Computing Machinery, 2021), 610–623, https://dl.acm.org/doi/10.1145/3442188.3445922.

32. Bietti, "From Ethics Washing to Ethics Bashing"; Green, "Contestation of Tech Ethics"; Young, Katell, and Krafft, "Confronting Power and Corporate Capture at the FAccT Conference."

33. De Waal and de Lange, "Introduction—the Hacker, the City and Their Institutions," 2.

34. Collins, *Intersectionality as Critical Social Theory*, 237–240.

35. See https://idatosabiertos.org.

36. Our website can be found at https://datoscontrafeminicidio.net. And our funding—which is always something important to ask about—is detailed in the acknowledgments section of this book.

37. I make this point explicitly because people have misunderstood our work as a project to collect and aggregate activist data. It is not. We have never asked activists for their data and do not intend to do so. See chapter 5 for activist-led efforts to aggregate and share data through Red Latinoamericana contra la Violencia de Género (Latin American Network against Gender Violence).

38. Our annual events are archived at https://datoscontrafeminicidio.net/en/edicion/2020-edition/ and https://datoscontrafeminicidio.net/en/2021-edition/.

39. The course was called Datos contra el Feminicidio: Teoría y práctica; see https://datoscontrafeminicidio.net/curso/.

40. Specifically, we began with Helena's list of other feminicide projects, which can be found on the Feminicidio Uruguay website: https://www.feminicidiouruguay.net/otros-sitios.

41. D'Ignazio et al., "Feminicide & Machine Learning"; Suárez Val, Martinez Cuba, Teng, and D'Ignazio, "Datos de Feminicidio, Trabajo Emocional y Autocuidado"; Suresh et al., "Towards Intersectional Feminist and Participatory ML"; D'Ignazio, "Human-Centred Computing and Feminicide Counterdata Production"; D'Ignazio et al., "Feminicide and Counterdata Production"; Suárez Val, D'Ignazio, and Fumega, "Data Against Feminicide"; Bhargava et al., "News as Data for Activists"; Fumega, "From Bias to Feminist AI."

42. Monárrez Fragoso, "Victims of Ciudad Juárez Feminicide"; Lagarde y de los Ríos, "Preface: Feminist Keys for Understanding Feminicide"; Fuentes, "(Re)reading the Boundaries and Bodies of Femicide"; Fuentes, "'Garbage of Society'"; PATH et al., *Strengthening Understanding of Femicide*; Lucchesi, "Mapping Geographies of Canadian Colonial Occupation."

43. Merry, *Seductions of Quantification*; Nelson, *Who Counts?*; Walklate et al., *Towards a Global Femicide Index*.

44. Merry, *Seductions of Quantification*, 45.

45. Shokooh Valle, "'How Will You Give Back?'"

46. Driver, *More or Less Dead*.

47. Dean, *Remembering Vancouver's Disappeared Women*.

CHAPTER 1

1. Annunziata et al., "El Caso de Argentina."

2. Belotti, Comunello, and Corradi, "Feminicidio and #NiUna Menos."

3. From Annunziata et al., "El Caso de Argentina"; and Chenou and Cepeda-Másmela, "#NiUnaMenos."

4. Russell, "Defining Femicide."

5. Radford and Russell, *Femicide*.

6. Activists and scholars are frequently confronted with questions about why it matters to disaggregate the murders of women. Dawn Wilcox, who runs Women Count USA, explains it like this: "I've had the question many times, 'what about men?' It matters when men are killed, too, so why should we count women's murders separately? My argument is that women's murders typically look very, very different than men's . . . these murders are especially brutal and vicious." Thus, differentiating by gender allows us to visibilize and disentangle patterns that are gendered. See Wilcox, "Invisible Women: Understanding the Scope of Lethal Male Violence against Women in the U.S."

7. Wright, "Public Women, Profit, and Femicide in Northern Mexico."

8. Tabuenca C., "From Accounting to Recounting."

9. Grupo de 8 Marzo translates to "group of March 8" (International Women's Day). Nuestras Hijas de Regreso a Casa translates to "bring our daughters home." Lourdes Portillo's documentary on the topic—*Señorita Extraviada* (2001)—showcases the lack of judicial and media response, the

victim-blaming, and the committed work of mothers and families to continue seeking justice. Other early work that describes the climate of impunity includes Benitez et al., *El Silencio Que La Voz de Todas Quiebra*; and Amnesty International's 2003 report *Mexico: Intolerable Killings*.

10. Chavez, *Primera Tormenta*.

11. Benitez et al., *El Silencio Que La Voz de Todas Quiebra*, 6.

12. In addition, Monárrez Fragoso developed the concept of *serial sexual feminicide* to describe those feminicides in which sexual violence (in the form of systematic torture, rape, mutilation, placement of the corpse) played a significant role. These have been prevalent in Ciudad Juárez since at least the 1990s. Monárrez Fragoso, "Feminicidio sexual serial en Ciudad Juárez 1993–2001."

13. Monárrez Fragoso, "La cultura del feminicidio en Ciudad Juárez, 1993–1999"; Monárrez Fragoso, "Crímenes en Ciudad Juárez, El feminicidio es el exterminio de la mujer en el patriarcado."

14. *Maquiladoras* are duty-free factories that proliferated in Northern Mexico following the signing of NAFTA. The 2006 documentary film *Maquilopolis* is a good starting point for learning more about them. They often employ women, who often end up in hazardous working conditions with few labor protections and low pay. Monárrez Fragoso, a sociologist by training, undertook one of the earliest academic projects to count feminicide in the context of Ciudad Juárez. To do so, she followed methods similar to the activists outlined in this book, and her monitoring project is ongoing (see appendix 1). Her methods and findings are described further in Monárrez Fragoso, "La cultura del feminicidio en Ciudad Juárez, 1993–1999"; Monárrez Fragoso, "Las Diversas Representaciones Del Feminicidio y Los Asesinatos de Mujeres En Ciudad Juárez, 1993–2005"; PATH et al., *Strengthening Understanding of Femicide*, 78–84.

15. Lagarde y de los Ríos, "Preface: Feminist Keys for Understanding Feminicide." Lagarde y de los Ríos's work on feminicide was pathbreaking and central to the concept's formulation in Latin American legislation and feminist movements. While her words here are strongly intersectional, in more recent years, Lagarde y de los Ríos has characterized the queer and trans rights movements as being a threat to women and to feminist movements, a position most often taken up by TERFs, transexclusionary radical feminists. It is disheartening and angering to witness this shift. I find myself traveling from deep admiration to profound disappointment. It isn't the first time someone I have admired turns out to be TERFy, and I'm sure it won't be the last. But here I will state my own position and hope that anyone leaning toward TERFdom might listen: Trans rights and queer rights and women's rights are not in competition—our fates and our liberation are bound together. If feminism fails trans women, trans people, and queer communities, then feminism fails all of us.

16. Lagarde y de los Ríos's work laid the groundwork for an important case before the Inter-American Court on Human Rights in 2009, called the Campo Algodonero case. The court ruled that the state failed to prevent, investigate, and prosecute disappearances and deaths of women killed in Ciudad Juarez.

17. See Carcedo, *No olvidamos ni aceptamos*; Carcedo Cabañas and Sagot Rodríguez, *Femicidio en Costa Rica 1990–1999*; Segato, "Cinco debates feministas."

18. Dawson, "Punishing Femicide"; Driver, *More or Less Dead*; Menjívar and Walsh, "Architecture of Feminicide."

19. García-Del Moral, "Transforming Feminicidio"; García-Del Moral, "Murders of Indigenous Women in Canada as Feminicides."

20. CONAMI and CHIRAPAQ, *Levantando nuestras voces por la paz y la seguridad de nuestros pueblos y continentes.*

21. Here and throughout the book, I do not translate the gender identity *travesti* because it is specific to the Latin American context. Originally a transphobic slur, *travesti* has been reappropriated by Latin American activists and used to reclaim the rights of gender nonconforming people. To learn more about the emerging areas of travesti art and theory, see cárdenas, *Poetic Operations*; Garriga-López et al., "Trans Studies En Las Américas."

22. Marco, "Social Movement Demands in Argentina and the Constitution of a 'Feminist People'"; Revilla Blanco, "Del ¡Ni Una Más! Al #NiUnaMenos: Movimientos de Mujeres y Feminismos En América Latina"; Alcaraz, "#NiUnaMenos."

23. Gago and Gutiérrez Aguilar, "Women Rising in Defense of Life."

24. Annunziata et al., "El Caso de Argentina."

25. *Ámbito*, "Raquel Vivanco Renunció al Ministerio de Mujeres Por 'No Encontrar Síntesis En Lo Que Desde Allí Se Debe Generar.'"

26. "Eleven Black Women: Why Did They Die? A Document of Black Feminism."

27. Williams, "#SayHerName"; African American Policy Forum, and Center for Intersectionality and Social Policy Studies, "Say Her Name"; Threadcraft, *Intimate Justice*.

28. Threadcraft, "North American Necropolitics and Gender."

29. Patricia Hill Collins has also written about Ida B. Wells's data activism in relation to violence as a "saturated site" where intersectional forms of domination become visible. See Collins, *Intersectionality as Critical Social Theory*, 160–167; Wells, "Red Record"; Brubaker, "Who Counts?"

30. See Mbembe, "Necropolitics"; Mbembe, *Necropolitics*.

31. Fuentes, "(Re)reading the Boundaries and Bodies of Femicide," 61.

32. Threadcraft, "North American Necropolitics and Gender," 566.

33. List, "Counting Women of Color"; Threadcraft, "Like Breonna Taylor, Black Women Are Often Killed in Private."

34. African American Policy Forum, and Center for Intersectionality and Social Policy Studies, "Say Her Name," 4.

35. Green-Gopher, "My Sister Was Murdered"; Nowell, "What to Know about Missing and Murdered Indigenous Persons Awareness Day"; Remarkable Women 2014, "Mona Woodward."

36. The woman's name is not spoken aloud or written here to honor the wishes of her family. Dean, *Remembering Vancouver's Disappeared Women*. As we will see throughout the book, there are diverse family wishes for both naming and counting victims.

37. Conn, "Women's Memorial March."

38. Culhane, "Their Spirits Live within Us," 593.

39. Simpson, "The State Is a Man."

40. Simpson, *As We Have Always Done*, 54; Perera and Pugliese, *Mapping Deathscapes*.

41. Deer, *The Beginning and End of Rape*.

42. Cabnal, *Feminismos diversos*.

43. I want to thank Geraldine Guerra Garcés for emphasizing this link between feminicide and dispossession based on her work with Indigenous women and communities in Ecuador.

44. Simpson, *As We Have Always Done*, 95–118.

45. García-Del Moral et al., "Femicide/Feminicide and Colonialism," 63.

46. George, Lucchesi, and Trillo, *To' Kee Skuy' Soo Ney-Wo-Chek'*; Sovereign Bodies, *To' Kee Skuy' Soo Ney-Wo-Chek'*; National Inquiry into Missing and Murdered Indigenous Women and Girls (Canada), *Reclaiming Power and Place*.

47. Lucchesi, "Nationwide Data Crisis."

48. From Lucchesi's feedback on the first draft of this manuscript.

49. I want to thank my collaborator Helena Suárez Val for our discussions on this point. Also see Manjoo, "United Nations Special Rapporteur on Violence against Women."

50. This table is reprinted from the original publication in the Harvard Dataverse, where each entry's sources and citations are documented. See D'Ignazio et al., "Table of Laws and Official Data about Feminicide."

51. Our team documented activist challenges to these laws in support of trans women in the notes that accompany the table at D'Ignazio et al., "Table of Laws."

52. United Nations, *Universal Declaration of Human Rights*.

53. See CDC, "Leading Causes of Death—Females—All Races and Origins—United States, 2018."

54. See CDC, "Leading Causes of Death—Females—Non-Hispanic Black—United States, 2018."

55. Puerto Rico, Ley Num. 40.

56. Laws include Savanna's Act and the Not Invisible Act, both passed in 2020, and Executive Order on Improving Public Safety and Criminal Justice for Native Americans and Addressing the Crisis of Missing or Murdered Indigenous People, issued by the president on November 15, 2021.

57. See US Department of Justice Consultation with Tribes, "Savanna's Act: Data Relevance and Access." Families and grassroots organizations have also criticized recent efforts such as Operation Lady Justice and the Not Invisible Commission for their failure to meaningfully and respectfully engage survivors, families and grassroots advocates. See Sovereign Bodies Institute, "Public Statement on Operation Lady Justice, from MMIWG & MMIP Grassroots Advocates."

58. Fumega, "Manos a La Obra."

59. Sovereign Bodies, *To' Kee Skuy' Soo Ney-Wo-Chek'*.

60. Data Cívica and Intersecta, *Datos para la Vida*.

61. United Nations High Commissioner for Human Rights, *Latin American Model Protocol*, 6.

62. Breña, "Las Feministas Saludan La Rectificación Del Fiscal Gertz Sobre El Tipo Penal de Feminicidio."

63. In the 2009 Campo Algodonero or Cotton Field case, the Inter-American Court of Human Rights used international frameworks of violence against women to condemn the state of Mexico for negligence in protecting the rights of women citizens of Ciudad Juárez.

64. Walklate et al., *Towards a Global Femicide Index*; Walklate and Fitz-Gibbon, "Re-imagining the Measurement of Femicide."

65. *Femicide Volume VII*; "Femicide Watch Initiative."

66. United Nations High Commissioner for Human Rights, *Latin American Model Protocol*.

67. ILDA, "ILDA Feminicide Data Standard (V2.0)."

68. Fumega, *Guide to Protocolize Processes of Femicide Identification*.

69. D'Ignazio, "5 Questions on Data and Feminicide with Silvana Fumega."

70. Monárrez Fragoso, "Las Diversas Representaciones Del Feminicidio y Los Asesinatos de Mujeres En Ciudad Juárez, 1993–2005," 357.

71. National Inquiry into Missing and Murdered Indigenous Women and Girls (Canada), *Reclaiming Power and Place*, 234.

72. Milan, "Data Activism as the New Frontier of Media Activism."

73. Milan and van der Velden, "Alternative Epistemologies of Data Activism"; Gutiérrez, *Data Activism and Social Change*; Milan and Gutierrez, "Technopolitics in the Age of Big Data."

74. Milan and van der Velden, "Alternative Epistemologies of Data Activism," 67.

75. Gray, Lämmerhirt, and Bounegru, *Changing What Counts*.

76. Renzi and Langlois, "Data Activism."

77. Pine and Liboiron, "Politics of Measurement and Action."

78. Pine and Liboiron, 3149.

79. Fregoso and Bejarano, *Terrorizing Women*, 25–27.

80. Suárez Val, "Caring, with Data."

81. Chenou and Cepeda-Másmela, "#NiUnaMenos."

82. Suárez Val, "Affect Amplifiers."

83. Lucchesi, "Mapping for Social Change."

84. Lucchesi, "Mapping Violence against Indigenous Women and Girls."

85. Carroll et al., "CARE Principles for Indigenous Data Governance."

86. Milan and Trere, "Big Data from the South(s)"; Couldry and Mejias, *Costs of Connection*; Dutta, "Whiteness, Internationalization, and Erasure."

87. Crooks, "Seeking Liberation."

88. Ricaurte, "Data Epistemologies, the Coloniality of Power, and Resistance."

89. Ricaurte.

90. Milan, "Counting, Debunking, Making, Witnessing, Shielding."

91. Fotopoulou, "Understanding Citizen Data Practices from a Feminist Perspective."

CHAPTER 2

1. Avilés and Rodríguez Reyes, *La Persistencia de La Indolencia*.

2. See Centro, *New Estimates*; Peoples Dispatch, "Puerto Rican Governor Ricardo Rosselló Resigns"; Alexandra, "Puerto Rico between Neoliberalism and 'Natural' Disasters."

3. Driver, *More or Less Dead*.

4. Kishore et al., "Mortality in Puerto Rico after Hurricane Maria."

5. Avilés and Rodríguez Reyes, *La Persistencia de La Indolencia*, 9.

6. Avilés and Rodríguez Reyes, 9.

7. Enders, *Applied Missing Data Analysis*; Fernstad, "To Identify What Is Not There."

8. Gargiulo, "Statistical Biases, Measurement Challenges, and Recommendations."

9. Dawson and Carrigan, "Identifying Femicide Locally and Globally."

10. Danner, Fort, and Young, "International Data on Women and Gender," 252; Boulding et al., *Handbook of International Data on Women*.

11. See kanarinka, "Missing Women, Blank Maps, and Data Voids"; Perez, *Invisible Women*.

12. Ọnụọha, "On Missing Data Sets."

13. Ọnụọha; D'Ignazio and Klein, *Data Feminism*.

14. Mills, *Racial Contract*, 1999.

15. Tuana, "Speculum of Ignorance," 11.

16. Bowleg, "'Master's Tools Will Never Dismantle the Master's House.'"

17. Carlson, "Data Silence in the Settler Archive." In addition, on data silences, literary scholar Michel-Rolph Trouillot has posited the idea of "archival silences" surrounding historical phenomena, such as the records that document the Haitian Revolution. Building on that work, Lauren Klein has outlined digital methods, including data visualization, to render visible those archival silences. Trouillot, *Silencing the Past*; Klein, "Image of Absence."

18. Davis, *The Uncounted*.

19. Tuana, "Speculum of Ignorance."

20. Thanks to Paola Ricaurte for pointing me to Bhattacharyya, "Epistemically Produced Invisibility."

21. Ọnụọha, "On Missing Data Sets."

22. Spade, *Normal Life*, 4.

23. Collins, *Black Feminist Thought*.

24. D'Ignazio and Klein, *Data Feminism*, 35–39.

25. Figueroa, "Tipificación de feminicidios y transfeminicidios."

26. Avilés and Rodríguez Reyes, *The Persistence of Indolence*, 9.

27. Fumega, "Manos a La Obra."

28. Sovereign Bodies, *To' Kee Skuy' Soo Ney-Wo-Chek'*; National Inquiry into Missing and Murdered Indigenous Women and Girls (Canada), *Reclaiming Power and Place*.

29. Menjívar and Walsh, "Architecture of Feminicide."

30. Segato, "Rita Segato"; Segato, *La guerra contra las mujeres*.

31. Sutherland et al., "Mediated Representations of Violence against Women"; Fairbairn and Dawson, "Canadian News Coverage of Intimate Partner Homicide"; Tiscareño-García and Miranda-Villanueva, "Victims and Perpetrators of Feminicide"; Wright, "Necropolitics, Narcopolitics, and Femicide"; Richards, Kirkland Gillespie, and Smith, "Exploring News Coverage of Femicide."

32. England, *Writing Terror on the Bodies of Women*. Lorena Fuentes has also written powerfully about "disposable women" in the context of Guatemala. Fuentes, "'Garbage of Society.'"

33. Instituto Patrícia Galvão, *Papel Social e Desafios Da Cobertura Sobre Feminicídio e Violência Sexual*.

34. Avilés and Rodríguez Reyes, *The Persistence of Indolence*.

35. Vanoli Imperiale, "El doble asesinato de las identidades transgénero."

36. List, "Counting Women of Color."

37. See Hopkins, "If I Am Taken, Will Anyone Look for Me?"

38. Based on search results from the *New York Times* website on January 30, 2023. A search for "Gabby Petito" returned nineteen results for the period from September 2021 to January 2023. Then I searched for "missing and murdered Indigenous women" and excluded Canada, resulting in eighteen entries about the United States, with the first mention from 2018.

39. Carlson, "Data Silence in the Settler Archive"; Barrowcliffe, "Closing the Narrative Gap"; O'Sullivan, "Lived Experience of Aboriginal Knowledges and Perspectives."

40. Gellman, "Landmark Femicide Case Fails to Fix El Salvador's Patriarchy."

41. Dalton and Thatcher, "What Does a Critical Data Studies Look like, and Why Do We Care?"

42. Currie et al., "Conundrum of Police Officer-Involved Homicides."

43. Meng and DiSalvo, "Grassroots Resource Mobilization through Counter-Data Action."

44. Crooks and Currie, "Numbers Will Not Save Us."

45. In particular, as I elaborate further in this book, Indigenous-led organizations and scholars explained that counterdata does not resonate for them because they do not see their work as a counterpoint to the settler state. There is a way in which counterhegemonic approaches still center the powerful people and institutions responsible for producing violence in the first place.

46. Segato, *Contra-pedagogías de la crueldad*.

47. Guerra Garces, "Algunas notas sobre el texto."

48. ILDA, Comunicando Datos, Data against Feminicide Course 2022.

49. Avilés and Rodríguez Reyes, *La Persistencia de La Indolencia*.

50. Deer, *Beginning and End of Rape*, 9.

51. Lauren Klein and I discuss the pitfalls of proof in D'Ignazio and Klein, *Data Feminism*, 57–59. See also Walcott, "Data or Politics?"

52. Merry, *Seductions of Quantification*, 46.

53. Datos Contra Feminicidio, "Challenges of Measuring, Comparing, and Standardizing 'global' Fem(in)icide Data."

54. See Walklate et al., *Towards a Global Femicide Index*. In a separate paper, Walklate and Fitz-Gibbon argue for "thick" femicide counts over "thin" femicide counts. Thick counts would be those that incorporate more context and indicators for each case, and they have a higher chance of getting at what the authors characterize as *slow femicide*—the violence that leads up to femicide and makes women's lives unlivable. Walklate and Fitz-Gibbon, "Re-imagining the Measurement of Femicide."

55. Ball, "Bigness of Big Data."

56. Keyes, "Counting the Countless."

57. Indeed, there is a good reason to be suspicious of data-driven targeting of minoritized groups by the state: it has happened a lot. See Seltzer and Anderson, "Dark Side of Numbers."

58. Walklate et al., *Towards a Global Femicide Index*, 68.

59. Here I want to thank Annita Lucchesi, Geraldina Guerra Garcés, Margaret Pearce, and Melissa Q. Teng in particular for their generative feedback and brainstorming with me on alternate framings.

60. Lagarde y de los Ríos, "Preface: Feminist Keys for Understanding Feminicide."

61. Simpson, "Ruse of Consent"; Tuck and Yang, "R-Words: Refusing Research."

62. D'Ignazio and Klein, *Data Feminism*, 26.

63. Segato, *Contra-pedagogías de la crueldad*.

64. Mobayed, "Kintsugi Method to Recount Data against Feminicide."

65. There is extensive work on the origins and meaning of the word *science* and, indeed, a whole field of STS devoted to exploring this in historic and contemporary relief. It is outside the scope of this book to engage with all of that literature, so I offer readers the pragmatic definition of science as "people seeking, systematizing and sharing knowledge" (Butler-Adam, "Weighty History and Meaning").

66. In particular, Collins describes how feminism, critical race theory, and decolonial theory all constitute "resistant knowledge projects" that arise from theory and practice and have strong presences both inside and outside academia. Collins, *Intersectionality as Critical Social Theory*, 87–121.

67. Helena's list of *mapeadoras* can be found on Feminicidio Uruguay's website: https://www .feminicidiouruguay.net/otros-sitios.

68. Data Against Feminicide has not published our list of feminicide data activist projects both to prevent enabling individual or systematic targeting of these groups and because we would want to obtain consent from the groups to be listed in any open database or directory. If you are interested in accessing the larger list of projects, contact us and we can have a conversation.

CHAPTER 3

1. Red Feminista Antimilitarista, "Historia."

2. Falquet, *Por las buenas o por las malas*; Segato, "Territory, Sovereignty, and Crimes of the Second State"; Federici, Gago, and Cavallero, *¿Quién Le Debe a Quién?*

3. See the glossary in chapter 8 for a definition of neoliberalism.

4. Red Feminista Antimilitarista, *Violencia Neoliberal Feminicida En Medellín.*

5. Unzúeta, "Articulaciones Feministas."

6. Martínez, "How Many More."

7. Lucchesi, "Inspirational Interview."

8. Néias—Observatório de Feminicídios Londrina, "Quem somos," https://www.observatorioneia .com/quem-somos.

9. Periodistas de a Pie (see https://periodistasdeapie.org.mx) covers many human rights issues in Mexico and works to build capacity and integrate a gender perspective into the reporting of journalists in its network.

10. Carol, "Personal Is Political"; Combahee River Collective, *Combahee River Collective Statement.*

11. Combahee River Collective; Sharpe et al., *Trying to Make the Personal Political.*

12. D'Ignazio et al., "'Personal Is Political.'"

13. Red Feminista Antimilitarista, *Paren La Guerra Contra Las Mujeres*, 8.

14. For a good summary of how this debate unfolded in the last quarter of the twentieth century, see Westmarland, "Quantitative/Qualitative Debate and Feminist Research"; Fonow and Cook, "Feminist Methodology"; Kelly, Regan, and Burton, "Defending the Indefensible?"

15. Merry, *Seductions of Quantification*.

16. There is now a wealth of scholarship on the harms of big data. For some starting points, see D'Ignazio and Klein, *Data Feminism*; Umoja Noble, *Algorithms of Oppression*; Crawford, Gray, and Miltner, "Critiquing Big Data"; Mattern, *City Is Not a Computer*; Walter and Andersen, *Indigenous Statistics*; Buolamwini and Gebru, "Gender Shades."

17. See, for example, the work of the Algorithmic Justice League, the Ban Facial Recognition coalition, and the Stop LAPD Spying Coalition.

18. See Pine and Liboiron, "Politics of Measurement and Action," 3149.

19. Stray, *Curious Journalist's Guide to Data*.

20. The designation of Hispanic as a US census category has an interesting history of its own. In 1930, the census included the category "Mexican" but dropped it in subsequent decades. The category of "Hispanic" was introduced in the 1980 census following advocacy from Mexican-American policy groups as well as political elites, who, Laura E. Gomez asserts, may have used it as an accommodationist counterpoint to "Chicano," which was perceived to be aligned with radical politics. See Gomez, "Birth of the 'Hispanic' Generation." Ultimately "Hispanic" was added as an ethnicity category, and information scholar Melanie Feinberg discusses how advocates tried to reform and simplify ethnic and racial categories in the 2020 census but were ultimately blocked by a right-wing government (whose interests were served by undercounting Hispanics). Although the 2020 census citizenship question received a great deal of press, Feinbaum asserts that the consequences of *not* reforming race and ethnic categories had similar effects as the citizenship question: the significant undercounting of Hispanic/Latinx people in the United States. See Feinberg, *Everyday Adventures with Unruly Data*, 125–126, 196.

21. Merry, *Seductions of Quantification*.

22. See Suárez Val, "Discordant Data," 54. Also, thanks to Susana Galen for pointing out that naming the violence is a long-standing feminist strategy, reaching back at least to second-wave feminists' conceptualization of marital rape and sexual harassment to describe practices that were normalized as routine aspects of a woman's life.

23. De Haan, "Violence as an Essentially Contested Concept"; cited in Krook, "Continuum of Violence."

24. Kelly, "Continuum of Sexual Violence."

25. Pine and Liboiron, "Politics of Measurement and Action," 3153; Liboiron, "Plastic Pollution."

26. See Alcaraz, "#NiUnaMenos."

27. Collins, *Intersectionality as Critical Social Theory*, 35–36.

28. Collins, "On Violence, Intersectionality and Transversal Politics."

29. This continues in a historical trajectory of women's and feminist groups' work going unacknowledged for their significant efforts to assemble data about gender-related violence, only to have it appropriated without attribution by others. For example, writer Roberto Bolaño's novel *2666* goes into graphic detail about murders of women and girls, which are widely known to

have been inspired by his research into the Ciudad Juárez crisis and communications with journalist Sergio Gonzalez. What is less widely known, according to scholar Julia E. Monárrez Fragoso, is that to create his "fictionalization," Bolano used the database assembled by the 8 de Marzo women's activist group, which contained detailed information about specific cases. He did this without attribution of the group's work, taking credit for creating "masterful" narratives of violence without acknowledging the labor and real people's lives that made it possible.

30. Ricaurte, "Data Epistemologies, the Coloniality of Power, and Resistance."

31. See the glossary in chapter 8 for definitions of systems of power such as patriarchy, settler colonialism, and racial capitalism.

32. Suresh, "Framework for Understanding Sources of Harm"; Gebru et al., "Datasheets for Datasets"; Gebru et al., "Documentation to Facilitate Communication"; Mitchell et al., "Model Cards for Model Reporting"; Peng, Mathur, and Narayanan, "Mitigating Dataset Harms Requires Stewardship."

33. Pine and Liboiron, "Politics of Measurement and Action."

34. Pickles, *History of Spaces*.

35. Porter, *Trust in Numbers*, 26–27, 206.

36. Scott, *Seeing Like a State*.

37. Benjamin, *Race after Technology*; Rich, *Blood, Bread, and Poetry*, 3.

38. These are discussed further in chapter 6. One key example from the Alianza is the multimedia mapping project Flores en el Aire (2022): https://www.spotlightinitiative.org/es/news/flores-en-el-aire-mapas-para-la-memoria-y-reparacion-de-victimas-del-femicidio.

39. See the toolkit in chapter 8 for more on data epistemologies.

40. DiSalvo, *Design as Democratic Inquiry*, 165.

CHAPTER 4

1. The Supreme Court of Argentina keeps a registry of femicides, which activists initially criticized for not including transfemicides, travesticidios, and cases where the perpetrator committed suicide. While they have reformed some of these counting practices, La Casa del Encuentro, Argentina's longest-running civil society femicide observatory, still considers the court's database to be significantly undercounting femicides compared with their own numbers. Ada Rico, president of the organization, stated to our team unequivocally, "No. We do not believe the official data."

2. The name of their observatory in Spanish is Observatorio MuMaLá: Mujeres, Disidencias, Derechos. Here I chose not to translate the concept of *disidencias* because there is not a completely corresponding concept in English. *Disidencias* names gender identities that are "dissident"— that is, that do not conform to cisgender and heterosexual norms and expectations. For more on Latin American theoretical approaches to disidencias, see Ortuño, "Teorías de la disidencia sexual." Although one could translate this in English as "LGBTQ+ people," the acronym style of

naming these identities in English does not communicate the same insurgent meaning as "dissident identities." Disidencias is something a bit closer to Kate Bornstein's framing of "gender outlaws"; see Bornstein, *Gender Outlaw*. That said, there is also not total agreement on the concept of disidencias across Latin American regions and political struggles either. For example, some members of the community reject its usage on the grounds that gender and sexual normativity itself is a fallacy.

3. *Compañeras* designates women members and *compañeres* designates nonbinary members of the collective. The *-es* ending for gendered language in Spanish is part of a movement, in Argentina and across Latin America, for the inclusion of nonbinary people in a language that has historically not acknowledged their existence. That said, the movement has faced right-wing backlash and bans. See Lankes, "In Argentina, One of the World's First Bans on Gender-Neutral Language."

4. Sovereign Bodies, *To' Kee Skuy' Soo Ney-Wo-Chek'*.

5. Sovereign Bodies Institute, *MMIWG2 & MMIP Organizing Toolkit*.

6. See the concept of the information ecosystem as outlined in "Why Information Matters," 10–17.

7. I'll take this moment to remind you that because the geographic scope of this project is broad—grassroots feminicide data activism in the Americas—this is not a comparative geographic analysis. For example, I'm not going to say "Here's what activists in Argentina said about missing data versus what those in Mexico said and what that reveals about those two places." The primary goal here is to look at common informatic challenges around feminicide data that surface for activists *despite* significant variation in geographic context. This is in line with Chandra Mohanty's notion of feminist solidarity that works by focusing on *commonalities across differences*—examining common struggles across diverse contexts while still attending to local differences and particularities; see Mohanty, "'Under Western Eyes' Revisited."

8. Google Alerts was used as a case discovery tool by many activists but often discarded because they encountered many irrelevant results. These experiences led to the co-design of an email alerts tool I describe further in chapter 7.

9. Crenshaw et al., *Say Her Name*.

10. D'Ignazio and Klein, *Data Feminism*, 149.

11. The Not Invisible Act (2019) had provisions for improving MMIW data collection and accessibility, along with directing the Department of Justice to develop better protocols to address missing and murdered Indigenous people.

12. Puig de la Bellacasa, "Matters of Care in Technoscience"; Suárez Val, Martinez Cuba, Teng, and D'Ignazio, "Datos de Feminicidio."

13. Suárez Val, "Affect Amplifiers."

14. Sun and Yin, "Opening up Mediation Opportunities."

15. West, Whittaker, and Crawford, *Discriminating Systems*.

16. US Bureau of Labor Statistics, "Labor Force Statistics from the Current Population Survey."

17. Duffy, *Making Care Count.*

18. Feinberg, *Everyday Adventures with Unruly Data*, 58.

19. Roberts, Sarah T., "Digital Refuse"; Irani, "Cultural Work of Microwork"; Miceli, Posada, and Yang, "Studying Up Machine Learning Data."

20. Roberts, *Behind the Screen.*

21. Posada, "Coloniality of Data Work"; Hao and Hernández, "How the AI Industry Profits from Catastrophe"; Miceli, Posada, and Yang, "Studying Up Machine Learning Data."

22. See Sambasivan et al., "'Everyone Wants to Do the Model Work'"; Muller and Strohmayer, "Forgetting Practices in the Data Sciences."

23. Rothschild et al., "Interrogating Data Work as a Community of Practice."

24. See also Lucchesi, "Mapping Violence against Indigenous Women and Girls."

25. Ricaurte, "Data Epistemologies, the Coloniality of Power, and Resistance."

26. Feinberg, *Everyday Adventures with Unruly Data*, 59.

27. For example, Sambasivan et al. talk about the ways that ignorance about upstream data limitations lead to *data cascades*—the magnification of harms downstream in the process. In the high-profile case involving Timnit Gebru and Margaret Mitchell from Google, researchers were censored and then fired for calling attention to the limitations of and significant potential harms of large language models. Sambasivan et al., "'Everyone Wants to Do the Model Work'"; Bender et al., "On the Dangers of Stochastic Parrots."

CHAPTER 5

1. Colectivo de Geografía Crítica del Ecuador, *Manifiesto Contra La Violencia Hacia Las Mujeres Desde La Geografía Crítica.*

2. Fundación ALDEA, *Feminicídios en Ecuador.*

3. Native Women's Association of Canada, *Voices of Our Sisters in Spirit*; Pearce, "Awkward Silence"; It Starts with Us, "Background."

4. Gartner, Dawson, and Crawford, *Woman Killing.*

5. Dawson, "Punishing Femicide."

6. D'Ignazio, "5 Questions on Data and Feminicide with Silvana Fumega."

7. Taillieu and Brownridge, "Violence against Pregnant Women."

8. Fundación ALDEA, *Feminicídios en Ecuador.*

9. George, Lucchesi, and Trillo, *To' Kee Skuy' Soo Ney-Wo-Chek'.*

10. This doesn't necessarily mean those categories are irrelevant, but it does reflect the difficulty in obtaining information to determine such categories. Whether a person was trafficked,

smuggled, or experienced FGM would be extremely difficult for activists to ascertain unless it was explicitly reported in the press.

11. Carcedo, *No olvidamos ni aceptamos*, 15; Sagot, "Violence against Women."

12. McLemore and D'Efilippo, "To Prevent Women from Dying in Childbirth, First Stop Blaming Them."

13. Trying to document indirect feminicides gets complicated when the official data are themselves missing. For example, see the Lost Mothers project about maternal mortality, discussed in chapter 8.

14. Bowker and Star, "Building Information Infrastructures for Social Worlds," 235.

15. Martin and Lynch, "Counting Things and People."

16. Collins, *Intersectionality as Critical Social Theory*, ch. 3.

17. Bowker and Star, *Sorting Things Out*; Bouk, Ackermann, and boyd, *Primer on Powerful Numbers*.

18. Helena Suárez Val's data frame is different from a data frame in the R programming language, but perhaps not as different as we might think. The R data frame operationalizes and arranges data into rows and columns, and Helena's data frame reflects on the implications of such arrangements, what impacts they enable, and what impacts they foreclose. Suárez Val, "Data Frames of Feminicide."

19. Grupo Guatemalteco de Mujeres, *Informe de Muertes Violentas de Mujeres-MVM En Guatemala*.

20. Bowker and Star, "Building Information Infrastructures for Social Worlds."

21. Other groups describe this violence as "Feminicide, stigmatized occupations" so that it could include strippers, escorts, and others who may not actually engage in sexual acts. And while the prostitution/sex work position is often depicted as a binary, there are also places where these two strands overlap and agree. A good primer complicating the sex worker/abolitionist binary can be found in Mackay, "Arguing against the Industry of Prostitution." For work on advocating for sex workers' rights, especially in relation to digital technology, see Kuo and Lee, *Dis/Organizing: How We Build Collectives beyond Institutions*.

22. Canadian Feminicide Observatory for Justice and Accountability, "Types of Femicide"; Canadian Femicide Observatory for Justice and Accountability, *Call It Femicide*; UNODOC, *Global Study on Homicide*, 36.

23. The exclusion of trans voices, and particularly trans women, from feminism is called *TERF feminism*, with TERF standing for *trans-exclusionary radical feminism*. Proponents of this position do not call themselves TERFs but rather use the term *gender critical*. Scholars have pointed out that TERF politics are bound up with gender essentialism and whiteness. For a further introduction, see Pearce, Erikainen, and Vincent, "TERF Wars."

24. Suárez Val, "Marcos de Datos de Feminicidio."

25. Caswell, "From Human Rights to Feminist Ethics."

26. Ruppert, Isin, and Bigio, "Data Politics," 1.

27. Mills, *Racial Contract*, 18.

28. Walter, "Conceptualizing and Theorizing the Indigenous Life."

29. Bouk, Ackermann, and boyd, *Primer on Powerful Numbers*.

30. Merry, *Seductions of Quantification*, 77.

31. Bailey, "Strategic Ignorance."

32. Suárez Val, "Caring, with Data."

33. Ruppert, Isin, and Bigio, "Data Politics," 2.

34. Lucchesi, "Mapping Violence against Indigenous Women and Girls," 390.

35. Ruppert, Isin, and Bigio, "Data Politics," 5.

36. Bold, Knowles, and Leach, "Feminist Memorializing and Cultural Countermemory." The domain of feminicide data activism is not unique in this regard. Another powerful contemporary project that foregrounds the use of data in the service of countermemory is COVID Black, discussed further in chapter 8.

37. Ruppert, Isin, and Bigio, "Data Politics," 2.

38. Hanna and Park, "Against Scale."

39. Tsing, "On Nonscalability," 506.

40. This resonates with the way in which Lauren Klein uses topic modeling to probe the multiple scales of invisible labor in abolitionist archives from the nineteenth century. See Klein, "Dimensions of Scale."

CHAPTER 6

1. See chapter 1, note 21, and chapter 4, note 2, for why I chose not to translate *travesti* and *disidencias*.

2. Kedar, *International Monetary Fund and Latin America*.

3. The formulation of these themes of activist data communication draws in part from a session Lauren and I co-organized with some of the authors of the Feminist Data Manifest-No, including Anita Chan, Tonia Sutherland, Marika Cifor, Patricia Garcia, T. L. Cowan, Lisa Nakamura, and Jasmine Rault, at the Allied Media Conference in 2020. The title of that workshop was "Feminist Data: Refuse, Reform, Reimagine, Revolt."

4. Wright, "Necropolitics, Narcopolitics, and Femicide." Other significant works about feminicide and necropolitics include Sagot Rodríguez, "Femicide as Necropolitics in Central America"; Fuentes, "(Re)reading the Boundaries and Bodies of Femicide."

5. Romeo, "Towards a Theory of Digital Necropolitics," 8–9.

6. Threadcraft, "North American Necropolitics and Gender."

7. Carlson, "Data Silence in the Settler Archive."

8. Snorton and Haritaworn, "Trans Necropolitics," 314.

9. Portal NE10, "#UmaPorUma."

10. Tronto and Fisher, "Toward a Feminist Theory of Caring."

11. See, for example, Chambers-Letson and Diaz, "Reparations."

12. Mobayed, "Kintsugi Method to Recount Data against Feminicide."

13. Otros Mapas, "Flores en el Aire."

14. See Janssen and Singh, "Data Intermediary."

15. In the case of the group that did not remove the person requested, it was because the perpetrator had been convicted of killing his wife and was serving time in jail. The children stated that they believed in their father's innocence, and the organization talked with them at length, on the phone and in person. Ultimately, the organization decided not to remove the name since the judicial system had ruled against the man.

16. Blanco and García, "Cortar el Hilo."

17. Blanco and García.

18. A short documentary about the stitching process is viewable at https://www.youtube.com/watch?v=Hu65ww5Ijrg.

19. Wendel, *Rwanda's Genocide Heritage*.

20. Members of the Red Chilena contra la Violencia hacia las Mujeres described to our team how they have faced misogynist pushback to their memorial when they have installed it in public spaces in Chile. Men have walked on top of it on purpose or told the women taking care of the memorial to "go back to your kitchens" or said of the killed women, "There must have been a reason they were killed." See http://www.nomasviolenciacontramujeres.cl/lazamiento-campana-regiones-2019/.

21. Suárez Val et al., "Data Artivism and Feminicide."

22. See No Estamos Todas at https://www.instagram.com/noestamostodas. In November 2022, Data Against Feminicide organized a panel discussion and workshop in which No Estamos Todas described more about their process; see https://datoscontrafeminicidio.net/visibilidades-arte-y-datos-de-feminicidio/.

23. As described in chapter 5, Dawn Wilcox, who runs Women Count USA, pays special attention to collecting (and sometimes retouching) the photos of women in her database. She created a video from all of the photos from her database in 2018 and set it to music. And in 2021, the African American Policy Forum partnered with Janelle Monae and other celebrities, plus the #SayHerName Mothers Network, to release a video—"Say Her Name (Hell You Talmbout)"—to honor the lives of Black women and protest their killings by police. Women & Girls Lost to Male Violence in 2018; Janelle Monáe, "Say Her Name (Hell You Talmbout)."

24. Asociación Civil La Casa del Encuentro and Beatriz Rico, *Por Ellas*.

25. Dean, *Remembering Vancouver's Disappeared Women*, 4.

26. Black, *REDress Project*.

27. Harjo, "Community Caretaking."

28. Red Feminista Antimilitarista, *Paren La Guerra Contra Las Mujeres*, 7–8.

29. "Presentes."

30. "Presentes."

31. "Argentina's Third Violent Transgender Death in a Month Sparks Call for Justice."

32. Such maps, their ethics, and their effects have been the subject of excellent scholarship by Annita Hetoevehotohke'e Lucchesi and Helena Suárez Val. See Suárez Val, "Affect Amplifiers"; Suárez Val, "Vibrant Maps"; Lucchesi, "Spatial Data and (De)colonization."

33. Walter, "Voice of Indigenous Data"; D'Ignazio and Klein, *Data Feminism*, 58–59.

34. Grupo Guatemalteco de Mujeres, *Informe de Muertes Violentas de Mujeres-MVM En Guatemala*.

35. Veronica Cúzco, representing Ecuador's National Institute of Statistics and Censuses, gave a public talk discussing this collaboration to the Data Against Feminicide course that Silvana, Helena, and I organized in Spring 2020. See ILDA, "Conversatorio.

36. This law is commonly referred to using the name of a child orphaned by feminicide in a high-profile case, but groups are trying to avoid the use of her name so as not to revictimize her. See the Data Against Feminicide event coproduced with the Interamerican Network against Femicide (RIAF): "More than Numbers: Voices of Relatives of Victims of Femicide," https://datoscontrafeminicidio.net/en/visibilities-art-and-feminicide-data/.

37. Bureau of Democracy, Human Rights, and Labor, *2021 Country Reports on Human Rights Practices*.

38. Gellman, "Landmark Femicide Case Fails to Fix El Salvador's Patriarchy."

39. Bowleg, "'Master's Tools Will Never Dismantle the Master's House'"; Krieger, "Data, 'Race,' and Politics."

40. Kennedy et al., "Work That Visualisation Conventions Do."

41. Fiscalía General de Justicia—Ciudad de México, "Atlas de Feminicidios de la Ciudad de México."

42. Pine and Liboiron, "Politics of Measurement and Action."

43. Here I use LGBTTTIQ+ because this is what Data Cívica uses. This acronym stands for lesbian, gay, bisexual, transgendered, transsexual, two-spirit, intersexed, queer and beyond.

44. Andrés Manuel López Obrador (a.k.a. AMLO), the Mexican president in 2020, has blamed feminicide on his predecessor's economic policy and on moral decay. He has repeatedly denounced feminist graffiti and protests, claimed that he is being unfairly targeted by the feminist movement, and is widely seen as not engaging seriously with the issue of feminicide. As Maricruz Ocampo, an activist quoted in the *Guardian* stated, "The message he's sending women is: I don't care." Agren, "'Message He's Sending Is I Don't Care.'"

45. There are, in fact, counterdata being assembled to document these actions. A group called the Restauradoras con Glitter (Preservationists with Glitter), comprised of preservationists associated

with Mexico's most prestigious museums, is working on assembling archives of feminist protests and documentation of graffiti on monuments. In some cases, they have intervened to delay
government cleaning so they can visually document feminist graffiti. Restauradoras con glitter,
Facebook page.

46. Buolamwini, "Facing the Coded Gaze."

47. See Cifor et al., "Feminist Data Manifest-No."

48. Lorde, "Master's Tools Will Never Dismantle the Master's House."

49. Lorde.

50. Crooks and Currie, "Numbers Will Not Save Us."

51. Lorde, "Master's Tools Will Never Dismantle the Master's House."

52. Carlson, "Data Silence in the Settler Archive," 99.

53. Vera et al., "When Data Justice and Environmental Justice Meet."

54. Muravyov, "Doubt to Be Certain."

55. Aguirre, "NO SON NÚMEROS, SON VIDAS QUE NOS ARREBATARON . . ."

56. Crooks and Currie, "Numbers Will Not Save Us"; Suárez Val, "Caring, with Data."

CHAPTER 7

1. See the glossary in chapter 8 for short definitions of systems of power like white supremacy,
settler colonialism, and patriarchy.

2. A great many authors have described the extractive tendencies of academic research in relation
to minoritized communities, and these insights have led to the development of whole research
methodologies such as participatory design (PD), which this case study discusses at length, as well
as participatory action research (PAR), community-based participatory action research (CBPR),
feminist participatory action research (FPAR), insurgent research, research justice, and more. If
this is a new perspective for you, a brilliant starting point is Linda Tuhiwai Smith's classic book,
Decolonizing Methodologies.

3. Muller, "Participatory Design"; Bødker et al., "Utopian Experience."

4. For a starting point for some of this work, see these articles: Ogbonnaya-Ogburu et al., "Critical Race Theory for HCI"; Rankin, Thomas, and Joseph, "Intersectionality in HCI"; Thomas et
al., "Discovering Intersectionality: Part 2"; Fox et al., "Imagining Intersectional Futures"; Bray
and Harrington, "Speculative Blackness"; Kumar et al., "Engaging Feminist Solidarity"; Alvarado
Garcia et al., "Decolonial Pathways."

5. Feminist HCI directly influenced me and Lauren Klein as we articulated the data feminism
principles (see chapter 8). Bardzell, "Feminist HCI." Another important HCI paper we drew from
was Dörk et al., "Critical InfoVis."

6. Dimond, Fiesler, and Bruckman, "Domestic Violence and Information Communication Technologies"; Gautam, Tatar, and Harrison, "Crafting, Communality, and Computing"; Haimson et

al., "Designing Trans Technology"; Scheuerman, Branham, and Hamidi, "Safe Spaces and Safe Places"; Sterling, "Designing for Trauma."

7. Dimond, "Feminist HCI for Real."

8. Silveira, dos Santos, and da Maia, "Estamos Juntas."

9. Shelby, "Technology, Sexual Violence, and Power-Evasive Politics," 558.

10. For more on carceral creep in HCI, see Kuo and Mohapatra, "Institutional Capture of Abolitionist Dissent"; Chordia, "Leveraging Transformative Justice."

11. Vigil-Hayes et al., "#Indigenous."

12. Strohmayer et al., "'We Come Together as One.'"

13. Crooks and Currie, "Numbers Will Not Save Us"; Cullen et al., "Intersectionality and Invisible Victims"; Ruse, "Experiences of Engagement and Detachment"; Suárez Val et al., "Monitoring, Recording, and Mapping Feminicide"; Suárez Val, Martinez Cuba, Teng, and D'Ignazio, "Datos de Feminicidio."

14. Alvarado García, Young, and Dombrowski, "On Making Data Actionable."

15. Krüger et al., "It Takes More Than One Hand to Clap."

16. This resonates with Donna Haraway's call to "stay with the trouble." And public scholar and designer Mushon Zer-Aviv exhorts us to use design to embrace (rather than eliminate) friction. Haraway, *Staying with the Trouble*; "Mushon Zer-Aviv Presents Friction & Flow at Better World X Design 2022."

17. See the introduction for a longer explanation of the origins and goals of Data Against Feminicide.

18. DiSalvo, *Design as Democratic Inquiry*; Dantec and DiSalvo, "Infrastructuring and the Formation of Publics in Participatory Design."

19. Asad, "Prefigurative Design as a Method for Research Justice."

20. For a good primer on how feminism and anticarceral approaches intersect, see Davis et al., *Abolition. Feminism. Now.* The abolition that the authors are talking about is the abolition of mass incarceration, not the abolition of sex work (which is often what abolition means in Latin American feminist discourse and is, paradoxically, a carceral approach to sex work that results in criminalizing its practitioners). For a practical primer on abolition, see Interrupting Criminalization, Project Nia, and Critical Resistance, *So Is This Actually an Abolitionist Proposal or Strategy?*

21. Lecher, "Police Are Looking to Algorithms to Predict Domestic Violence."

22. Honeywell, "Honeywell Wins Bengaluru Safe City Project."

23. Radhika Radhakrishnan, PhD student and research affiliate in the Data + Feminism Lab, has written about the ways that the Safe City project surveils and controls women's bodies. Radhakrishnan, "Cost of Safety." In the US context, Andrea Ritchie details the long history of sexual assault and violence perpetrated by law enforcement against Black women, gender nonconforming people, and women of color. Ritchie, *Invisible No More*.

24. Shokooh Valle, "'How Will You Give Back?'"; Stengers, *Invention of Modern Science*, 90.

25. Bødker and Kyng, "Participatory Design that Matters," 10.

26. De Castro Leal, Strohmayer, and Krüger, "On Activism and Academia," 6.

27. In fact, Helena often cites the "Friendship as Method" paper to describe how we work together. See Tillmann-Healy, "Friendship as Method."

28. Asad, "Prefigurative Design as a Method for Research Justice."

29. I'm not asserting that everyone could or should use start-up funds or leftover grant money. This happened to be a tactic available to us that we used to secure ourselves time and flexibility for exploration and relationship-building while still being able to compensate interviewees and participants as well as staff and students working on the project. If you are reading this and you are a funder, please consider investing in long-term relationship building and unrestricted operational funds over funding individual projects and novelties! For more on the structural limitations imposed by the nonprofit industrial complex, see INCITE! Women of Color against Violence, *The Revolution Will Not Be Funded*.

30. Okun, "White Supremacy Culture."

31. Interviewees were paid seventy-five dollars or the equivalent in their home currency, either to themselves personally or their organization, per their preference. Participants in the co-design process were paid hourly at twenty-five dollars per hour. Participants in the pilot study in which groups used our tools and provided feedback on them were paid $200 per organization.

32. Smith, *Decolonizing Methodologies*, xi.

33. It takes a village to raise a software tool. Please see the acknowledgments for all the staff, students, and activists who contributed to the tools.

34. Roberts et al., "Media Cloud."

35. A team of five people, including myself, Helena, and three graduate students—Angeles Martinez Cuba, Mariel García-Montes, and Harini Suresh—labeled the first sets of training data in Spanish and English. The process of reading articles about brutal violence was intense for all of us. We held space for discussing our emotional reactions and any stories that stayed with us. I am grateful to scholar and UX researcher Julia DeCook for speaking with us about secondary trauma at one of those sessions. The HCI community has recently begun to discuss researcher trauma and systematize resources for researchers engaging emotionally demanding topics through a workshop: Feuston et al., "Researcher Wellbeing and Best Practices in Emotionally Demanding Research."

36. D'Ignazio et al., "Feminicide & Machine Learning."

37. See https://datoscontrafeminicidio.net/en/2021-edition/.

38. Abebe et al., "Roles for Computing in Social Change."

39. We have not made these datasets public to date because we see potential for misuse and misinterpretation of ML models about feminicide by actors who have less expertise than the activists in feminicide and in the flawed information ecosystem surrounding feminicide, as well as fewer

ethical commitments to and relations with the people described by these news stories. That said, we encourage projects to get in touch with us if they are interested in using our training data and we would be happy to talk.

40. Sovereign Bodies, *To' Kee Skuy' Soo Ney-Wo-Chek'*.

41. D'Ignazio and Klein, *Data Feminism*, 149.

42. Harini Suresh presented this paper at the FAccT 2022 conference, and it received a Distinguished Student Paper Award. Suresh et al., "Towards Intersectional Feminist and Participatory ML."

43. There are many ways in which participation and participatory methods can themselves be exploitative and undemocratic. For example, by requiring enormous amounts of time and free labor from communities, by imposing inappropriate cultural norms of participation from outside a community, by requiring people to show up at inconvenient times to have a basic say in their political future (e.g., voting in the United States), by operating as window-dressing for decisions that have already been made by people in power, by profiteering on people's participatory contributions (e.g., social media), and more. See, for example, Terranova, "Free Labor"; Cooke and Kothari, *Participation*; Busch and Palmås, *Corruption of Co-Design*.

44. This is why the widely cited "Gender Shades" paper by Joy Buolamwini and Timnit Gebru used the Fitzpatrick skin type classification system from the field of dermatology (a skin color scale) rather than racial classification to demonstrate bias in facial recognition technologies. Buolamwini and Gebru, "Gender Shades." For more on race as a technology, see the work of Falguni Sheth, Wendy Chun, Beth Coleman and Ruha Benjamin: Sheth, "Technology of Race"; Chun, "Introduction: Race and/as Technology"; Coleman, "Race as Technology"; Benjamin, "Catching Our Breath."

45. Thank you to Annita Lucchesi for making this point during the review process.

46. The same can be said about gender—which is also a social, political, and historical system of classification. There is a growing body of work that is challenging technologies that try to undertake gender inference from names, photos, and voices, as these are often based on harmful, trans-exclusionary, binary (and empirically wrong!) notions of gender. See Keyes, "Misgendering Machines"; Scheuerman et al., "How We've Taught Algorithms to See Identity"; Albert and Delano, "Sex Trouble."

47. De Castro Leal, Strohmayer, and Krüger, "On Activism and Academia"; Bødker and Kyng, "Participatory Design That Matters."

48. Sloane et al., "Participation Is Not a Design Fix for Machine Learning."

49. Bødker and Kyng, "Participatory Design That Matters." Also see the group called The Maintainers, who advocate for putting a focus on the care and maintenance of technical and informatic systems: https://themaintainers.org/about/.

50. Irani and Silberman, "From Critical Design to Critical Infrastructure."

51. The authors emphasize that deploying these technologies is not about technosolutionism for trans people: "We, and many of the participants in our study, are well aware that technology

cannot solve the systemic problems that trans people face, but that technology can make small inroads to making trans lives more livable." Haimson et al., "How Transgender People and Communities Were Involved in Trans Technology Design Processes."

52. Ghoshal, Mendhekar, and Bruckman, "Toward a Grassroots Culture of Technology Practice," 7.

53. Ethan Zuckerman runs the Initiative for Digital Public Infrastructure at the University of Massachusetts Amherst: https://publicinfrastructure.org/. See also Zuckerman, "Case for Digital Public Infrastructure."

54. I want to thank James Scott-Brown for pointing me to Oliver's work. Oliver, Savičić, and Vasiliev, *Critical Engineering Manifesto*; Oliver, "36C3—Server Infrastructure for Global Rebellion."

55. Barendregt et al., "Defund Big Tech, Refund Community."

56. Bødker and Kyng, "Participatory Design That Matters," 19.

CHAPTER 8

1. Although this quote ended up circulating on the internet as the work of one person—Lilla Watson—Watson herself describes it as the outcome of a collective process, and she desired that it be credited as "Aboriginal activists group, Queensland, 1970s."

2. Suárez Val, "Datos Discordantes."

3. Adapted from D'Ignazio and Klein, *Data Feminism*.

4. Adapted from UnLeading, "Ableism."

5. Adapted from Murrey, "Colonialism."

6. Adapted from Vera et al., "When Data Justice and Environmental Justice Meet."

7. Drawn from Tuck and Yang, "Decolonization Is Not a Metaphor."

8. From the Challenging White Supremacy Workshop: https://www.cwsworkshop.org.

9. Nenquino, "Mapeo Territorial Waorani."

10. McLemore and D'Efilippo, "To Prevent Women from Dying in Childbirth, First Stop Blaming Them."

BIBLIOGRAPHY

Abebe, Rediet, Solon Barocas, Jon Kleinberg, Karen Levy, Manish Raghavan, and David G. Robinson. "Roles for Computing in Social Change." In *Proceedings of the 2020 Conference on Fairness, Accountability, and Transparency*, 252–260. New York: Association for Computing Machinery, 2020.

Agren, David. "'The Message He's Sending Is I Don't Care': Mexico's President Criticized for Response to Killings of Women." *The Guardian*, February 21, 2020. https://www.theguardian.com /world/2020/feb/21/mexico-femicide-crisis-amlo-response.

Aguirre, Victoria. "NO SON NÚMEROS, SON VIDAS QUE NOS ARREBATARON . . ." Mumalá—Mujeres de la Matria Latinoamericana, Facebook, March 23, 2022. https://www.facebook.com /MumalaNacional/posts/288753676753283.

Albert, Kendra, and Maggie Delano. "Sex Trouble: Sex/Gender Slippage, Sex Confusion, and Sex Obsession in Machine Learning Using Electronic Health Records." *Patterns* 3, no. 8 (August 12, 2022). https://doi.org/10.1016/j.patter.2022.100534.

Alcaraz, Maria Florencia. "#NiUnaMenos: Politicising the Use of Technologies." *Feminist Talk* (blog), September 4, 2017. https://www.genderit.org/feminist-talk/special-edition-niunamenos -politicising-use-technologies.

Alexandra, Zoe. "Puerto Rico between Neoliberalism and 'Natural' Disasters." *Peoples Dispatch*, February 4, 2020. https://peoplesdispatch.org/2020/02/04/puerto-rico-between-neoliberalism-and -natural-disasters/.

Alvarado Garcia, Adriana, Juan F. Maestre, Manuhuia Barcham, Marilyn Iriarte, Marisol Wong-Villacres, Oscar A. Lemus, Palak Dudani, Pedro Reynolds-Cuéllar, Ruotong Wang, and Teresa Cerratto Pargman. "Decolonial Pathways: Our Manifesto for a Decolonizing Agenda in HCI Research and Design." In *Extended Abstracts of the 2021 CHI Conference on Human Factors in Computing Systems*, edited by Yoshifumi Kitamura, Aaron Quigley, Katherine Isbister, and Takeo Igarashi, 1–9. New York: Association for Computing Machinery, 2021.

Alvarado García, Adriana, Alyson L. Young, and Lynn Dombrowski. "On Making Data Actionable: How Activists Use Imperfect Data to Foster Social Change for Human Rights Violations in Mexico." *Proceedings of the ACM on Human–Computer Interaction* 1, no. CSCW (December 6, 2017): article 19, 1–19. https://doi.org/10.1145/3134654.

Alvarez, Sonia E., Claudia de Lima Costa, Veronica Feliu, Rebecca Hester, Norma Klahn, and Millie Thayer. *Translocalities/Translocalidades: Feminist Politics of Translation in the Latin/a Américas*. Durham, NC: Duke University Press, 2014.

Ámbito. "Raquel Vivanco Renunció al Ministerio de Mujeres Por 'No Encontrar Síntesis En Lo Que Desde Allí Se Debe Generar.'" *Ámbito*, March 3, 2022. https://www.ambito.com/politica/mujeres/raquel-vivanco-renuncio-al-ministerio-no-encontrar-sintesis-lo-que-alli-se-debe-generar-n5384709.

Annunziata, Rocío, Emilia Arpini, Tomás Gold, and Bárbara Zeifer. "El Caso de Argentina." In *Activismo político en la era de internet*, edited by B. Sorj and S. Fausto, 35–110. São Paulo: Ediçoes Plataforma Democrática, 2016.

Anti-Eviction Mapping Project. *Counterpoints: A San Francisco Bay Area Atlas of Displacement & Resistance*. Oakland, CA: PM Press, 2021.

"Argentina's Third Violent Transgender Death in a Month Sparks Call for Justice." *The Guardian*, October 14, 2015. https://www.theguardian.com/world/2015/oct/14/argentina-transgender-women-violence-cristina-fernandez.

Asad, Mariam. "Prefigurative Design as a Method for Research Justice." *Proceedings of the ACM on Human–Computer Interaction* 3, no. CSCW (November 2019): article 200, 1–18. https://doi.org/10.1145/3359302.

Asociación Civil La Casa del Encuentro and Ada Beatriz Rico. *Por Ellas: 10 Años de Informes de Femicidios en Argentina*. Buenos Aires: Asociación Civil La Casa del Encuentro, 2020. https://www.porellaslibro.com/libros/porellas2020-spanish.pdf.

Avilés, Luis A., and Luis Emmanuel Rodríguez Reyes. *La Persistencia de La Indolencia: Feminicidios En Puerto Rico 2014–2018*. San Juan: Proyecto Matria and Kilómetro 0, 2019.

Avilés, Luis A., and Luis Emmanuel Rodríguez Reyes. *The Persistence of Indolence: Feminicides in Puerto Rico 2014–1018*. Translated by Eïrïc R. Durändal-Stormcrow. San Juan: Matria Project and Kilómetro 0, 2019.

Bailey, Alison. "Strategic Ignorance." In *Race and Epistemologies of Ignorance*, edited by Shannon Sullivan and Nancy Tuana, 77–94. Albany: State University of New York Press, 2007.

Ball, Patrick. "The Bigness of Big Data: Samples, Models, and the Facts We Might Find When Looking at Data." In *The Transformation of Human Rights Fact-Finding*, edited by Philip Alston and Sarah Knuckey, 425–440. Oxford: Oxford University Press, 2015.

Bardzell, Shaowen. "Feminist HCI: Taking Stock and Outlining an Agenda for Design." In *CHI '10: Proceedings of the SIGCHI Conference on Human Factors in Computing Systems*, 1301–1310. Atlanta: Association for Computing Machinery, 2010. https://dl.acm.org/doi/10.1145/1753326.1753521.

Barendregt, Wolmet, Christoph Becker, EunJeong Cheon, Andrew Clement, Pedro Reynolds-Cuéllar, Douglas Schuler, and Lucy Suchman. "Defund Big Tech, Refund Community." Tech Otherwise, February 5, 2021. https://doi.org/10.21428/93b2c832.e0100a3f.

Barrowcliffe, Rose. "Closing the Narrative Gap: Social Media as a Tool to Reconcile Institutional Archival Narratives with Indigenous Counter-Narratives." *Archives and Manuscripts* 49, no. 3 (September 2, 2021): 151–166. https://doi.org/10.1080/01576895.2021.1883074.

Belotti, Francesca, Francesca Comunello, and Consuelo Corradi. "Feminicidio and #NiUna Menos: An Analysis of Twitter Conversations during the First 3 Years of the Argentinean Movement." *Violence against Women* 27, no. 8 (2021): 1035–1063.

Bender, Emily M., Timnit Gebru, Angelina McMillan-Major, and Shmargaret Shmitchell. "On the Dangers of Stochastic Parrots: Can Language Models Be Too Big?" In *FAccT '21: Proceedings of the 2021 ACM Conference on Fairness, Accountability, and Transparency*, 610–623. New York: Association for Computing Machinery, 2021. https://dl-acm-org.libproxy.mit.edu/doi/10.1145/3442188.3445922.

Benítez, Rohry, Adriana Candia, Patricia Cabrera, Guadelupe De la Mora, Josefina Martínez, and Isabel Velázquez. *El Silencio Que La Voz deTodas Quiebra: Mujeres y Víctimas de Ciudad Juárez*. Chihuahua: Ediciones de Azar, 1999.

Benjamin, Ruha. "Catching Our Breath: Critical Race STS and the Carceral Imagination." *Engaging Science, Technology, and Society* 2 (2016): 145–156.

Benjamin, Ruha. *Race after Technology: Abolitionist Tools for the New Jim Code*. Cambridge, MA: Polity Press, 2019.

Bhargava, Rahul, Harini Suresh, Amelia Lee Doğan, Wonyoung So, Helena Suárez Val, and Catherine D'Ignazio. "News as Data for Activists: A Case Study in Feminicide Counterdata Production." Paper presented at Computation + Journalism Conference, Brown Institute at Columbia, New York, June 11, 2022. https://api-web.dataplusfeminism.mit.edu/uploads/a1df3e3809e843c4a6a0d86e5b579889.pdf.

Bhattacharyya, Sayan. "Epistemically Produced Invisibility." In *Global Debates in the Digital Humanities*, edited by Paola Ricaurte, Sukanta Chaudhuri, and Domenico Fiormonte, 3–14. Minneapolis: University of Minnesota Press, 2022. https://muse.jhu.edu/pub/23/oa_edited_volume/book/100081.

Bietti, Elettra. "From Ethics Washing to Ethics Bashing: A View on Tech Ethics from within Moral Philosophy." In *Proceedings of the 2020 Conference on Fairness, Accountability, and Transparency*, 210–219. New York: Association for Computing Machinery, 2020. https://doi.org/10.1145/3351095.3372860.

Black, Jaime. *The REDress Project*. Accessed August 5, 2023. https://www.jaimeblackartist.com/exhibitions/.

Blanco, Marby, and Alejandra García. "Cortar el Hilo Artist Statement." March 2020.

Bødker, Susanne, Pelle Ehn, John Kammersgaard, Morten Kyng, and Yngve Sundblad. "A Utopian Experience." In *Computers and Democracy: A Scandinavian Challenge*, edited by Gro Bjerknes, Pelle Ehn, and Morten Kyng, 251–278. Aldershot: Avebury, 1987.

Bødker, Susanne, and Morten Kyng. "Participatory Design that Matters—Facing the Big Issues." *ACM Transactions on Computer-Human Interaction* 25, no. 1 (February 2018): 4:1–4:31. https://doi.org/10.1145/3152421.

Bold, Christine, Ric Knowles, and Belinda Leach. "Feminist Memorializing and Cultural Countermemory: The Case of Marianne's Park." *Signs: Journal of Women in Culture and Society* 28, no. 1 (2002): 125–148. https://doi.org/10.1086/340905.

Bornstein, Kate. *Gender Outlaw: On Men, Women, and the Rest of Us*. Revised and updated. New York: Vintage Books, 2016.

Bouk, Dan, Kevin Ackermann, and danah boyd. *A Primer on Powerful Numbers: Selected Readings in the Social Study of Public Data and Official Numbers*. New York: Data and Society, February 2022. https://datasociety.net/wp-content/uploads/2022/03/APrimerOnPowerfulNumbers_032022.pdf.

Boulding, Elise, Shirley A. Nuss, Dorothy L. Carson, and Michael A. Greenstein. *Handbook of International Data on Women*. Beverly Hills, CA: Sage Publications, 1976.

Bowker, Geoffrey C., and Susan Leigh Star. "Building Information Infrastructures for Social Worlds—The Role of Classifications and Standards." In *Community Computing and Support Systems*, edited by T. Ishida, 231–248. Berlin: Springer, 1998.

Bowker, Geoffrey C., and Susan Leigh Star. *Sorting Things Out: Classification and Its Consequences*. Cambridge, MA: MIT Press, 1999.

Bowleg, Lisa. "'The Master's Tools Will Never Dismantle the Master's House': Ten Critical Lessons for Black and Other Health Equity Researchers of Color." *Health Education & Behavior* 48, no. 3 (June 1, 2021): 237–249. https://doi.org/10.1177/10901981211007402.

Bray, Kirsten, and Christina Harrington. "Speculative Blackness: Considering Afrofuturism in the Creation of Inclusive Speculative Design Probes." In *DIS '21: Proceedings of the 2021 ACM Designing Interactive Systems Conference*, 1793–1806. New York: Association for Computing Machinery, 2021. https://doi.org/10.1145/3461778.3462002.

Breña, Carmen Morán. "Las Feministas Saludan La Rectificación Del Fiscal Gertz Sobre El Tipo Penal de Feminicidio." *EL PAIS*, August 14, 2020. https://elpais.com/mexico/2020-08-14/las-feministas-saludan-la-rectificacion-del-fiscal-gertz-sobre-el-tipo-penal-de-feminicidio.html.

Broussard, Meredith. *Artificial Unintelligence: How Computers Misunderstand the World*. Cambridge, MA: MIT Press, 2018.

Broussard, Meredith. *More Than a Glitch: What Everyone Needs to Know About Making Technology Antiracist and Accessible to All*. Cambridge, MA: MIT Press, 2023.

Brubaker, Anne M. "Who Counts? Urgent Lessons from Ida B. Wells's Radical Statistics." *American Quarterly* 74, no. 2 (2022): 265–293. https://doi.org/10.1353/aq.2022.0019.

Buolamwini, Joy. "Facing the Coded Gaze with Evocative Audits and Algorithmic Audits." PhD diss., Massachusetts Institute of Technology, 2022. https://dspace.mit.edu/handle/1721.1/143396.

Buolamwini, Joy, and Timnit Gebru. "Gender Shades: Intersectional Accuracy Disparities in Commercial Gender Classification." In *Conference on Fairness, Accountability and Transparency: FAT 2018*, edited by Sorelle A. Friedler and Christo Wilson, 1–15. New York: ACM, 2018.

Bureau of Democracy, Human Rights, and Labor. *2021 Country Reports on Human Rights Practices: El Salvador*. Washington, DC: US Department of State, 2021. https://www.state.gov/reports/2021-country-reports-on-human-rights-practices/el-salvador/.

Busch, Otto von, and Karl Palmås. *The Corruption of Co-Design: Political and Social Conflicts in Participatory Design Thinking*. New York: Routledge, 2023.

Butler-Adam, John. "The Weighty History and Meaning behind the Word 'Science.'" The Conversation, October 1, 2015. http://theconversation.com/the-weighty-history-and-meaning-behind-the-word-science-48280.

Cabnal, Lorena. *Feminismos diversos: el feminismo comunitario*. Barcelona: ACSUR Las Segovias, 2010. https://porunavidavivible.files.wordpress.com/2012/09/feminismos-comunitario-lorena-cabnal.pdf.

Canadian Femicide Observatory for Justice and Accountability. *Call It Femicide—Understanding Gender-Related Killings of Women and Girls in Canada 2018*. Social Sciences and Humanities Research Council of Canada, CSSLRV, Canada Excellence Research Chairs, 2018. https://www.femicideincanada.ca/callitfemicide.pdf.

Canadian Feminicide Observatory for Justice and Accountability. "Types of Femicide." Accessed August 5, 2023. https://www.femicideincanada.ca/about/types.

Carcedo, Ana, ed. *No olvidamos ni aceptamos: Femicidio en Centroamérica, 2000–2006*. San José, Costa Rica: Associación Centro Feminista de Información y Acción, 2010.

Carcedo Cabañas, Ana, and Montserrat Sagot Rodríguez. *Femicidio en Costa Rica 1990–1999*. Instituto Nacional de las Mujeres (INAMU), 2000.

cárdenas, micha. *Poetic Operations: Trans of Color Art in Digital Media*. Durham, NC: Duke University Press, 2022.

Carlson, Bronwyn. "Data Silence in the Settler Archive." In *Mapping Deathscapes: Digital Geographies of Racial and Border Violence*, edited by Suvendrini Perera, 84–105. New York: Routledge, 2022.

Carmi, E. "A Feminist Critique to Digital Consent." *Seminar.net* 17, no. 2 (August 31, 2021). https://doi.org/10.7577/seminar.4291.

Carol, Hanisch. "The Personal Is Political." In *Notes from the Second Year: Women's Liberation, Major Writings of the Radical Feminists*, 76–78. Pamphlet. 1970.

Carroll, Stephanie Russo, Ibrahim Garba, Oscar L. Figueroa-Rodríguez, Jarita Holbrook, Raymond Lovett, Simeon Materechera, Mark Parsons, et al. "The CARE Principles for Indigenous Data Governance." *Data Science Journal* 19, no. 43 (2020): 1–12. https://doi.org/10.5334/dsj-2020-043.

Caswell, Michelle, and Marika Cifor. "From Human Rights to Feminist Ethics: Radical Empathy in Archives." *Archivaria* 81 (Spring 2016): 23–43. https://escholarship.org/uc/item/0mb9568h.

CDC. "From the CDC—Leading Causes of Death—Females—All Races and Origins—United States, 2018." Centers for Disease Control and Prevention, last reviewed September 27, 2019. https://www.cdc.gov/women/lcod/2018/all-races-origins/index.htm#all-ages.

CDC. "From the CDC—Leading Causes of Death—Females—Non-Hispanic Black—United States, 2018." Centers for Disease Control and Prevention, last reviewed March 3, 2022. https://www.cdc.gov/women/lcod/2018/nonhispanic-black/index.htm.

Centro. *New Estimates: 135,000+ Post-Maria Puerto Ricans Relocated to Stateside.* New York: Centro, March 2018. https://centropr-archive.hunter.cuny.edu/sites/default/files/data_sheets/PostMaria-NewEstimates-3-15-18.pdf.

CEPAL. "Prevenir el feminicidio: Una tarea prioritaria para la sociedad en su conjunto." Infographic. CEPAL, United Nations, November 11, 2016. https://oig.cepal.org/es/infografias/prevenir-feminicidio-tarea-prioritaria-la-sociedad-su-conjunto.

Chambers-Letson, Joshua, and Robert G. Diaz, eds. "Reparations." Special issue, *Women & Performance: A Journal of Feminist Theory* 16, no. 2 (2006).

Chavez, Susan. *Primera Tormenta: Poemas De Susan Chavez* (blog), Accessed September 14, 2023. https://primeratormenta.blogspot.com/.

Chenou, Jean-Marie, and Carolina Cepeda-Másmela. "#NiUnaMenos: Data Activism from the Global South." *Television & New Media* 20, no. 4 (May 1, 2019): 396–411. https://doi.org/10.1177/1527476419828995.

Chordia, Ishita. "Leveraging Transformative Justice in Organizing Collective Action towards Community Safety." In *CHI EA '22: Extended Abstracts of the 2022 CHI Conference on Human Factors in Computing Systems*, article 50, 1–4. New York: Association for Computing Machinery, 2022. https://doi.org/10.1145/3491101.3503820.

Chun, Wendy Hui Kyong. "Introduction: Race and/as Technology; or, How to Do Things with Race." *Camera Obscura* 24, no. 1 (2009): 7–35.

Cifor, M., P. Garcia, T. L. Cowan, J. Rault, T. Sutherland, A. Chan, J. Rode, A. L. Hoffmann, N. Salehi, and L. Nakamura. *Feminist Data Manifest-No* (website). Accesed September 14, 2023. https://www.manifestno.com/.

Colectivo de Geografía Crítica de Ecuador. *Manifiesto Contra La Violencia Hacia Las Mujeres Desde La Geografía Crítica.* Colectivo de Geografía Crítica de Ecuador, August 8, 2017. https://geografiacriticaecuador.org/wp-content/uploads/2017/08/Manifiesto-geogr%C3%A1fico-contra-violencia-hacia-las-mujeres-FINAL.pdf.

Coleman, Beth. "Race as Technology." *Camera Obscura* 24, no. 1 (2009): 176–207.

Collins, Patricia Hill. *Black Feminist Thought: Knowledge, Consciousness, and the Politics of Empowerment.* 2nd ed., rev. New York: Routledge, 2002.

Collins, Patricia Hill. *Intersectionality as Critical Social Theory.* Durham, NC: Duke University Press, 2019.

Collins, Patricia Hill. "On Violence, Intersectionality and Transversal Politics." *Ethnic and Racial Studies* 40, no. 9 (2017): 1450–1475. https://doi.org/10.1080/01419870.2017.1317827.

Combahee River Collective. *The Combahee River Collective Statement*. 1977. https://www.blackpast .org/african-american-history/combahee-river-collective-statement-1977/.

Conn, Heather. "Women's Memorial March." *The Canadian Encyclopedia*, January 22, 2020. https://www.thecanadianencyclopedia.ca/en/article/women-s-memorial-march.

Cooke, Bill, and Uma Kothari, eds. *Participation: The New Tyranny?* London: Zed Books, 2001.

Couldry, Nick, and Ulises Mejias. *The Costs of Connection: How Data Is Colonizing Human Life and Appropriating It for Capitalism*. Stanford, CA: Stanford University Press, 2019.

Crawford, Kate, Mary L. Gray, and Kate Miltner. "Critiquing Big Data: Politics, Ethics, Epistemology." *International Journal of Communication* 8 (2014): 1663–1672.

Crenshaw, Kimberlé W., Andrea J. Ritchie, Rachel Anspach, Rachel Gilmer, and Luke Harris. *Say Her Name: Resisting Police Brutality against Black Women*. New York: Columbia Law School, Center for Intersectionality and Social Policy Studies, 2015. https://scholarship.law.columbia.edu/faculty _scholarship/3226.

Criado Perez, Caroline. *Invisible Women: Data Bias in a World Designed for Men*. New York: Abrams Press, 2019.

Crooks, Roderic. "Seeking Liberation: Surveillance, Datafication, and Race." *Surveillance & Society* 20, no. 4 (December 16, 2022): 413–419. https://doi.org/10.24908/ss.v20i4.15983.

Crooks, Roderic, and Morgan Currie. "Numbers Will Not Save Us: Agonistic Data Practices." *Information Society* 37 (2021): 1–19. https://doi.org/10.1080/01972243.2021.1920081.

Culhane, Dara. "Their Spirits Live within Us: Aboriginal Women in Downtown Eastside Vancouver Emerging into Visibility." *American Indian Quarterly* 27, no. 3–4 (2003): 593–606.

Cullen, Patricia, Myrna Dawson, Jenna Price, and James Rowlands. "Intersectionality and Invisible Victims: Reflections on Data Challenges and Vicarious Trauma in Femicide, Family and Intimate Partner Homicide Research." *Journal of Family Violence* 36 (2021): 619–628. https://doi .org/10.1007/s10896-020-00243-4.

Currie, Morgan, Britt S. Paris, Irene Pasquetto, and Jennifer Pierre. "The Conundrum of Police Officer-Involved Homicides: Counter-Data in Los Angeles County." *Big Data & Society* 3, no. 2 (December 1, 2016). https://doi.org/10.1177/2053951716663566.

Dalton, Craig, and Jim Thatcher. "What Does a Critical Data Studies Look Like, and Why Do We Care? Seven Points for a Critical Approach to 'Big Data.'" *Society and Space* 29 (May 12, 2014). https://www.societyandspace.org/articles/what-does-a-critical-data-studies-look-like-and-why -do-we-care.

Danner, Mona, Lucía Fort, and Gay Young. "International Data on Women and Gender: Resources, Issues, Critical Use." *Women's Studies International Forum* 22, no. 2 (March–April 1999): 249–259. https://doi.org/10.1016/S0277-5395(99)00011-4.

Data Cívica and Intersecta. *Datos para la Vida*. Mexico City: Data Cívica, Intersecta, January 2022. https://s3.us-east-1.amazonaws.com/media.datacivica.org/assets/pdf/InformeDatosParaLaVida 2022.pdf.

Datos Contra Feminicidio. "The Challenges of Measuring, Comparing, and Standardizing 'global' Fem(in)icide Data." Video, April 12, 2021. https://www.youtube.com/watch?v=n4PQvqeZ7 M4&t=3s.

Davis, Angela Y., Gina Dent, Erica R. Meiners, and Beth E. Richie. *Abolition. Feminism. Now*. Chicago: Haymarket Books, 2022.

Davis, Sara L. M. *The Uncounted: Politics of Data in Global Health*. Cambridge: Cambridge University Press, 2020.

Dawson, Myrna. "Punishing Femicide: Criminal Justice Responses to the Killing of Women over Four Decades." *Current Sociology* 64, no. 7 (2016): 996–1016. https://doi.org/10.1177/0011392 115611192.

Dawson, Myrna, and Michelle Carrigan. "Identifying Femicide Locally and Globally: Understanding the Utility and Accessibility of Sex/Gender-Related Motives and Indicators." *International Sociological Association* 69, no. 5 (September 2021), 682–704. https://journals.sagepub.com/doi /abs/10.1177/0011392120946359.

Dean, Amber. *Remembering Vancouver's Disappeared Women: Settler Colonialism and the Difficulty of Inheritance*. Toronto: University of Toronto Press, 2015.

de Castro Leal, Débora, Angelika Strohmayer, and Max Krüger. "On Activism and Academia: Reflecting Together and Sharing Experiences among Critical Friends." In *Proceedings of the 2021 CHI Conference on Human Factors in Computing Systems*, 1–18. New York: Association for Computing Machinery, 2021. https://dl.acm.org/doi/10.1145/3411764.3445263.

Deer, Sarah. *The Beginning and End of Rape: Confronting Sexual Violence in Native America*. Minneapolis: University of Minnesota Press, 2015.

De Haan, Willem. "Violence as an Essentially Contested Concept." In *Violence in Europe*, edited by Sophie Body-Gendrot and Pieter Spierenburg, 27–40. New York: Springer, 2009.

de Waal, Martijn, and Michiel de Lange. "Introduction—the Hacker, the City and Their Institutions: From Grassroots Urbanism to Systemic Change." In *The Hackable City: Digital Media and Collaborative City-Making in the Network Society*, edited by Michiel de Lange and Martijn de Waal, 1–22. Singapore: Springer, 2018.

D'Ignazio, Catherine. "5 Questions on Data and Feminicide with Silvana Fumega." *Data Feminism* (blog), July 9, 2020. https://medium.com/data-feminism/5-questions-on-data-and-feminicide-with -silvana-fumega-318526a7c107.

D'Ignazio, Catherine. "Human-Centred Computing and Feminicide Counterdata Production." In *The Routledge International Handbook on Femicide and Feminicide*, edited by Myrna Dawson and Saide Mobayed Vega, 529–541. London: Routledge, 2023.

D'Ignazio, Catherine, Isadora Cruxên, Helena Suárez Val, Angeles Martinez Cuba, Mariel García-Montes, Silvana Fumega, Harini Suresh, and Wonyoung So. "Feminicide and Counterdata Production: Activist Efforts to Monitor and Challenge Gender-Related Violence." *Patterns* 3, no. 7 (July 8, 2022). https://doi.org/10.1016/j.patter.2022.100530.

D'Ignazio, Catherine, Angeles Martinez Cuba, Alessandra Jungs de Almeida, and Valentina Pedroza Munoz. "Table of Laws and Official Data about Feminicide." Dataset, Harvard Dataverse, March 24, 2023. https://doi.org/10.7910/DVN/VQKIPJ.

D'Ignazio, Catherine, and Lauren F. Klein. *Data Feminism*. Cambridge, MA: MIT Press, 2020.

D'Ignazio, Catherine, Rebecca Michelson, Alexis Hope, Josephine Hoy, Jennifer Roberts, and Kate Krontiris. "'The Personal Is Political': Hackathons as Feminist Consciousness Raising." *Proceedings of the ACM on Human–Computer Interaction* 4, no. CSCW2 (2020): article 150, 1–23. https://doi .org/10.1145/3415221.

D'Ignazio, Catherine, Helena Suárez Val, Silvana Fumega, Harini Suresh, Isadora Cruxên, Wonyoung So, Angeles Martinez Cuba, and Mariel Garcia-Montes. "Feminicide & Machine Learning: Detecting Gender-Based Violence to Strengthen Civil Sector Activism." Paper presented at Mechanism Design for Social Good (MD4SG '20), August 17, 2020. http://www.kanarinka.com /wp-content/uploads/2021/01/DIgnazio-et-al.-2020-Feminicide-Machine-Learning-Detecting -Gender-ba.pdf.

Dimond, Jill P. "Feminist HCI for Real: Designing Technology in Support of a Social Movement." PhD diss., Georgia Institute of Technology, December 2012. https://www.proquest.com /docview/1288414343/abstract/9EB92BA69BA747A8PQ/1.

Dimond, Jill P., Casey Fiesler, and Amy S. Bruckman. "Domestic Violence and Information Communication Technologies." *Interacting with Computers* 23, no. 5 (September 2011): 413–421. https://doi.org/10.1016/j.intcom.2011.04.006.

DiSalvo, Carl. *Design as Democratic Inquiry: Putting Experimental Civics into Practice*. Cambridge, MA: MIT Press, 2022.

Dörk, Marian, Patrick Feng, Christopher Collins, and Sheelagh Carpendale. "Critical InfoVis: Exploring the Politics of Visualization." In *CHI '13 Extended Abstracts on Human Factors in Computing Systems*, 2189–2198. New York: Association for Computing Machinery, 2013. https://doi .org/10.1145/2468356.2468739.

Driver, Alice. *More or Less Dead: Feminicide, Haunting, and the Ethics of Representation in Mexico*. Tucson: University of Arizona Press, 2015.

Drucker, Johanna. "Visualization." In *Uncertain Archives: Critical Keywords for Big Data*, edited by Nanna Bonde Thylstrup, Daniela Agostinho, Annie Ring, Catherine D'Ignazio, and Kristin Veel, 561–569. Cambridge, MA: MIT Press, 2021.

Duffy, Mignon. *Making Care Count: A Century of Gender, Race, and Paid Care Work*. Rutgers University Press, 2011.

Dutta, Mohan J. "Whiteness, Internationalization, and Erasure: Decolonizing Futures from the Global South." *Communication and Critical/Cultural Studies* 17, no. 2 (2020): 228–235. https://doi .org/10.1080/14791420.2020.1770825.

ECMIA, CONAMI, and CHIRAPAQ. *Levantando nuestras voces por la paz y la seguridad de nuestros pueblos y continentes—VIII Encuentro Continental de Mujeres Indígenas de Las Américas*. Lima: CHIRAPAQ,

May 2021. http://ecmia.org/index.php/publicaciones/244-viii-encuentro-continental-de-mujeres
-indigenas-de-las-americas.

Edenfield, Avery C. "Queering Consent: Design and Sexual Consent Messaging." *Communication Design Quarterly* 7, no. 2 (August 26, 2019): 50–63. https://doi.org/10.1145/3358931.3358938.

"Eleven Black Women: Why Did They Die? A Document of Black Feminism." *Radical America* 13, no. 6 (1979): 41–50.

Enders, Craig K. *Applied Missing Data Analysis*. 2nd ed. New York: Guilford Press, 2022. https://www.guilford.com/books/Applied-Missing-Data-Analysis/Craig-Enders/9781462549863.

England, Sarah. *Writing Terror on the Bodies of Women: Media Coverage of Violence against Women in Guatemala*. Lanham: Lexington Books, 2018.

Eubanks, Virginia. *Automating Inequality: How High-Tech Tools Profile, Police, and Punish the Poor*. New York: St. Martin's Press, 2018.

Fairbairn, Jordan, and Myrna Dawson. "Canadian News Coverage of Intimate Partner Homicide: Analyzing Changes over Time." *Feminist Criminology* 8, no. 3 (July 1, 2013): 147–176. https://doi.org/10.1177/1557085113480824.

Falquet, Jules. *Por las buenas o por las malas: las mujeres en la globalización*. Bogota: Colección General, Serie Estudios de Género, 2018. http://julesfalquet.com/wp-content/uploads/2018/06/Por-la-buenas-o-por-las-malas-PDF-integral.pdf.

Falquet, Jules. "Violence against Women and (De-)colonization of the 'Body-Territory.'" In *The Globalization of Gender*, edited by Iona Cîrstocea, Delphine Lacombe, and Elisabeth Marteu, 81–101. London: Routledge, 2019.

Federici, Silvia, Verónica Gago, and Lucía Cavallero, eds. *¿Quién Le Debe a Quién? Ensayos Transnacionales de Desobediencia Financiera*. Buenos Aires: Tinta Limón, 2021. https://tintalimon.com.ar/public/x2aajtw1zgx4u2c3gkvgowfl4se5/quien%20le%20debe%20a%20quien.pdf.

Feinberg, Melanie. *Everyday Adventures with Unruly Data*. Cambridge, MA: MIT Press, 2022.

Femicide Volume VII: Establishing a Femicide Watch in Every Country. Vienna: Academic Council on the United Nations System (ACUNS) Vienna Liaison Office, 2017.

"Femicide Watch Initiative." United Nations Human Rights, Office of the High Commissioner, Accessed August 8, 2023. https://www.ohchr.org/en/issues/women/srwomen/pages/femicidewatch.aspx.

Fernstad, S. J. "To Identify What Is Not There: A Definition of Missingness Patterns and Evaluation of Missing Value Visualization." *Information Visualization* 18, no. 2 (2019): 230–250. https://doi.org/10.1177/1473871618785387.

Feuston, Jessica L., Arpita Bhattacharya, Nazanin Andalibi, Elizabeth A. Ankrah, Sheena Erete, Mark Handel, Wendy Moncur, Sarah Vieweg, and Jed R. Brubaker. "Researcher Wellbeing and Best Practices in Emotionally Demanding Research." In *Extended Abstracts of the 2022 CHI Conference on Human Factors in Computing Systems*, article 72: 1–6. New York: Association for Computing Machinery, 2022. https://doi.org/10.1145/3491101.3503742.

Figueroa, Naylie Vélez. "Tipificación de feminicidios y transfeminicidios: Visibiliza la violencia de género." *Todas*, August 27, 2021. https://www.todaspr.com/tipificacion-de-feminicidios-y-trans feminicidios-visibiliza-la-violencia-de-genero/.

Fiscalía General de Justicia—Ciudad de México. "Atlas de Feminicidios de la Ciudad de México." GeoAtlas. Accessed October 7, 2022. https://atlasfeminicidios.fgjcdmx.gob.mx/index.html.

Fonow, Mary Margaret, and Judith A. Cook. "Feminist Methodology: New Applications in the Academy and Public Policy." *Signs: Journal of Women in Culture and Society* 30, no. 4 (2005): 2211–2236. https://doi.org/10.1086/428417.

Fotopoulou, Aristea. "Understanding Citizen Data Practices from a Feminist Perspective: Embodiment and the Ethics of Care." In *Citizen Media and Practice: Currents, Connections, Challenges*, edited by Hilde C. Stephansen and Emiliano Treré, 227–242. London: Routledge, 2019.

Fox, Sarah, Amanda Menking, Stephanie Steinhardt, Anna Lauren Hoffmann, and Shaowen Bardzell. "Imagining Intersectional Futures: Feminist Approaches in CSCW." In *CSCW '17 Companion: Companion of the 2017 ACM Conference on Computer Supported Cooperative Work and Social Computing*, 387–393. New York: Association for Computing Machinery, 2017. https://doi.org /10.1145/3022198.3022665.

Fregoso, Rosa-Linda, and Cynthia Bejarano. *Terrorizing Women: Feminicide in the Americas*. Durham, NC: Duke University Press, 2010.

Fuentes, Lorena. "'The Garbage of Society': Disposable Women and the Socio-Spatial Scripts of Femicide in Guatemala." *Antipode* 52, no. 6 (2020): 1667–1687. https://doi.org/10.1111/anti.12669.

Fuentes, Lorena. "(Re)reading the Boundaries and Bodies of Femicide: Exploring Articulations within the Discursive Economy of Gendered Violence in 'Post War' Guatemala." PhD diss., Birkbeck, University of London, April 2016.

Fumega, Silvana. "Data against Feminicide" In *From Bias to Feminist AI*. <A+> Alliance, 2021. https://feministai.pubpub.org/pub/data-against-feminicide/release/1.

Fumega, Silvana. *Guide to Protocolize Processes of Femicide Identification for Later Registration*. ILDA, August 5, 2019. https://doi.org/10.5281/ZENODO.3360671.

Fumega, Silvana. "Manos a La Obra: Primer Taller de Estandarización de Datos Sobre Femicidios En Argentina." ILDA, June 25, 2018. https://idatosabiertos.org/en/manos-a-la-obra-primer -taller-de-estandarizacion-de-datos-sobre-femicidios-en-argentina/.

Fundación ALDEA. *Feminicídios en Ecuador: Realidades ocultas, datos olvidados e invisibilizados*. Quito: Fundación ALDEA, 2022. https://nube.fundacionaldea.org/index.php/s/QqmLd25Ex596g36.

Gago, Verónica, and Raquel Gutiérrez Aguilar. "Women Rising in Defense of Life: Tracing the Revolutionary Flows of Latin American Women's Many Uprisings." *NACLA Report on the Americas* 50, no. 4 (October 31, 2018): 364–368. https://doi.org/10.1080/10714839.2018.1550978.

Gallardo, Adriana. "How We Collected Nearly 5,000 Stories of Maternal Harm." ProPublica, March 20, 2018. https://www.propublica.org/article/how-we-collected-nearly-5-000-stories-of -maternal-harm.

García-Del Moral, Paulina. "The Murders of Indigenous Women in Canada as Feminicides: Toward a Decolonial Intersectional Reconceptualization of Femicide." *Signs: Journal of Women in Culture and Society* 43, no. 4 (June 2018): 929–954. https://doi.org/10.1086/696692.

García-Del Moral, Paulina. "Transforming Feminicidio: Framing, Institutionalization and Social Change." *Current Sociology* 64, no. 7 (November 1, 2016): 1017–1035. https://doi.org/10.1177/0011392115618731.

García-Del Moral, Paulina, Dolores Figueroa Romero, Patricia Torres Sandoval, and Laura Hernández Pérez. "Femicide/Feminicide and Colonialism." In *The Routledge International Handbook on Femicide and Feminicide*, edited by Myrna Dawson and Saide Mobayed Vega, 60–69. London: Routledge, 2023.

Gargiulo, Maria. "Statistical Biases, Measurement Challenges, and Recommendations for Studying Patterns of Femicide in Conflict." *Peace Review* 34, no. 2 (April 3, 2022): 163–176. https://doi.org/10.1080/10402659.2022.2049002.

Garriga-López, Claudia Sofia, Denilson Lopes, Cole Rizki, and Juana Maria Rodriguez, eds. "Trans Studies En Las Américas." Special issue, *TSQ: Transgender Studies Quarterly* 6, no. 2 (2019). https://www.dukeupress.edu/trans-studies-en-las-americas.

Gartner, Rosemary, Myrna Dawson, and Maria Crawford. *Woman Killing: Intimate Femicide in Ontario, 1974–1994*. Women We Honour Action Committee, 1994. https://femicideincanada.ca/what-is-femicide/history/woman-killing-intimate-femicide-in-ontario-1974-1994/.

Gautam, Aakash, Deborah Tatar, and Steve Harrison. "Crafting, Communality, and Computing: Building on Existing Strengths to Support a Vulnerable Population." In *Proceedings of the 2020 CHI Conference on Human Factors in Computing Systems*, 1–14. New York: Association for Computing Machinery, 2020. https://doi.org/10.1145/3313831.3376647.

Gebru, Timnit, Jamie Morgenstern, Briana Vecchione, Jennifer Wortman Vaughan, Hanna Wallach, Hal Daumé III, and Kate Crawford. "Datasheets for Datasets." *Communications of the ACM* 64, no. 12 (December 2021): 86–92. https://doi.org/10.1145/3458723.

Gellman, Mneesha. "Landmark Femicide Case Fails to Fix El Salvador's Patriarchy." *Globe Post*, February 6, 2020. https://theglobepost.com/2020/02/06/femicides-el-salvador/.

George, Blythe, Annita Lucchesi, and Chelsea Trillo. *To' Kee Skuy' Soo Ney-wo-chek': I Will See You Again in a Good Way Year 2 Progress Report on Missing and Murdered Indigenous Women, Girls, and Two Spirit People in Northern California*. Yurok Tribal Court and Sovereign Bodies Institute, July 2021. https://2a840442-f49a-45b0-b1a1-7531a7cd3d30.filesusr.com/ugd/6b33f7_d7e4c0de2a434f6e9d4b1608a0648495.pdf.

Ghoshal, Sucheta, Rishma Mendhekar, and Amy Bruckman. "Toward a Grassroots Culture of Technology Practice." *Association for Computing Machinery* 4, no. CSCW1 (May 29, 2020): article 54, 1–28. https://doi.org/10.1145/3392862.

Gomez, Laura E. "The Birth of the 'Hispanic' Generation: Attitudes of Mexican-American Political Elites toward the Hispanic Label." *Latin American Perspectives* 19, no. 4 (1992): 45–58.

Gramsci, Antonio. *Prison Notebooks*. Translated by Joseph A. Buttigieg. New York: Columbia University Press, 2011.

Gray, Jonathan, Danny Lämmerhirt, and Liliana Bounegru. *Changing What Counts: How Can Citizen-Generated and Civil Society Data Be Used as an Advocacy Tool to Change Official Data Collection?* Rochester, NY: SSRN, March 3, 2016. https://doi.org/10.2139/ssrn.2742871.

Green, Ben. "The Contestation of Tech Ethics: A Sociotechnical Approach to Technology Ethics in Practice." *Journal of Social Computing* 2, no. 3 (September 2021): 209–225. https://doi.org/10.23919/JSC.2021.0018.

Green-Gopher, Deborah. "My Sister Was Murdered." LinkedIn, November 24, 2017. https://www.linkedin.com/pulse/my-sister-murdered-deborah-green/.

Grupo Guatemalteco de Mujeres. *Informe de Muertes Violentas de Mujeres-MVM En Guatemala*. Grupo Guatemalteco de Mujeres—GGM, 2018. https://ggm.org.gt/wp-content/uploads/2018/10/Informe-de-muertes-violentas-de-mujeres-LOW.pdf.

Guerra Garces, Geraldina. "Algunas notas sobre el texto." Personal communication, January 22, 2023.

Gutiérrez, Miren. *Data Activism and Social Change*. Cham, Switzerland: Palgrave Macmillan, 2018.

Guyan, Kevin. *Queer Data: Using Gender, Sex and Sexuality Data for Action*. Bloomsbury Publishing, 2022.

Haimson, Oliver L., Dykee Gorrell, Denny L. Starks, and Zu Weinger. "Designing Trans Technology: Defining Challenges and Envisioning Community-Centered Solutions." In *Proceedings of the 2020 CHI Conference on Human Factors in Computing Systems*, 1–13. New York: Association for Computing Machinery, 2020. https://doi.org/10.1145/3313831.3376669.

Haimson, Oliver L., Kai Nham, Hibby Thach, and Aloe DeGuia. "How Transgender People and Communities Were Involved in Trans Technology Design Processes." In *Proceedings of the 2023 CHI Conference on Human Factors in Computing Systems*, 1–16. New York: Association for Computing Machinery, 2023. https://oliverhaimson.com/PDFs/HaimsonHowTransgenderPeople.pdf.

Hanna, Alex, and Tina M. Park. "Against Scale: Provocations and Resistances to Scale Thinking." Paper presented at the Resistance AI Workshop at the Conference on Neural Information Processing Systems (NeurIPS), 2020. https://doi.org/10.48550/arXiv.2010.08850.

Hao, Karen, and Andrea Paola Hernández. "How the AI Industry Profits from Catastrophe." *MIT Technology Review*, April 20, 2022. https://www.technologyreview.com/2022/04/20/1050392/ai-industry-appen-scale-data-labels/.

Haraway, Donna J. *Staying with the Trouble: Making Kin in the Chthulucene*. Durham, NC: Duke University Press, 2016.

Harjo, Laura. "Community Caretaking: Decolonizing Indigenous Anti-violence Work." Virtual event, March 31, 2022. https://www.umb.edu/news_events_media/events/community_caretaking_decolonizing_indigenous_anti_violence_work.

Honeywell. "Honeywell Wins Bengaluru Safe City Project." Press release, October 20, 2021. https://www.honeywell.com/content/dam/honeywellbt/en/documents/downloads/india-hail /announcements/press-release-honeywell-wins-bengaluru-safe-city-project.pdf.

Hopkins, Ruth. "If I Am Taken, Will Anyone Look for Me?" *The Cut*, October 1, 2021. https:// www.thecut.com/2021/10/what-about-the-missing-women-who-look-like-me.html.

ILDA. Comunicando Datos. Data Against Feminicide Course 2022. Video, May 2, 2022. https:// www.youtube.com/watch?v=3bTckDI3H0c.

ILDA. "Conversatorio: Haciendo Datos Desafíos y Aprendizajes Desde El Activismo, El Periodismo y El Estado." Video, June 7, 2022. https://www.youtube.com/watch?v=pB7RBJO117w.

ILDA. "ILDA Feminicide Data Standard (V2.0)." Dataset. Accessed August 9, 2023. https://docs .google.com/spreadsheets/d/0Bz4bv9Y3kT3Tc01LZlFZVVZma0UzUE92bUxZcVBLMHZqcTlr /edit?resourcekey=0-RA5_9Qai1U65WDrY1d-VwQ&usp=embed_facebook.

INCITE! Women of Color Against Violence. *The Revolution Will Not Be Funded: Beyond the Non-Profit Industrial Complex*. Durham, NC: Duke University Press, 2017.

Instituto Patrícia Galvão. *Papel Social e Desafios Da Cobertura Sobre Feminicídio e Violência Sexual*. São Paulo: Instituto Patrícia Galvão, 2019.

Interrupting Criminalization, Project Nia, and Critical Resistance. *So Is This Actually an Abolitionist Proposal or Strategy? A Collection of Resources to Aid in Evaluation and Reflection*. Interrupting Criminalization, Project Nia, and Critical Resistance, 2022.

Irani, Lilly. "The Cultural Work of Microwork." *New Media & Society* 17, no. 5 (2015): 720–739. https://doi.org/10.1177/1461444813511926.

Irani, Lilly, and M. Six Silberman. "From Critical Design to Critical Infrastructure: Lessons from Turkopticon." *Interactions* 21, no. 4 (July 1, 2014): 32–35. https://doi.org/10.1145/2627392.

It Starts with Us. "Background." Accessed February 13, 2023. http://itstartswithus-mmiw.com /about/.

Janssen, Heleen, and Jatinder Singh. "Data Intermediary." *Internet Policy Review* 11, no. 1 (March 30, 2022). https://policyreview.info/glossary/data-intermediary.

kanarinka. "Missing Women, Blank Maps, and Data Voids: What Gets Counted Counts." *MIT Center for Civic Media* (blog), March 22, 2016. https://civic.mit.edu/index.html%3Fp=1153 .html.

Kedar, Claudia. *The International Monetary Fund and Latin America: The Argentine Puzzle in Context*. Philadelphia: Temple University Press, 2013.

Keilty, Patrick, ed. *Queer Data Studies*. Feminist Technosciences. University of Washington Press, 2023.

Kelly, Liz. "The Continuum of Sexual Violence." In *Women, Violence and Social Control*, edited by Jalna Hanmer and Mary Maynard, 46–60. London: Palgrave Macmillan, 1987. https://doi.org /10.1007/978-1-349-18592-4_4.

Kelly, Liz, Linda Regan, and Sheila Burton. "Defending the Indefensible? Quantitative Methods and Feminist Research." In *Working Out: New Directions for Women's Studies*, edited by Hilary Hinds, Ann Phoenix, and Jackie Stacey, 149–161. London: Psychology Press, 1992.

Kennedy, Helen, Rosemary Lucy Hill, Giorgia Aiello, and William Allen. "The Work That Visualisation Conventions Do." *Information, Communication and Society* 19, no. 6 (2016): 715–735.

Keyes, Os. "Counting the Countless." *Real Life*, April 8, 2019. https://reallifemag.com/counting-the-countless/.

Keyes, Os. "The Misgendering Machines: Trans/HCI Implications of Automatic Gender Recognition." *Proceedings of the ACM on Human–Computer Interaction* 2, no. CSCW (November 1, 2018): article 88, 1–22. https://doi.org/10.1145/3274357.

Kishore, Nishant, Domingo Marqués, Ayesha Mahmud, Mathew V. Kiang, Irmary Rodriguez, Arlan Fuller, Peggy Ebner, et al. "Mortality in Puerto Rico after Hurricane Maria." *New England Journal of Medicine* 379, no. 2 (July 12, 2018): 162–170. https://doi.org/10.1056/NEJMsa1803972.

Klein, Lauren F. "Dimensions of Scale: Invisible Labor, Editorial Work, and the Future of Quantitative Literary Studies." *PMLA* 135, no. 1 (January 2020): 23–39. https://doi.org/10.1632/pmla.2020.135.1.23.

Klein, Lauren F. "The Image of Absence: Archival Silence, Data Visualization, and James Hemings." *American Literature* 85, no. 4 (December 1, 2013): 661–688. https://doi.org/10.1215/00029831-2367310.

Krieger, Nancy. "Data, 'Race,' and Politics: A Commentary on the Epidemiological Significance of California's Proposition 54." *Journal of Epidemiology & Community Health* 58, no. 8 (August 1, 2004): 632–633. https://doi.org/10.1136/jech.2003.018549.

Krook, Mona Lena. "A Continuum of Violence." In *Violence against Women in Politics*. Online edition. Oxford Academic Books, 2020. http://academic.oup.com/book/36672/chapter/321695317.

Krüger, Max, Anne Weibert, Débora de Castro Leal, Dave Randall, and Volker Wulf. "It Takes More Than One Hand to Clap: On the Role of 'Care' in Maintaining Design Results." In *Proceedings of the 2021 CHI Conference on Human Factors in Computing Systems*, 1–14. New York: Association for Computing Machinery, 2021. https://doi.org/10.1145/3411764.3445389.

Kumar, Neha, Naveena Karusala, Azra Ismail, Marisol Wong-Villacres, and Aditya Vishwanath. "Engaging Feminist Solidarity for Comparative Research, Design, and Practice." In *Proceedings of the ACM on Human–Computer Interaction* 3, no. CSCW (November 7, 2019): article 167, 1–24. https://doi.org/10.1145/3359269.

Kuo, Rachel, and Lorelei Lee. *Dis/Organizing: How We Build Collectives beyond Institutions: A Non-comprehensive Community Toolkit and Report.* 2021. https://hackinghustling.org/research-2/disorganizing-toolkit/.

Kuo, Rachel, and Mon Mohapatra. "The Institutional Capture of Abolitionist Dissent: Ending Genres of Police Science." *Interactions* 28, no. 6 (2021): 30–31. https://doi.org/10.1145/3491135.

Lagarde y de los Ríos, Marcela. "Preface: Feminist Keys for Understanding Feminicide: Theoretical, Political, and Legal Construction." In *Terrorizing Women: Feminicide in the Americas*, edited by

Rosa-Linda Fregoso and Cynthia Bejarno, xi–xxvi. Durham, NC: Duke University Press, 2010. http://www.jstor.org/stable/j.ctv11smqfd.3.

Lankes, Ana. "In Argentina, One of the World's First Bans on Gender-Neutral Language." *New York Times*, July 20, 2022. https://www.nytimes.com/2022/07/20/world/americas/argentina-gender-neutral-spanish.html.

Lecher, Colin. "Police Are Looking to Algorithms to Predict Domestic Violence." The Markup, June 29, 2022. https://themarkup.org/the-breakdown/2022/06/29/police-are-looking-to-algorithms-to-predict-domestic-violence.

Le Dantec, Christopher A., and Carl DiSalvo. "Infrastructuring and the Formation of Publics in Participatory Design." *Social Studies of Science* 43, no. 2 (April 1, 2013): 241–264. https://doi.org/10.1177/0306312712471581.

Lee, Una, and Dann Toliver. *Building Consentful Tech* (zine). Una Lee and Dann Toliver, 2017. http://www.consentfultech.io/wp-content/uploads/2019/10/Building-Consentful-Tech.pdf.

Leurs, Koen. "Feminist Data Studies: Using Digital Methods for Ethical, Reflexive and Situated Socio-Cultural Research." *Feminist Review* 115, no. 1 (March 1, 2017): 130–154. https://doi.org/10.1057/s41305-017-0043-1.

Ley General de Acceso de las Mujeres a una Vida Libre de Violencia. Congreso General De Los Estados Unidos Mexicanos. 2007. https://www.diputados.gob.mx/LeyesBiblio/pdf/LGAMVLV.pdf.

Liboiron, Max. "Plastic Pollution: Terrible & Charismatic Waste." Peabody Museum of Archaeology & Ethnology, 2012. https://discardstudies.com/2012/01/26/plastic-pollution-terrible-charismatic-waste-upcoming-talks-by-max-liboiron/.

List, Monica. "Counting Women of Color: Being Angry about 'Missing White Woman Syndrome' Is Not Enough." *MSU Bioethics* (blog), October 25, 2021. https://msubioethics.com/2021/10/25/counting-women-of-color-missing-white-woman-syndrome-list/.

Lorde, Audre. "The Master's Tools Will Never Dismantle the Master's House." (Comments at the "The personal and the political panel," Second Sex Conference, New York, September 29, 1979.) In *Sister Outsider: Essays and Speeches*, 110–114. Berkeley, CA: Crossing Press, 1984. https://monoskop.org/images/b/be/Lorde_Audre_1983_1984_The_Masters_Tools_Will_Never_Dismantle_the_Masters_House.pdf.

Lucchesi, Annita. "A Nationwide Data Crisis: Missing and Murdered Indigenous Women and Girls." Webinar, Native Alliance Against Violence, January 24, 2019. https://www.youtube.com/watch?v=Urtj2QriZg8.

Lucchesi, Annita Hetoevehotohke'e. "Inspirational Interview: Annita Lucchesi, Executive Director, Sovereign Bodies Institute—Part I." Pixel Project, July 28, 2019. https://www.thepixelproject.net/2019/07/28/inspirational-interview-annita-lucchesi-executive-director-sovereign-bodies-institute-part-i/.

Lucchesi, Annita Hetoevehotohke'e. "Mapping for Social Change: A Cartography and Community Activism in Mobilizing against Colonial Gender Violence." *Mapping Meaning, the Journal*, no. 2 (2018): 14–22.

Lucchesi, Annita Hetoevehotohke'e. "Mapping Geographies of Canadian Colonial Occupation: Pathway Analysis of Murdered Indigenous Women and Girls." *Gender, Place & Culture* 26, no. 6 (June 3, 2019): 868–887. https://doi.org/10.1080/0966369X.2018.1553864.

Lucchesi, Annita Hetoevehotohke'e. "Mapping Violence against Indigenous Women and Girls: Beyond Colonizing Data and Mapping Practices." *ACME: An International Journal for Critical Geographies* 21, no. 4 (May 5, 2022): 389–398.

Lucchesi, Annita Hetoevehotohke'e. "Spatial Data and (De)colonization: Incorporating Indigenous Data Sovereignty Principles into Cartographic Research." *Cartographica: The International Journal for Geographic Information and Geovisualization* 55, no. 3 (2020): 163–169.

Mackay, Finn. "Arguing against the Industry of Prostitution: Beyond the Abolitionist versus Sex Worker Binary." *Feminist Current* (blog), June 24, 2013. https://www.feministcurrent.com/2013/06/24/arguing-against-the-industry-of-prostitution-beyond-the-abolitionist-versus-sex-worker-binary/.

Manjoo, Rashida. "United Nations Special Rapporteur on Violence against Women, Its Causes and Consequences." In *Prevention, Investigation and Prosecution of Gender-Related Killings of Women and Girls*, 1–6. Bangkok: UNODC, 2014.

Marco, Garciela Di. "Social Movement Demands in Argentina and the Constitution of a 'Feminist People.'" In *Beyond Civil Society: Activism, Participation, and Protest in Latin America*, edited by Sonia E. Alvarez, Jeffrey W. Rubin, Millie Thayer, Gianpaolo Baiocchi, and Agustín Laó-Montes, 122–140. Durham, NC: Duke University Press, 2017.

Martin, Aryn, and Michael Lynch. "Counting Things and People: The Practices and Politics of Counting." *Social Problems* 56, no. 2 (2009): 243–266. https://doi.org/10.1525/sp.2009.56.2.243.

Martínez, Aranzazú Ayala. "How Many More, the Project That Seeks to Make Feminicides in Bolivia Visible." LADO B, November 24, 2016. https://www.ladobe.com.mx/2016/11/cuantas-mas-el-proyecto-que-busca-visibilizar-a-los-feminicidios-en-bolivia/.

Mattern, Shannon. *A City Is Not a Computer: Other Urban Intelligences*. Princeton, NJ: Princeton University Press, 2021.

Mbembe, Achille. "Necropolitics." Translated by Libby Meintjes. *Public Culture* 15, no. 1 (2003): 11–40.

Mbembe, Achille. *Necropolitics*. Durham, NC: Duke University Press, 2019.

McHugh, Jess. "Opinion | Why Aren't Women in the U.S. Also Protesting against Femicide?" *Washington Post*, March 10, 2020. https://www.washingtonpost.com/opinions/2020/03/10/why-arent-women-us-also-protesting-against-femicide/.

McLemore, Monica R., and Valentina D'Efilippo. "To Prevent Women from Dying in Childbirth, First Stop Blaming Them." *Scientific American*, May 1, 2019. https://www.scientificamerican.com/article/to-prevent-women-from-dying-in-childbirth-first-stop-blaming-them/.

Mejía Berdeja, Ricardo, and Ricardo Monreal Ávila. Iniciativa que reforma el artículo 325 del código penal federal, Pub. L. No. 3723-V. March 7, 2013. https://oig.cepal.org/sites/default/files/2013_mex_refcodpenal.pdf.

Meltis Vejar, Mónica, Martha Muñoz Aristizabal, Marcela Zendejas Lasso de la Vega, Ixchel Cisneros Soltero, and Nora Hinojo Escamilla. Interview with Colectiva SJF by Jimena Acosta, Catherine D'Ignazio, and Melissa Teng. Zoom video, June 14, 2022.

Meng, Amanda, and Carl DiSalvo. "Grassroots Resource Mobilization through Counter-Data Action." *Big Data & Society* 5, no. 2 (2018). https://doi.org/10.1177/2053951718796862.

Menjívar, Cecilia, and Shannon Drysdale Walsh. "The Architecture of Feminicide: The State, Inequalities, and Everyday Gender Violence in Honduras." *Latin American Research Review* 52, no. 2 (August 2017): 221–240. https://doi.org/10.25222/larr.73.

Merry, Sally. *The Seductions of Quantification: Measuring Human Rights, Gender Violence, and Sex Trafficking.* Chicago: University of Chicago Press, 2016.

Mexico: Intolerable Killings: 10 Years of Abductions and Murders in Ciudad Juárez and Chihuahua. Amnesty International, 2003. https://www.amnesty.org/en/documents/amr41/027/2003/en/.

Mhlambi, Sabelo. *From Rationality to Relationality: Ubuntu as an Ethical and Human Rights Framework for Artificial Intelligence Governance.* Carr Center for Human Rights Policy Discussion Paper Series, no. 2020-009 (Spring 2020). https://carrcenter.hks.harvard.edu/files/cchr/files/ccdp_2020-009_sabelo_b.pdf.

Miceli, Milagros, Julian Posada, and Tianling Yang. "Studying Up Machine Learning Data: Why Talk about Bias When We Mean Power?" *Proceedings of the ACM on Human–Computer Interaction* 6, no. GROUP (January 14, 2022): article 34, 1–14. https://doi.org/10.1145/3492853.

Milan, Stefania. "Counting, Debunking, Making, Witnessing, Shielding: What Critical Data Studies Can Learn from Data Activism During the Pandemic." In *New Perspectives in Critical Data Studies*, edited by Andreas Hepp, Juliane Jarke, and Leif Kramp, 445–467. Cham, Switzerland: Palgrave Macmillan, 2022.

Milan, Stefania. "Data Activism as the New Frontier of Media Activism." In *Media Activism in the Digital Age*, edited by Victor W. Pickard and Guobin Yang, 151–163. London: Routledge, 2017.

Milan, Stefania, and Miren Gutiérrez. "Technopolitics in the Age of Big Data: The Rise of Proactive Data Activism in Latin America." In *Networks, Movements and Technopolitics in Latin America*, edited by Francisco Sierra Caballero and Tommaso Gravante, 95–109. Cham, Switzerland: Springer International Publishing, 2018. https://doi.org/10.1007/978-3-319-65560-4.

Milan, Stefania, and Emiliano Trere. "Big Data from the South(s): Beyond Data Universalism." *Television & New Media* 20, no. 4 (2019): 319–335. https://doi.org/10.1177/1527476419837739.

Milan, Stefania, and Lonneke van der Velden. "The Alternative Epistemologies of Data Activism." *Digital Culture and Society* 2, no. 2 (2016): 57–74. http://doi.org/10.14361/dcs-2016-0205.

Mills, Charles W. *The Racial Contract.* Ithaca, NY: Cornell University Press, 1999.

Mitchell, Margaret, Simone Wu, Andrew Zaldivar, Parker Barnes, Lucy Vasserman, Ben Hutchinson, Elena Spitzer, Inioluwa Deborah Raji, and Timnit Gebru. "Model Cards for Model Reporting." In *FAT* '19: Proceedings of the Conference on Fairness, Accountability, and Transparency*, 220–229. New York: Association for Computing Machinery, 2019. https://doi.org/10.1145/3287560.3287596.

MIT Media Lab. "City Science Summit Guadalajara—the Power of WITHOUT." Video, 1:53:10, October 7, 2020. https://www.youtube.com/watch?v=x-iy-J9vH0o&t=3854s.

Mobayed, Saide. "Recontar Feminicidios. Activismo de Datos En México." In *25N Día Internacional de La Eliminación de La Violencia Contra La Mujer*, edited by Alethia Fernández de la Reguera and Fabiola de Lachica Huerta, 53–65. Mexico City: UNAM, 2022. https://archivos.juridicas.unam.mx/www/bjv/libros/15/7157/2.pdf.

Mobayed, Saide. "The Kintsugi Method to Recount Data against Feminicide." Paper presented at the 4S/ESOCITE 2022 conference, Cholula, México, December 9, 2022.

Mohanty, Chandra Talpade. "'Under Western Eyes' Revisited: Feminist Solidarity through Anticapitalist Struggles." *Signs* 28, no. 2 (2003): 499–535. https://doi.org/10.1086/342914.

Monáe, Janelle. "Say Her Name (Hell You Talmbout)." Official lyric video, September 23, 2021. https://www.youtube.com/watch?v=kQbeUN-IfyQ.

Monárrez Fragoso, Julia E. "The Victims of Ciudad Juárez Feminicide: Sexually Fetishized Commodities." In *Terrorizing Women: Feminicide in the Americas*, edited by Rosa-Linda Fregoso and Cynthia Bejarano, translated by Sara Koopman, 59–69. Durham, NC: Duke University Press, 2010. https://doi.org/10.1215/9780822392644-004.

Monárrez Fragoso, Julia E. "Crímenes en Ciudad Juárez, El feminicidio es el exterminio de la mujer en el patriarcado." Interview by Graciela Atencio. Accessed January 18, 2023. https://www.jornada.com.mx/2003/09/01/articulos/61_juarez_monarrez.htm.

Monárrez Fragoso, Julia E. "Feminicidio sexual serial en Ciudad Juárez 1993–2001." *Debate Feminista* 25 (April 1, 2002): 279–305. https://doi.org/10.22201/cieg.2594066xe.2002.25.642.

Monárrez Fragoso, Julia E. "Serial Sexual Femicide in Ciudad Juarez, 1993–2001." *A Journal of Chicano Studies* 28, no. 2 (2003): 153–178.

Monárrez Fragoso, Julia Estela. "La cultura del feminicidio en Ciudad Juárez, 1993–1999." *Frontera norte* 12, no. 23 (June 2000): 87–117.

Monárrez Fragoso, Julia E. "Las Diversas Representaciones Del Feminicidio y Los Asesinatos de Mujeres En Ciudad Juárez, 1993–2005." In *Violencia Contra Las Mujeres e Inseguridad Ciudadana En Ciudad Juárez*, vol. 2, by Julia Estela Monárrez Fragoso, Luis E. Cervera Gómez, César M. Fuentes Flores, and Rodolfo Rubio Salas. Ciudad Juárez: El Colegio de la Frontera Norte; Miguel Ángel Porrúa, 2010.

Monroe-White, Thema. "Emancipatory Data Science: A Liberatory Framework for Mitigating Data Harms and Fostering Social Transformation." In *SIGMIS-CPR '21: Proceedings of the 2021 Computers and People Research Conference*, 23–30. New York: Association for Computing Machinery, 2021. https://doi.org/10.1145/3458026.3462161.

Muller, Michael J. "Participatory Design: The Third Space in HCI." In *The Human–Computer Interaction Handbook: Fundamentals, Evolving Technologies and Emerging Applications*, 1051–1068. Hillsdale, NJ: L. Erlbaum Associates, 2002.

Muller, Michael, and Angelika Strohmayer. "Forgetting Practices in the Data Sciences." In *CHI '22: Proceedings of the 2022 CHI Conference on Human Factors in Computing Systems*, 1–19. New York: Association for Computing Machinery, 2022. https://doi.org/10.1145/3491102.3517644.

Muravyov, Dmitry. "Doubt to Be Certain: Epistemological Ambiguity of Data in the Case of Grassroots Mapping of Traffic Accidents in Russia." *Social Movement Studies*, October 5, 2022. https://doi.org/10.1080/14742837.2022.2128327.

Murrey, Amber. "Colonialism." In *Encyclopedia of Human Geography*, edited by Audrey Kobayashi, 315–326. Amsterdam: Elsevier, 2020. https://www.academia.edu/39187665/Colonialism_Ency clopedia_of_Human_Geography_2nd_ed_.

"Mushon Zer-Aviv Presents Friction & Flow at Better World X Design 2022." Video, RISD, November 9, 2022. https://www.youtube.com/watch?v=J2uV8DByF9g.

National Inquiry into Missing and Murdered Indigenous Women and Girls (Canada). *Reclaiming Power and Place: The Final Report of the National Inquiry into Missing and Murdered Indigenous Women and Girls*. 2019. https://www.mmiwg-ffada.ca/final-report/.

Native Women's Association of Canada. *Voices of Our Sisters in Spirit: A Report to Families and Communities*. 2nd ed. Ottawa: Native Women's Association of Canada, 2009.

NE10. "#UmaPorUma." Accessed August 9, 2023. http://produtos.ne10.uol.com.br/umaporuma /index.php.

Nelson, Diane M. *Who Counts?: The Mathematics of Death and Life after Genocide*. Durham, NC: Duke University Press, 2015.

Nenquino, Opi. "Mapeo Territorial Waorani." Revista de La Universidad de Mexico, July 2018. https://www.revistadelauniversidad.mx/articles/15aa78e7-d712-4ae5-8daf-9431390313bd/mapeo -territorial-waorani.

Nowell, Cecilia. "What to Know about Missing and Murdered Indigenous Persons Awareness Day." *Washington Post*, May 5, 2022. https://www.washingtonpost.com/nation/2022/05/05 /missing-murdered-indigenous-persons-awareness-day/.

Ogbonnaya-Ogburu, Ihudiya Finda, Angela D. R. Smith, Kentaro Toyama, and Alexandra To. "Critical Race Theory for HCI." In *Proceedings of the 2020 CHI Conference on Human Factors in Computing Systems*, 1–16. New York: Association for Computing Machinery, 2020. https://doi .org/10.1145/3313831.3376392.

Okun, Tema. "White Supremacy Culture." dRworks. Accessed August 9, 2023. https://www.white supremacyculture.info/uploads/4/3/5/7/43579015/okun_-_white_sup_culture_2020.pdf.

Oliver, Julian. "36C3—Server Infrastructure for Global Rebellion." Video, 1:03:11. Media.ccc.de, December 27, 2019. https://www.youtube.com/watch?v=I_O3zj3p52A.

Oliver, Julian, Gordan Savičić, and Danja Vasiliev. *The Critical Engineering Manifesto*. Berlin: Critical Engineering Working Group, 2011–2021. https://criticalengineering.org/.

O'Neil, Cathy. *Weapons of Math Destruction: How Big Data Increases Inequality and Threatens Democracy*. New York, Crown, 2016.

ONU Mujeres, INMUJERES, and CONAVIM. *Violencia Feminicida En México: Aproximaciones y Ten-dencias*. Mexico City: ONU Mujeres, December 2020. http://cedoc.inmujeres.gob.mx/documentos _download/ViolenciaFeminicidaMX-V8.pdf.

Ọnụọha, Mimi. "On Missing Data Sets." GitHub, January 24, 2018. https://github.com/Mimi Onuoha/missing-datasets.

Ortiz, Lu, and Daragh Brady. "One Woman Is Behind the Most Up-to-Date Interactive Map of Femicides in Mexico." Global Voices, June 15, 2017. https://globalvoices.org/2017/06/15/one -woman-is-behind-the-most-up-to-date-interactive-map-of-femicides-in-mexico/.

Ortuño, Gabriela González. "Teorías de la disidencia sexual: De contextos populares a usos elitis-tas. La teoría queer en América latina frente a las y los pensadores de disidencia sexogenérica." *De Raíz Diversa. Revista Especializada en Estudios Latinoamericanos* 3, no. 5 (2016): 179–200. https:// doi.org/10.22201/ppela.24487988e.2016.5.58507.

O'Sullivan, Sandy. "A Lived Experience of Aboriginal Knowledges and Perspectives: How Cul-tural Wisdom Saved My Life." In *Practice Wisdom: Values and Interpretations*, edited by Joy Higgs, 107–112. Leiden: Brill, 2019. https://research.usc.edu.au/esploro/outputs/bookChapter/A-lived -experience-of-Aboriginal-knowledges-and-perspectives-How-cultural-wisdom-saved-my-life /99451434202621.

Otros Mapas. "Flores en el Aire." PNUD. Accessed February 22, 2023. https://www.otrosmapas .org/flores-en-el-aire.

PATH (Program for Appropriate Technology in Health), *InterCambios* (Inter-American Alliance for the Prevention of Gender-based Violence, MRC (Medical Research Council of South Africa), and WHO (World Health Organization). *Strengthening Understanding of Femicide: Using Research to Gal-vanize Action and Accountability*. 2009. https://media.path.org/documents/GVR_femicide_rpt.pdf.

Pearce, Maryanne. "An Awkward Silence: Missing and Murdered Vulnerable Women and the Canadian Justice System." PhD diss., University of Ottawa, 2013. https://doi.org/10.20381/ruor -3344.

Pearce, Ruth, Sonja Erikainen, and Ben Vincent. "TERF Wars: An Introduction." *Sociological Review* 68, no. 4 (July 1, 2020): 677–698. https://doi.org/10.1177/0038026120934713.

Peng, Kenny, Arunesh Mathur, and Arvind Narayanan. "Mitigating Dataset Harms Requires Stewardship: Lessons from 1000 Papers." Paper presented at the Conference on Neural Informa-tion Processing Systems (NeurIPS), Track on Datasets and Benchmarks, 2021. https://arxiv.org /pdf/2108.02922.pdf?tpcc=nleyeonai.

Peoples Dispatch. "Puerto Rican Governor Ricardo Rosselló Resigns after a Week of Massive Protests." Peoples Dispatch, July 25, 2019. https://peoplesdispatch.org/2019/07/25/puerto-rican -governor-ricardo-rossello-resigns-after-a-week-of-massive-protests/.

Perera, Suvendrini, and Joseph Pugliese, eds. *Mapping Deathscapes: Digital Geographies of Racial and Border Violence*. London: Routledge, 2021.

Pickles, John. *A History of Spaces: Cartographic Reason, Mapping and the Geo-Coded World*. London: Routledge, 2003.

Pine, Kathleen, and Max Liboiron. "The Politics of Measurement and Action." In *Proceedings of the 33rd Annual ACM Conference on Human Factors in Computing Systems*, 3147–3156. New York: Association for Computing Machinery, 2015. https://doi.org/10.1145/2702123.2702298.

Porter, Theodore M. *Trust in Numbers: The Pursuit of Objectivity in Science and Public Life.* Princeton, NJ: Princeton University Press, 1995.

Posada, Julian. "The Coloniality of Data Work: Power and Inequality in Outsourced Data Production for Machine Learning." PhD diss., University of Toronto, 2022. https://tspace.library .utoronto.ca/bitstream/1807/126388/1/Posada_Gutierrez_Julian__Alberto_202211_PhD_thesis.pdf.

"Presentes: Periodismo de Géneros, Diversidad, y Derechos Humanos Desde América Latina." Presentes. Accessed August 9, 2023. https://agenciapresentes.org/nosotres/.

Puerto Rico. Ley Num. 40. August 27, 2021. https://openstates.org/pr/bills/2021-2024/PS130/.

Puig de la Bellacasa, María. "Matters of Care in Technoscience: Assembling Neglected Things." *Social Studies of Science* 41, no. 1 (2011): 85–106.

Radford, Jill, and Diana E. H. Russell. *Femicide: The Politics of Woman Killing.* Woodbridge, CT: Twayne, 1992.

Radhakrishnan, Radhika. "The Cost of Safety. Surveillance Will Not Keep Women Safe." *Smashboard*, January 14, 2021. https://smashboard.org/the-cost-of-safety-surveillance-will-not-keep -women-safe/.

Rankin, Yolanda A., Jakita O. Thomas, and Nicole M. Joseph. "Intersectionality in HCI: Lost in Translation." *Interactions* 27, no. 5 (2020): 68–71.

Red Feminista Antimilitarista. "Historia." Accessed May 31, 2021. https://www.redfeministaanti militarista.org/nosotras/historia.

Red Feminista Antimilitarista. *Paren La Guerra Contra Las Mujeres.* Medellín: Editores Publicidad, 2019. https://redfeministaantimilitarista.org/images/documentos/Revista_Paren_La_Guerra_Contra _Las_Mujeres_RFA_2019.pdf.

Red Feminista Antimilitarista. *Violencia Neoliberal Feminicida En Medellín.* Medellín: Red Juvenil Feminista y Antimilitarista, 2014. https://www.redfeministaantimilitarista.org/repositorio/public aciones?task=callelement&format=raw&item_id=49&element=0f25a93b-d6ac-449e-b003 -402116058020&method=download.

Remarkable Women 2014. "Mona Woodward." Remarkable Women 2014, October 22, 2013. https://remarkablewomen2014.wordpress.com/2013/10/22/mona-woodward/.

Renzi, Alessandra, and Ganaele Langlois. "Data Activism." In *Compromised Data: New Paradigms in Social Media Theory and Methods*, edited by Greg Elmer, Joanna Redden, and Ganaele Langlois, 202–225. London: Bloomsbury Academic, 2015.

Restauradoras con glitter. Facebook. Accessed August 9, 2023. https://www.facebook.com/restau radoras.glitterMX/.

Revilla Blanco, Marisa. "Del ¡Ni Una Más! Al #NiUnaMenos: Movimientos de Mujeres y Feminismos En América Latina." *Política y Sociedad* 56, no. 1 (2019): 47–67. https://doi.org/10.5209 /poso.60792.

Ricaurte, Paola. "Data Epistemologies, the Coloniality of Power, and Resistance." *Television & New Media* 20, no. 4 (May 1, 2019): 350–365. https://doi.org/10.1177/1527476419831640.

Rich, Adrienne. *Blood, Bread, and Poetry: Selected Prose, 1979–1985*. New York: Norton, 1986.

Richards, Tara N., Lane Kirkland Gillespie, and M. Dwayne Smith. "Exploring News Coverage of Femicide: Does Reporting the News Add Insult to Injury?" *Feminist Criminology* 6, no. 3 (2011): 178–202. https://doi.org/10.1177/1557085111409919.

Ritchie, Andrea J. *Invisible No More: Police Violence Against Black Women and Women of Color*. Boston: Beacon Press, 2017.

Roberts, Hal, Rahul Bhargava, Linas Valiukas, Dennis Jen, Momin M. Malik, Cindy Sherman Bishop, Emily B. Ndulue, et al. "Media Cloud: Massive Open Source Collection of Global News on the Open Web." In *Proceedings of the Fifteenth International AAAI Conference on Web and Social Media (ICWSM 2021)*, 1034–1045. Palo Alto, CA: AAAI Press, 2021. https://ojs.aaai.org/index.php/ICWSM/article/download/18127/17930/21622.

Roberts, Sarah T. *Behind the Screen: Content Moderation in the Shadows of Social Media*. New Haven, CT: Yale University Press, 2021.

Roberts, Sarah T. "Digital Refuse: Canadian Garbage, Commercial Content Moderation and the Global Circulation of Social Media's Waste." *Media Studies Publications* 14 (2016). https://ir.lib.uwo.ca/commpub/14/.

Romeo, Francesca M. "Towards a Theory of Digital Necropolitics." PhD diss., University of California Santa Cruz, 2021. https://escholarship.org/uc/item/1059d63h.

Rothschild, Annabel, Amanda Meng, Carl DiSalvo, Britney Johnson, Ben Rydal Shapiro, and Betsy DiSalvo. "Interrogating Data Work as a Community of Practice." *Proceedings of the ACM on Human–Computer Interaction* 6, no. CSCW2 (November 11, 2022): article 307, 1–28. https://doi.org/10.1145/3555198.

Ruppert, Evelyn, Engin Isin, and Didier Bigio. "Data Politics." *Big Data & Society* 4, no. 2 (2017). https://doi.org/10.1177/2053951717717749.

Ruse, Jamie-Leigh. "Experiences of Engagement and Detachment When Counting the Dead for Menos Días Aquí, a Civilian-Led Count of the Dead of Mexico's Drugs War." *Journal of Latin American Cultural Studies* 25, no. 2 (April 2016): 215–236.

Russell, Diana. "Defining Femicide." Speech given at the UN Symposium on Femicide: A Global Issue that Demands Action. Vienna, Austria: November 2012. https://www.dianarussell.com/defining-femicide-.html.

Russell, Legacy. *Glitch Feminism: A Manifesto*. London: Verso, 2020.

Sagot Rodríguez, Montserrat. "Violence against Women: Contributions from Latin America." In *The Oxford Hanbook of the Sociology of Latin America*, edited by Xóchitl Bada and Liliana Rivera-Sánchez, 520–539. New York: Oxford University Press, 2020. https://doi.org/10.1093/oxfordhb/9780190926557.013.32.

Sagot Rodríguez, Montserrat. "Femicide as Necropolitics in Central America." *Labrys Estudos Feministas* 24 (2013): 1–26. https://www.labrys.net.br/labrys24/feminicide/monserat.htm.

Salguero, María. "Yo te nombro: El Mapa de los Feminicidios en México" [I name you: Map of Feminicides in Mexico]. *Yo te nombro: El Mapa de los Feminicidios en México* (blog). Accessed January 4, 2022. http://mapafeminicidios.blogspot.com/p/inicio.html.

Salmenrón Arroyo, Christina, Lydiette Carrión Rivera, and Isabel Montoya Ramos. *Un Manual Urgente Para La Cobertura de Violencia Contra Las Mujeres y Feminicidios En México*. Iniciativa Spotlight ONU Mujeres, 2021. https://mexico.unwomen.org/sites/default/files/Field%20Office%20Mexico/Documentos/Publicaciones/2021/MANUAL%20PERIODISTAS-SPOTLIGHT.pdf.

Sambasivan, Nithya, Shivani Kapania, Hannah Highfill, Diana Akrong, Praveen Paritosh, and Lora Aroyo. "'Everyone Wants to Do the Model Work, Not the Data Work': Data Cascades in High-Stakes AI." In *Proceedings of the 2021 CHI Conference on Human Factors in Computing Systems*, 1–15. New York: Association for Computing Machinery, 2021. https://doi.org/10.1145/3411764.3445518.

Scheuerman, Morgan Klaus, Stacy M. Branham, and Foad Hamidi. "Safe Spaces and Safe Places: Unpacking Technology-Mediated Experiences of Safety and Harm with Transgender People." *Proceedings of the ACM on Human–Computer Interaction* 2, no. CSCW (November 1, 2018): 1–27. https://doi.org/10.1145/3274424.

Scheuerman, Morgan Klaus, Kandrea Wade, Caitlin Lustig, and Jed R. Brubaker. "How We've Taught Algorithms to See Identity: Constructing Race and Gender in Image Databases for Facial Analysis." *Proceedings of the ACM on Human–Computer Interaction* 4, no. CSCW1 (May 29, 2020): article 58, 1–35. https://doi.org/10.1145/3392866.

Scott, James C. *Seeing Like a State: How Certain Schemes to Improve the Human Condition Have Failed*. New Haven, CT: Yale University Press, 1998.

Segato, Rita Laura. "Las Nuevas Formas de La Guerra y El Cuerpo de Las Mujeres." *Sociedade e Estado* 29, no. 2 (May 2014): 341–371.

Segato, Rita Laura. "Cinco debates feministas: Temas para una reflexión divergente sobre la violencia contra las mujeres." In *La guerra contra las mujeres*, 629–653. Mexico City: Bonilla Artigas Editores, 2016. https://www.torrossa.com/es/resources/an/4535908.

Segato, Rita Laura. *Contra-pedagogías de la crueldad*. Buenos Aires: Prometeo Libros, 2018.

Segato, Rita Laura. *La guerra contra las mujeres*. Madrid: Traficantes de Sueños, 2016.

Segato, Rita Laura. "Rita Segato: 'Los femicidios se repiten porque se muestran como un espectáculo.'" Interview by Ailín Trepiana, LMNeuquén, September 6, 2019. https://www.lmneuquen.com/rita-segato-los-femicidios-se-repiten-porque-se-muestran-como-un-espectaculo-n649114.

Segato, Rita Laura. "Territory, Sovereignty, and Crimes of the Second State: The Writing on the Body of Murdered Women." In *Terrorizing Women*, edited by Rosa-Linda Fregoso and Cynthia Bejarano, translated by Sara Koopman, 70–92. Durham, NC: Duke University Press, 2010. https://doi.org/10.1215/9780822392644-005.

Seltzer, William, and Margo Anderson. "The Dark Side of Numbers: The Role of Population Data Systems in Human Rights Abuses." *Social Research* 68, no. 2 (2001): 481–513.

Semple, Kirk, and Paulina Villegas. "The Grisly Deaths of a Woman and a Girl Shock Mexico and Test Its President." *New York Times*, February 19, 2020. https://www.nytimes.com/2020/02/19/world/americas/mexico-violence-women.html.

Sharpe, Lori, Jane Ginsburg, Gail Gordon, Mariame Kaba, and Jacqui Shine. *Trying to Make the Personal Political: Feminism and Consciousness-Raising*. Chicago: Half Letter Press, 2017.

Shelby, Renee. "Technology, Sexual Violence, and Power-Evasive Politics: Mapping the Anti-Violence Sociotechnical Imaginary." *Science, Technology, & Human Values* 48, no. 3 (2023): 552–581. https://doi.org/10.1177/01622439211046047.

Sheth, Falguni. "The Technology of Race." *Radical Philosophy Review* 7, no. 1 (2004): 77–98.

Shokooh Valle, Firuzeh. "'How Will You Give Back?': On Becoming a Compañera as a Feminist Methodology from the Cracks." *Journal of Contemporary Ethnography* 50, no. 6 (2021): 835–861.

Silveira, Giulia Bordignon, Daniela Scherer dos Santos, and Gabriela Felten da Maia. "Estamos Juntas: Expert System to Support the Identification and Denunciation of Violence against Women." In *Proceedings of the XIV Brazilian Symposium on Information Systems*, 1–9. New York: Association for Computing Machinery, 2018. https://doi.org/10.1145/3229345.3229381.

Simpson, Audra. "The Ruse of Consent and the Anatomy of 'Refusal': Cases from Indigenous North America and Australia." *Postcolonial Studies* 20, no. 1 (January 2, 2017): 18–33. https://doi.org/10.1080/13688790.2017.1334283.

Simpson, Audra. "The State Is a Man: Theresa Spence, Loretta Saunders and the Gender of Settler Sovereignty." *Theory & Event* 19, no. 4 (2016). https://www.proquest.com/docview/1866315122?parentSessionId=RULkBTqia9qjKxrLHQ438oCcxze3O3WzRmhqPAyNEII%3D&pq-origsite=primo.

Simpson, Leanne Betasamosake. *As We Have Always Done: Indigenous Freedom through Radical Resistance*. 3rd ed. Minneapolis: University of Minnesota Press, 2017.

Sloane, Mona, Emanuel Moss, Olaitan Awomolo, and Laura Forlano. 2022. "Participation Is Not a Design Fix for Machine Learning." In *EAAMO '22: Equity and Access in Algorithms, Mechanisms, and Optimization*, article 1, 1–6. New York: Association for Computing Machinery, 2022. //doi.org/10.1145/3551624.3555285.

Smith, Linda Tuhiwai. *Decolonizing Methodologies: Research and Indigenous Peoples*. 3rd ed. London: Bloomsbury Publishing, 2021.

Snorton, C. Riley, and Jin Haritaworn. "Trans Necropolitics: A Transnational Reflection on Violence, Death, and the Trans of Color Afterlife." In *The Transgender Studies Reader Remix*, edited by Susan Stryker and Dylan M. Blackston, 305–316. London: Routledge, 2013.

Sovereign Bodies Institute. *To' Kee Skuy' Soo Ney-Wo-Chek': I Will See You Again in a Good Way Progress Report #MMIWG2*. Sovereign Bodies, July 2020. https://2a840442-f49a-45b0-b1a1-7531a7cd3d30.filesusr.com/ugd/6b33f7_c7031acf738f4f05a0bd46bf96486e58.pdf.

Sovereign Bodies Institute. *MMIWG2 & MMIP Organizing Toolkit.* Accessed June 2, 2022. https://www.sovereign-bodies.org/_files/ugd/6b33f7_2585fecaf9294450a595509cb701e7af.pdf.

Sovereign Bodies Institute. "Public Statement on Operation Lady Justice, from MMIWG & MMIP Grassroots Advocates." May 2021. https://www.sovereign-bodies.org/_files/ugd/6b33f7_1c1b448 93a2e4385a8314b53e31ea4be.pdf.

Spade, Dean. *Normal Life: Administrative Violence, Critical Trans Politics, and the Limits of Law.* Rev. ed. Durham, NC: Duke University Press, 2015. https://doi.org/10.1515/9780822374794.

Stengers, Isabelle. *The Invention of Modern Science.* Translated by Daniel W. Smith. Vol. 19. Minneapolis: University of Minnesota Press, 2000.

Sterling, S. Revi. "Designing for Trauma: The Roles of ICTD in Combating Violence against Women (VAW)." In *Proceedings of the Sixth International Conference on Information and Communications Technologies and Development*, vol. 2, 159–162. New York: Association for Computing Machinery, 2013. https://doi.org/10.1145/2517899.2517908.

Stray, Jonathan. *The Curious Journalist's Guide to Data.* Tow Center for Digital Journalism, GitBook, March 2016. https://towcenter.gitbooks.io/curious-journalist-s-guide-to-data/content/footnotes/.

Strohmayer, Angelika, Janis Lena Meissner, Alexander Wilson, Sarah Charlton, and Laura Mcintyre. "'We Come Together as One . . . and Hope for Solidarity to Live on': On Designing Technologies for Activism and the Commemoration of Lost Lives." In *Proceedings of the 2020 ACM Designing Interactive Systems Conference*, 87–100. New York: Association for Computing Machinery, 2020.

Suárez Val, Helena. "Affect Amplifiers: Feminist Activists and Digital Cartographies of Feminicide." In *Networked Feminisms: Activist Assemblies and Digital Practices*, edited by Shana MacDonald, Brianna I. Wiens, Michelle MacArthur, and Milena Radzikowska, 163–186. Lanham MD: Lexington Books, 2021.

Suárez Val, Helena. "Caring, with Data: An Exploration of the Affective Politicality of Feminicide Data." PhD diss., University of Warwick, 2023.

Suárez Val, Helena. "Data Frames of Feminicide: An Ontological Reconstruction and Critical Analysis of Activist and Government Datasets of Gender-Related Murders of Women." MA thesis, University of Warwick, 2019.

Suárez Val, Helena. "Datos Discordantes. Información Pública Sobre Femicidio En Uruguay." *Mundos Plurales* 7, no. 1 (2020): 53–78. https://doi.org/10.17141/mundosplurales.1.2021.3937.

Suárez Val, Helena. "Marcos de Datos de Feminicidio. Reconstrucción ontológica y análisis crítico de dos datasets de asesinatos de mujeres por razones de género." *Informatio. Revista del Instituto de Información de la Facultad de Información y Comunicación* 26, no. 1 (May 31, 2021): 313–346. https://doi.org/10.35643/Info.26.1.15.

Suárez Val, Helena. "Vibrant Maps. Exploring the Reverberations of Feminist Digital Mapping." *Inmaterial. Diseño, Arte y Sociedad* 3, no. 5 (2018): 113–139.

Suárez Val, Helena, Catherine D'Ignazio, Jimena Acosta, Melissa Q. Teng, and Silvana Fumega. "Data Artivism and Feminicide." Forthcoming in "Big Data & AI in Latin America," special issue, *Big Data & Society* (2023).

Suárez Val, Helena, Catherine D'Ignazio, and Silvana Fumega. "Data Against Feminicide." Forthcoming in *Making a Difference! Novel Research Methods in the Datafied Society*, edited by Mirko Schäfer and Tracey P. Lauriault, 2024.

Suárez Val, Helena, Sonia Madrigal, Ivonne Ramírez, and María Salguero. "Monitoring, Recording, and Mapping Feminicide—Experiences from Mexico and Uruguay." In *Femicide Volume XII Living Victims of Femicide*, 67–73. ACUNS, 2019. https://www.researchgate.net/publication/337592833_Monitoring_recording_and_mapping_feminicide-Experiences_from_Mexico_and_Uruguay.

Suárez Val, Helena, Angeles Martinez Cuba, Melissa Q. Teng, and Catherine D'Ignazio. "Datos de Feminicidio, Trabajo Emocional y Autocuidado." *4S Backchannels Blog* (blog), June 13, 2022. https://medium.com/data-feminism-lab-mit/datos-de-feminicidio-trabajo-emocional-y-auto cuidado-fa0ba5daeb2b.

Sun, Yu, and Siyuan Yin. "Opening up Mediation Opportunities by Engaging Grassroots Data: Adaptive and Resilient Feminist Data Activism in China." *New Media & Society*, June 2, 2022. https://doi.org/10.1177/14614448221096806.

Suresh, Harini. "A Framework for Understanding Sources of Harm throughout the Machine Learning Life Cycle." In *Equity and Access in Algorithms, Mechanisms, and Optimization*, 1–9. New York: Association for Computing Machinery, 2021. https://doi.org/10.1145/3465416.3483305.

Suresh, Harini, Rajiv Movva, Amelia Lee Dogan, Rahul Bhargava, Isadora Cruxen, Angeles Martinez Cuba, Guilia Taurino, Wonyoung So, and Catherine D'Ignazio. "Towards Intersectional Feminist and Participatory ML: A Case Study in Supporting Feminicide Counterdata Collection." In *Proceedings of the 2022 ACM Conference on Fairness, Accountability, and Transparency*, 667–678. New York: Association for Computing Machinery, 2022. https://dl.acm.org/doi/10.1145/3531146.3533132.

Sutherland, Georgina, Patricia Easteal, Kate Holland, and Cathy Vaughan. "Mediated Representations of Violence against Women in the Mainstream News in Australia." *BMC Public Health* 19 (May 3, 2019). https://doi.org/10.1186/s12889-019-6793-2.

Tabuenca C., María Socorro. "From Accounting to Recounting: Esther Chávez Cano and the Articulation of Advocacy, Agency, and Justice on the US-Mexico Border." In *Mexican Public Intellectuals*, edited by Debra A. Castillo and Stuart A. Day, 139–161. New York: Palgrave Macmillan, 2014. https://doi.org/10.1057/9781137392299_7.

Taillieu, Tamara L., and Douglas A. Brownridge. "Violence against Pregnant Women: Prevalence, Patterns, Risk Factors, Theories, and Directions for Future Research." *Aggression and Violent Behavior* 15, no. 1 (2010): 14–35. https://doi.org/10.1016/j.avb.2009.07.013.

Taylor, John, and Tahu Kukutai, eds. *Indigenous Data Sovereignty: Toward an Agenda*. Canberra: ANU Press, 2016.

Taylor, Keeanga-Yamahtta. "Until Black Women Are Free, None of Us Will Be Free." *New Yorker*, July 20, 2020. https://www.newyorker.com/news/our-columnists/until-black-women-are-free-none-of-us-will-be-free.

Terranova, Tiziana. "Free Labor: Producing Culture for the Digital Economy." *Social Text* 18, no. 2 (2000): 33–58.

Thomas, Jakita O., Neha Kumar, Alexandra To, Quincy Brown, and Yolanda A. Rankin. "Discovering Intersectionality: Part 2: Reclaiming Our Time." *Interactions* 28, no. 4 (2021): 72–75.

Threadcraft, Shatema. *Intimate Justice: The Black Female Body and the Body Politic*. Oxford: Oxford University Press, 2016.

Threadcraft, Shatema. "Like Breonna Taylor, Black Women Are Often Killed in Private—Even When It's by Police." *Washington Post*, March 22, 2022. https://www.washingtonpost.com/politics/2022/03/22/invisible-black-women-death-race-gender-disadvantage/.

Threadcraft, Shatema. "North American Necropolitics and Gender: On #BlackLivesMatter and Black Femicide." *South Atlantic Quarterly* 116, no. 3 (2017): 553–579. https://doi.org/10.1215/00382876-3961483.

Tillmann-Healy, Lisa M. "Friendship as Method." *Qualitative Inquiry* 9, no. 5 (2003): 729–749.

Tiscareño-García, Elizabeth, and Óscar-Mario Miranda-Villanueva. "Victims and Perpetrators of Feminicide in the Language of the Mexican Written Press." *Comunicar* 28, no. 63 (2020): 51–60.

Tronto, Joan C., and Berenice Fisher. "Toward a Feminist Theory of Caring." In *Circle of Care*, edited by E. Abel and M. Nelson, 36–54. New York: SUNY Press, 1990.

Trouillot, Michel-Rolph. *Silencing the Past: Power and the Production of History*. 2nd rev. ed. Boston: Beacon Press, 2015.

Tsing, Anna Lowenhaupt. "On Nonscalability: The Living World Is Not Amenable to Precision-Nested Scales." *Common Knowledge* 18, no. 3 (2012): 505–524.

Tuana, Nancy. "The Speculum of Ignorance: The Women's Health Movement and Epistemologies of Ignorance." *Hypatia* 21, no. 3 (2006): 1–19.

Tuck, Eve, and K. Wayne Yang. "Decolonization Is Not a Metaphor." *Decolonization: Indigeneity, Education & Society* 1, no. 1 (2012): 1–40.

Tuck, Eve, and K. Wayne Yang. "R-Words: Refusing Research." In *Humanizing Research: Decolonizing Qualitative Inquiry with Youth and Communities*, edited by D. Paris and M. T. Winn, 223–247. Thousand Oaks, CA: Sage, 1987. https://doi.org/10.4135/9781544329611/.

Umoja Noble, Safiya. *Algorithms of Oppression: How Search Engines Reinforce Racism*. New York: NYU Press, 2018.

United Nations. *Universal Declaration of Human Rights*. 2015. https://www.un.org/en/udhrbook/pdf/udhr_booklet_en_web.pdf.

United Nations High Commissioner for Human Rights. *Latin American Model Protocol for the Investigation of Gender-Related Killings of Women (Femicide/Feminicide)*. United Nations Entity for Gender Equality and the Empowerment of Women (UN Women), 2014. https://lac.unwomen.org/en/digiteca/publicaciones/2014/10/modelo-de-protocolo.

UnLeading. "Ableism." Accessed March 30, 2023. https://www.yorku.ca/edu/unleading/ableism/.

UNODOC. *Global Study on Homicide—Gender-Related Killing of Women and Girls*. United Nations Office on Drugs and Crime, 2019. https://www.unodc.org/documents/data-and-analysis/GSH2018 /GSH18_Gender-related_killing_of_women_and_girls.pdf.

Unzúeta, Mónica. "Articulaciones Feministas: Contemporary Bolivian Feminisms and the Struggle against Gender Violence." Washington University in St. Louis: Center for the Humanities, May 13, 2020. https://humanities.wustl.edu/news/articulaciones-feministas-contemporary-bolivian -feminisms-and-struggle-against-gender-violence.

US Bureau of Labor Statistics. "Labor Force Statistics from the Current Population Survey." Last modified January 25, 2023. https://www.bls.gov/cps/cpsaat11.htm.

US Department of Justice Consultation with Tribes. "Savanna's Act: Data Relevance and Access." June 17–18, 2021. https://www.justice.gov/d9/fieldable-panel-panes/basic-panes/attachments/2022 /01/27/tribal_and_agency_feedback_on_savannas_act_consultation_on_data_relevance_and _access.pdf.

Vanoli Imperiale, Sofía Carolina. "El doble asesinato de las identidades transgénero: Un análisis crítico del discurso de la prensa policial en Uruguay." PhD diss., Universidad ORT Uruguay, 2014.

Vera, Lourdes A., Dawn Walker, Michelle Murphy, Becky Mansfield, Ladan Mohamed Siad, and Jessica Ogden. "When Data Justice and Environmental Justice Meet: Formulating a Response to Extractive Logic through Environmental Data Justice." *Information, Communication and Society* 22, no. 7 (2019): 1012–1028. https://doi.org/10.1080/1369118X.2019.1596293.

Vigil-Hayes, Morgan, Marisa Duarte, Nicholet Deschine Parkhurst, and Elizabeth Belding. "#Indigenous: Tracking the Connective Actions of Native American Advocates on Twitter." In *Proceedings of the 2017 ACM Conference on Computer Supported Cooperative Work and Social Computing*, 1387–1399. New York: Association for Computing Machinery, 2017. https://doi.org/10.1145 /2998181.2998194.

Walcott, Rinaldo. "Data or Politics? Why the Answer Still Remains Political." Royal Society of Canada, November 12, 2020. https://rsc-src.ca/en/covid-19/impact-covid-19-in-racialized -communities/data-or-politics-why-answer-still-remains.

Walklate, Sandra, and Kate Fitz-Gibbon. "Re-imagining the Measurement of Femicide: From 'Thin' Counts to 'Thick' Counts." *Current Sociology* 71, no. 1 (April 12, 2022). https://doi.org /10.1177/00113921221082698.

Walklate, Sandra, Kate Fitz-Gibbon, Jude McCulloch, and JaneMaree Maher. *Towards a Global Femicide Index: Counting the Costs*. London: Routledge, 2019. https://doi.org/10.4324/9781138393134.

Walter, Maggie. "The Voice of Indigenous Data: Beyond the Markers of Disadvantage." *Griffith Review* 60 (2018): 256–263.

Walter, Maggie. "Conceptualizing and Theorizing the Indigenous Life." In *The Oxford Handbook of Indigenous Sociology*, edited by M. Walter, A. A. Gonzales, and R. Henr, 12–30. New York: Oxford University Press, 2023. http://ecite.utas.edu.au/151035.

Walter, Maggie, and Chris Andersen. *Indigenous Statistics*. London: Routledge, 2013.

Wells, Ida B. "A Red Record. Tabulated Statistics and Alleged Causes of Lynchings in the United States, 1892-1893-1894." Chicago: Donohue & Henneberry, 1894. https://digitalcollections.nypl .org/items/510d47df-8dbd-a3d9-e040-e00a18064a99.

Wendel, Delia Duong Ba. *Rwanda's Genocide Heritage: Between Justice and Sovereignty*. Durham, NC: Duke University Press, 2024.

West, S. M., M. Whittaker, and K. Crawford. *Discriminating Systems: Gender, Race and Power in AI*. AI Now Institute, April 2019. https://ainowinstitute.org/publication/discriminating-systems -gender-race-and-power-in-ai-2.

Westmarland, Nicole. "The Quantitative/Qualitative Debate and Feminist Research: A Subjective View of Objectivity." *Forum Qualitative Sozialforschung* 2, no. 1 (2001). https://www.qualitative -research.net/index.php/fqs/article/view/974/2124.

Why Information Matters: A Foundation for Resilience. Internews and the Rockefeller Foundation, May 2015. https://www.unhcr.org/innovation/wp-content/uploads/2017/10/150513-Internews _WhyInformationMatters.pdf.

Wilcox, Dawn. "Invisible Women: Understanding the Scope of Lethal Male Violence against Women in the U.S." Video, RIAF, January 16, 2022. https://www.youtube.com/watch?v =YUwlLEdc9CQ.

Williams, Sherri. "#SayHerName: Using Digital Activism to Document Violence against Black Women." *Feminist Media Studies* 16, no. 5 (September 2, 2016): 922–925. https://doi.org/10.1080 /14680777.2016.1213574.

"Women & Girls Lost to Male Violence In 2018." Video, Dawn Wilcox, July 16, 2018. https:// www.youtube.com/watch?v=S9P70qU3cyw.

Wright, Melissa W. "Public Women, Profit, and Femicide in Northern Mexico." *South Atlantic Quarterly* 105, no. 4 (2006): 681–698. https://doi.org/10.1215/00382876-2006-003.

Wright, Melissa W. "Necropolitics, Narcopolitics, and Femicide: Gendered Violence on the Mexico-U.S. Border." *Signs: Journal of Women in Culture and Society* 36, no. 3 (March 1, 2011): 707–731. https://doi.org/10.1086/657496.

Young, Meg, Michael Katell, and P. M. Krafft. "Confronting Power and Corporate Capture at the FAccT Conference." In *Proceedings of the 2022 ACM Conference on Fairness, Accountability, and Transparency*, 1375–1386. New York: Association for Computing Machinery, 2022. https://doi .org/10.1145/3531146.3533194.

Zuboff, Shoshana. *The Age of Surveillance Capitalism: The Fight for a Human Future at the New Frontier of Power*. New York: PublicAffairs, 2019.

Zuckerman, Ethan. "The Case for Digital Public Infrastructure." *Knight First Amendment Institute* (blog), January 17, 2020. https://knightcolumbia.org/content/the-case-for-digital-public -infrastructure.

NAME INDEX

The letter *f* following a page number denotes a figure; the letter *t* following a page number denotes a table.

SUBJECT INDEX

The letter *f* following a page number denotes a figure; the letter *t* following a page number denotes a table.